新编Android应用开发从入门到精通

何福贵/等编著

前 言

本书基于当前最新的 Android Studio 2. 3 版本和 Android SDK，从 Android 发展的前沿角度出发，展示了 Android 开发的最新相关知识内容。通过本书的学习，您将掌握实用的移动终端开发基础知识和应用技能，精通 Android 项目开发技术，从而能够胜任应用程序的实际开发任务，为培养综合应用能力铺平了道路。

全书共 12 章，以 Android 项目开发的视角，循序渐进地讲解并展示了 Android 项目开发过程的主要流程，具体如下。

第 1 章介绍了 Android 的开发环境，包括两种环境的搭建方法，以及 Eclipse 项目到 Android Studio 项目的转化方法，完成开发前的准备工作。

第 2 章介绍了 Android 软件项目开发的整体流程及 Android 开发过程中的代码规范，让读者对 Android 项目开发形成整体的了解。

第 3 章介绍了 Android 界面设计，包括布局、控件和 Activity，以及新的设计方法。

第 4 章对 Android 应用程序的各组成部分进行了深入讲解，包括事件处理机制、Android 多线程、Android 广播组件、后台服务 Service、AsyncTask、Handler 等。

第 5 章针对 Android 界面的设计，介绍了一些更复杂和高级的界面设计方法，包括 Android 的一些新控件的使用方法。通过本章的学习，读者将能够设计出更美观的界面。

第 6 章对 Android 常用的数据持久化方案进行了详细讲解，包括 SharedPreferences 存储、SQLite 数据库操作和最新的 LitePal 数据库操作等。

第 7 章介绍了与 Android 相关的动画技术，包括绘图动画、Drawable 动画、矢量动画等基本的图形类和二维动画，以及 Open GL ES 三维动画。

第 8 章介绍了 Android 音视频的操作方法，包括 Android 系统类的实现方法，并介绍了被 Android 开发者广泛应用的基于 FFmpeg 开发并开源的轻量级视频播放器 Ijkplayer。

第 9 章介绍了 Android 的权限机制，讲解了 JSON 格式数据的构造和解析方法。

第 10 章介绍了 Android 目前应用最广泛的无线通信技术，包括 WiFi、蓝牙和 NFC。不光展示了这三种技术的应用方法，还提供了对应的实际项目。

第 11 章介绍了 Android 的开源库和开源项目，包括一些典型 Android 开源库的获取和使用方法，一些典型 Android 开源项目的功能，以及获取 Android 开源资源的方法。

第 12 章介绍了应用程序的托管和发布方法。

总体来说，本书具有如下特点。

(1) 面向项目。按照实际项目的特点进行编写，以项目为主线进行内容讲解。

(2) 面向前沿。立足于 Android 发展的前沿角度，使用最新的开发环境。

(3) 有序分类。对知识进行了科学编排，使每一章既具有独立性，整体上又具有完整性。

本书由何福贵主要编写，其他参与编写人员还包括闫秀珍、何小波等。由于编写时间仓促，作者水平有限，书中疏漏和错误之处在所难免，望广大专家、读者提出宝贵意见。

目　录

第四章　应用程序的构成部件

第五章　界面设计更进一步——UI 高级设计

第六章　数据持久化方案

第七章　让界面动起来——Android 动画

第八章　更丰富的应用——Android 多媒体

第九章　连接到远方——Android 网络开发

第十章　更方便的通信——Android 无线通信

第十一章　Android 的开源库和开源项目

第十二章　Android 应用程序托管和发布

第一章

Android开发环境

Android 系统是目前市场占有率最高的移动操作系统。由于 Google 的开放政策，任何手机厂商和个人都能免费获取 Android 操作系统的源码，并且可以自由地使用和定制。如今，不管在哪里，都可以看到 Android 手机。今天的 Android 世界可谓欣欣向荣，让我们一起走进 Android 的世界吧。

1.1 Android 开发环境简介

工欲善其事，必先利其器，开发每一种应用程序，必须先准备集成开发环境（Integrated Development Environment，IDE）。目前 Android 应用程序的开发环境有两类：（1）Eclipse + ADT + Android SDK + JDK；（2）Android Studio + JDK。随着 Android Studio 2.2 的发布，Android 应用程序的开发环境已从 Eclipse 转向了 Android Studio。目前已经发布了 Android Studio 2.4 Preview 版和 Android Studio 2.3 正式版。Android 开发环境可以搭建在目前任何一种主流系统（Mac、Windows、Linux）上。

1.2 基于 Eclipse 的开发环境

Eclipse 是比较常见的开发 Android 应用程序的环境之一，到目前为止，大部分 Android 应用程序是基于 Eclipse 开发的。

Android 开发可以使用 Windows XP、Windows Vista、Mac OS、Linux 等操作系统平台。Android 开发所需要的工具为：JDK + Eclipse + Android SDK + ADT，在下载这些工具时，要依据不同的操作系统下载不同的版本。下面列出了 Android 开发工具版本的匹配关系，如表 1-1 所示。

表 1-1　Android 开发工具的版本对应关系

Android SDK 版本	JDK 版本	Eclipse 版本	ADT 版本
Android 6.0	Java 1.8 or higher	EclipseJuno（Version 4.2.1）or higher	ADT 23.0.7
Android4.2.2（SDK Tools r21.1.）	Java 1.6 or higher	Eclipse Helios（Version 3.6.2）or higher	ADT 21.1.0 ADT 21.0.1
Android4.2（SDK Tools r20.0.3）	Java 1.6 or higher	Eclipse Helios（Version 3.6.2）or higher	ADT 20.0.3
Android4.1（SDK Tools r20.0.1）	Java 1.6 or higher	Eclipse Helios（Version 3.6.2）or higher	ADT 20.0.2

（续）

Android SDK 版本	JDK 版本	Eclipse 版本	ADT 版本
Android4. 0. 3 （SDK Tools r18）	Java 1. 6 or higher	Eclipse Helios （Version 3. 6. 2） or higher	ADT 18. 0. 0
Android4. 0. 3 （SDK Tools r17）	Java 1. 6 or higher	Eclipse Helios （Version 3. 6. 2） or higher	ADT 17. 0. 0
Android4. 0. 3 （SDK Tools r16）	Java 1. 6 or higher	Eclipse Helios （Version 3. 6） or higher	ADT 16. 0. 0
Android4. 0 （SDK Tools r15）	Java 1. 6 or higher	Eclipse3. 3 或者 3. 4	ADT 15. 0. 1
Android3. 0 （SDK Tools r10）	Java 1. 6 or higher	Eclipse3. 3 或者 3. 4	. ADT 10. 0. 0
Android2. 3 （SDK Tools r8）	Java 1. 6 or higher	Eclipse3. 3 或者 3. 4	ADT 8. 0. 0

1. 2. 1　开发环境的搭建

本节介绍如何搭建开发环境。

1. JDK 的安装

JDK 即 Java Development Kit，简单来说，JDK 是面向开发人员的 Java SDK，它提供了 Java 的开发环境和运行环境。因为 Android 应用程序是面向 Java 的应用程序，Android 开发语言使用的是 Java，所以要安装 JDK 开发工具包（JDK 的下载地址读者可自行查询，此处不再赘述），该工具包的下载界面如图 1-1 所示。

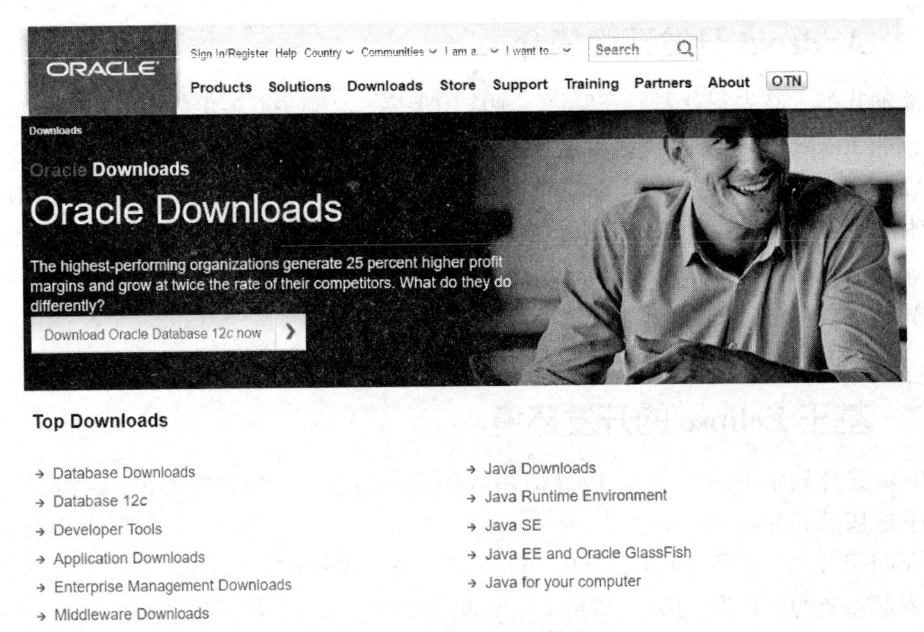

图 1-1　Java JDK 的下载界面

Java 有三个类型版本，便于软件开发人员、服务提供商和设备生产商针对特定的市场进行开发。

（1）Java SE（Java Platform，Standard Edition）。Java SE 以前称为进行开发和部署 J2SE。它允许对在桌面、服务器、嵌入式环境和实时环境中使用的 Java 应用程序。Java SE 包含了支持 Java Web 服务开发的类，并为 Java Platform，Enterprise Edition（Java EE）提供基础。

（2）Java EE（Java Platform，Enterprise Edition）。这个版本以前称为 J2EE。企业版本帮助开发和部署可移植、健壮、可伸缩且安全的服务器端 Java 应用程序。

(3) Java ME（Java Platform，Micro Edition）。这个版本以前称为 J2ME。Java ME 为在移动设备和嵌入式设备（比如手机、PDA、电视机顶盒和打印机）上运行的应用程序提供健壮且灵活的环境。

目前最新的版本为 JDK 8u121，在下载之前要选择 JDK 运行的平台，目前 JDK 可运行的平台有：Linux、Mac OS、Solaris SPARC、Window 四种，而 Windows 平台又分为 32 位和 64 位两种，本书选择 64 位 SE 版本，对应的文件为 jdk- 8u111-windows-x64. exe，在下载完成之后，安装过程如图 1-2 所示。

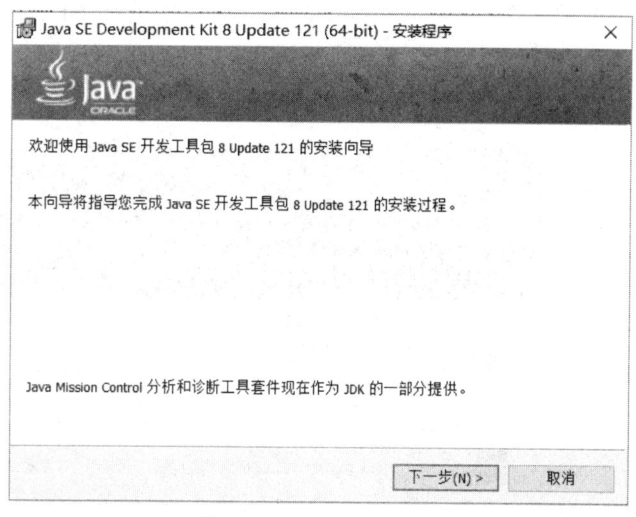

图 1-2　JDK 的安装过程

2. Eclipse

Eclipse 是一个开放源代码的、基于 Java 的可扩展开发平台。就其本身而言，它只是一个框架和一组服务，用于通过插件组件构建开发环境。Eclipse 的初始下载界面如图 1-3 所示。之后单击 Download Packages 链接，进入下一界面。

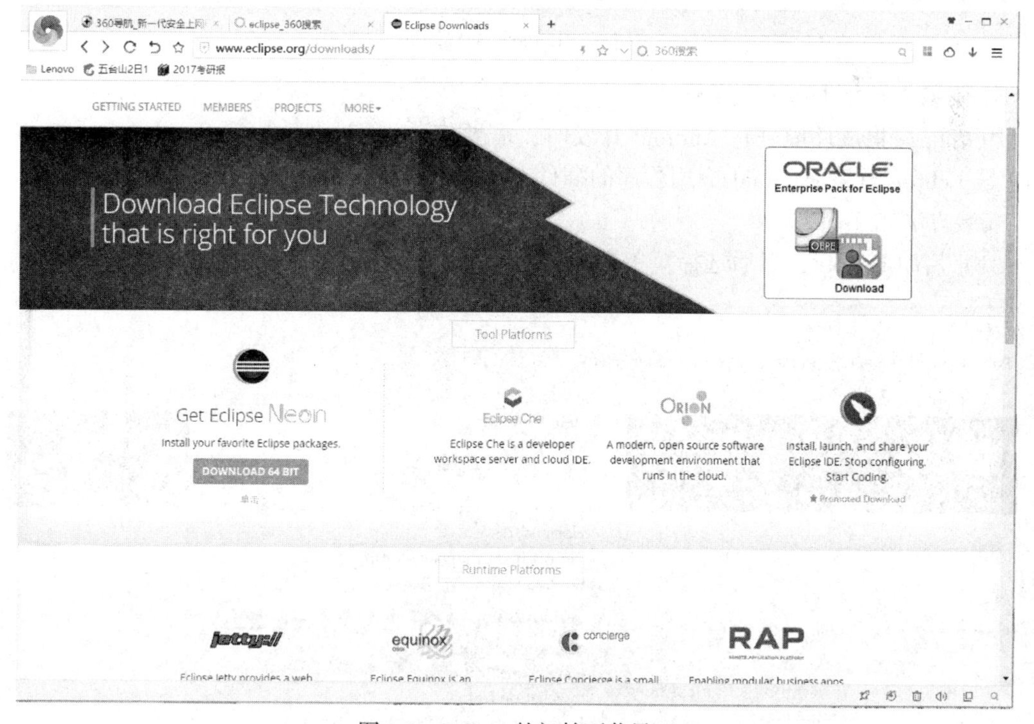

图 1-3　Eclipse 的初始下载界面

在如图 1-4 所示的界面中选择 Eclipse IDE for Java EE Developers，选择 64 bit 进行下载，

下载的文件名为“eclipse-jee-neon-2-win32-x86_64. zip”。

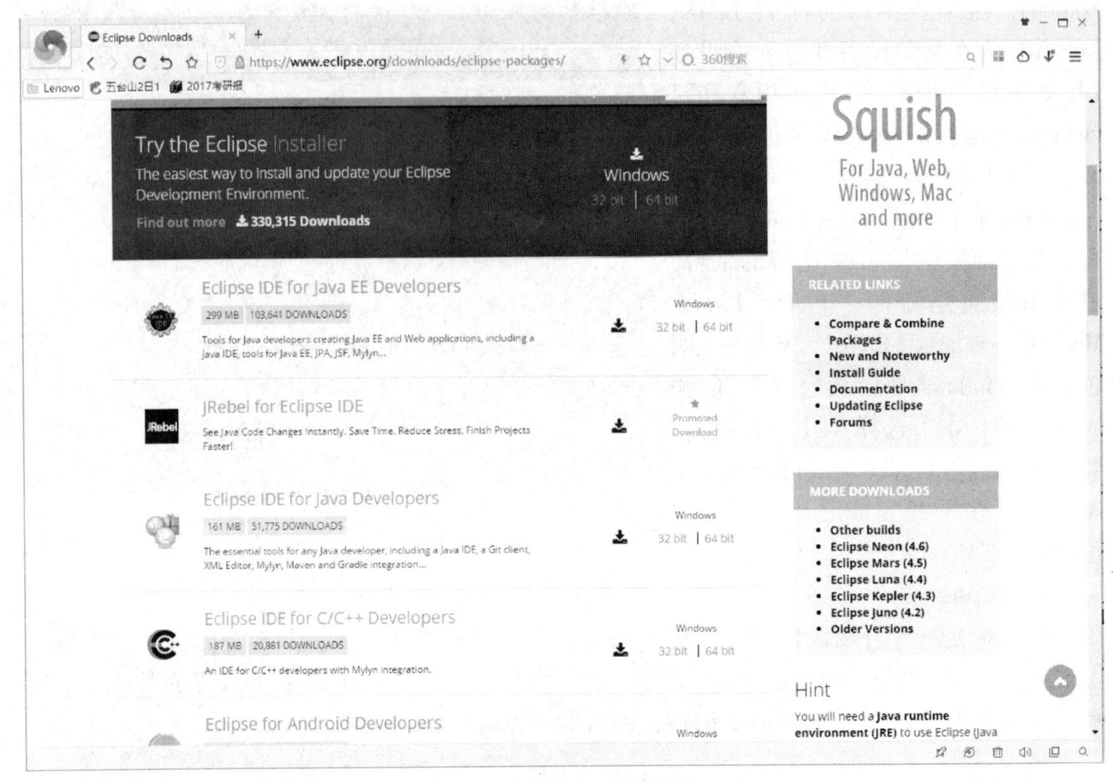

图 1-4　Eclipse 下载界面

3. 安装 Android　ADT 插件

用 Eclipse 集成环境进行 Android 开发时，需要安装 ADT（Android Development Tools）。ADT 是 Eclipse 开发 Android 应用程序的插件。

安装方法如下。

（1）启动 Eclipse，然后选择菜单栏中 Help ＞Install New Software 命令，如图 1-5 所示。

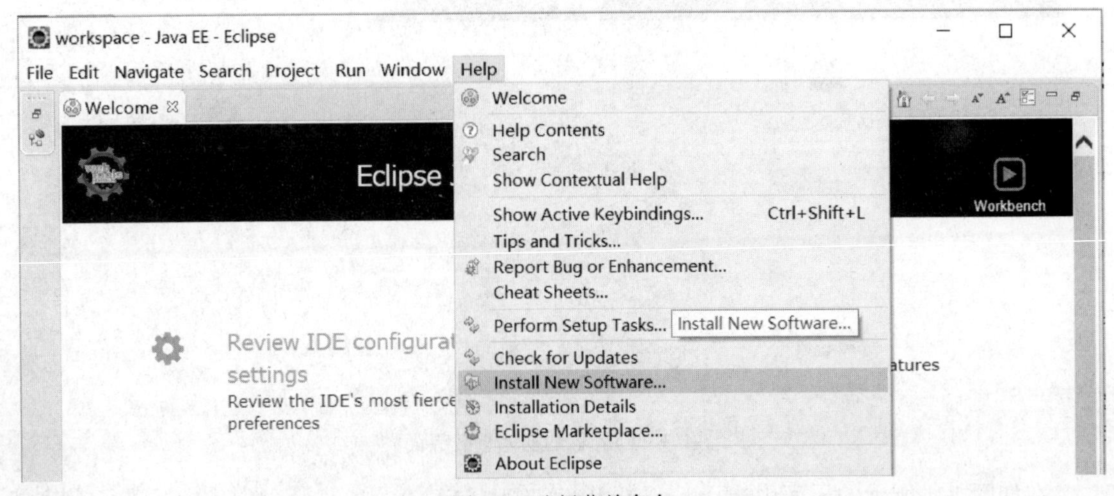

图 1-5　选择菜单命令

（2）在弹出的窗口中，单击 Add 按钮，在弹出的对话框中添加存储库，输入 ADT 的名称，以及位置：https://dl-ssl. google. com/android/eclipse/，单击 OK 按钮，如图 1-6 所示。

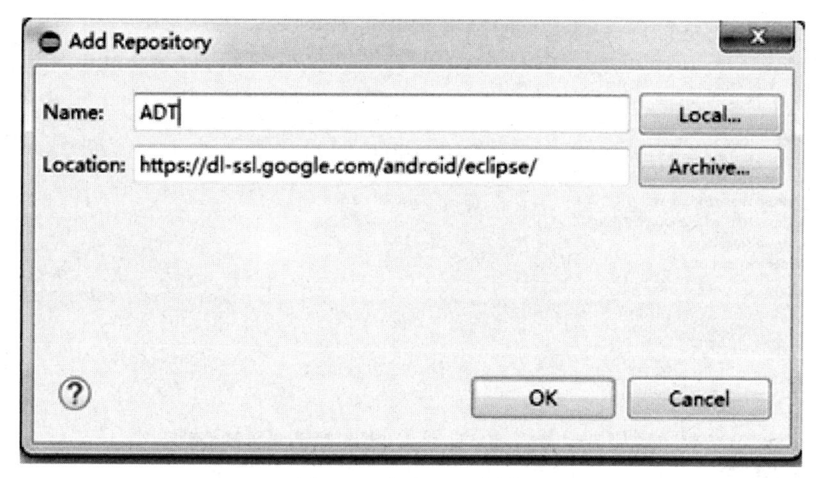

图 1-6　安装 ADT 插件

下载完 ADT 软件后，选择菜单栏中 Help ＞Install New Software 命令，在弹出的窗口中，单击 Add 按钮，然后单击 Archive 按钮，选中 ADT 文件，如图 1-7 所示。

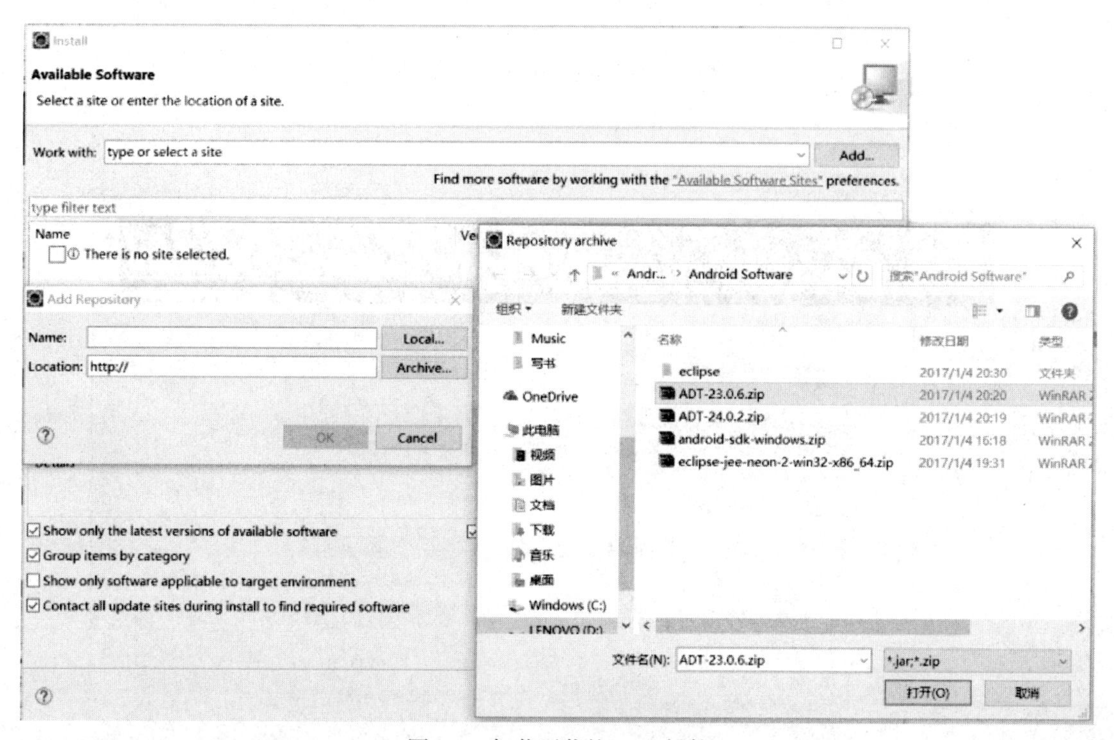

图 1-7　加载下载的 ADT 插件

（3）打开 ADT 插件后，选择 Developer Tools 选项，如图 1-8 所示。

（4）单击 Next 按钮，开始安装 ADT-23. 0. 6，直到安装完成，如图 1-9 所示。

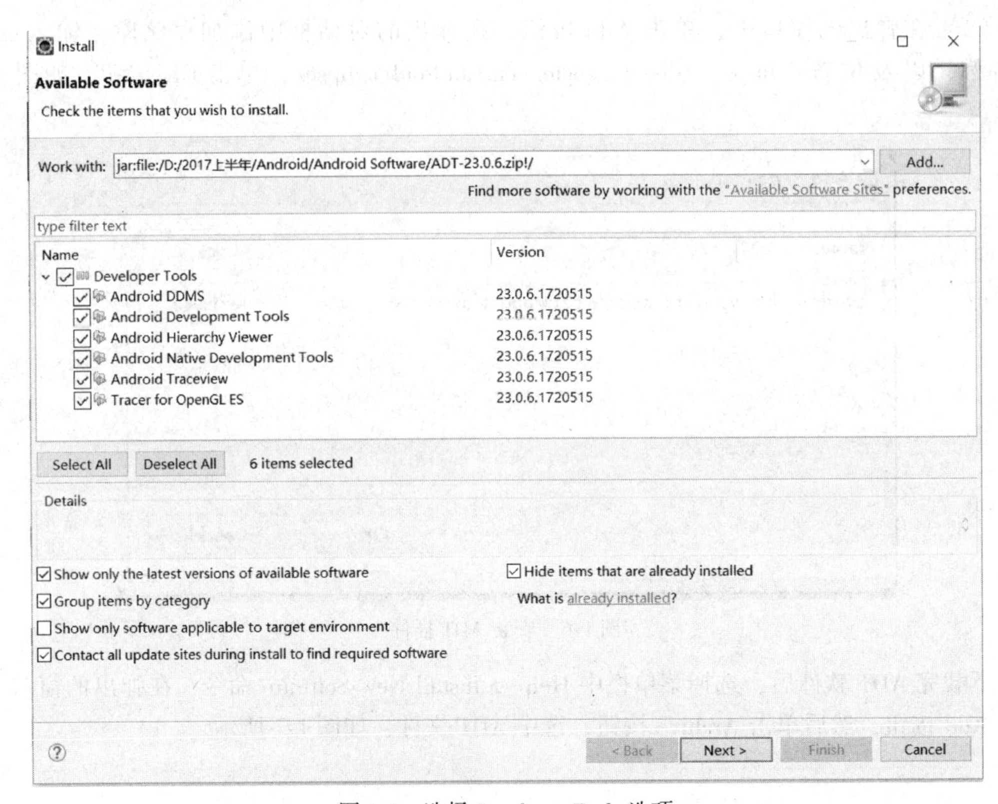

图 1-8　选择 Developer Tools 选项

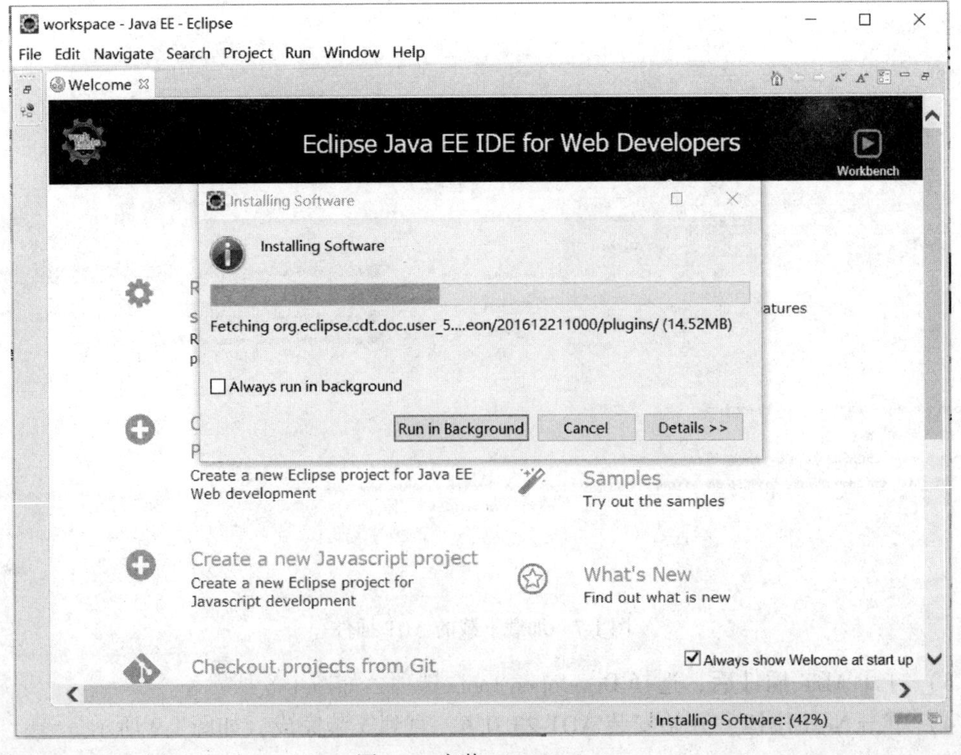

图 1-9　安装 ADT-23. 0. 6

4. Android SDK

Android SDK 可从 Android 的官网下载，选择的 Android SDK 版本应与 ADT 的版本对应，前面安装的 ADT 为“ADT-23. 0. 6. zip”，那么对应的版本也要是 23 以上的 Android SDK，本书选择的文件为“android-sdk-windows-r24-updated. 7z”，将压缩文件“android-sdk-windows-r24-updated. 7z”解压，可以看到下载的文件和文件夹，如图 1-10 所示。

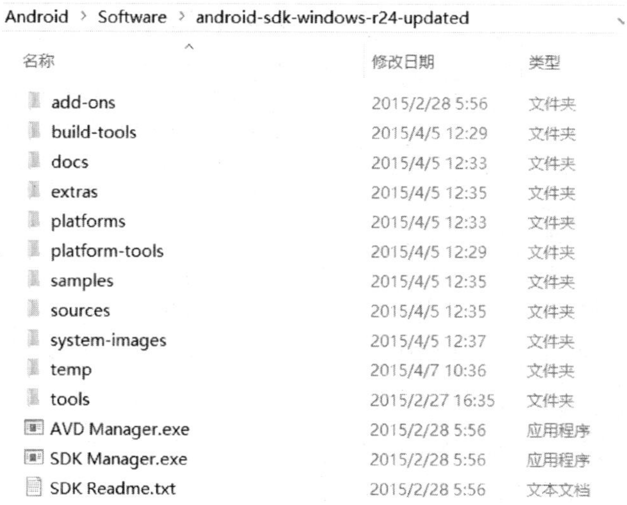

Android > Software > android-sdk-windows-r24-updated

名称	修改日期	类型
add-ons	2015/2/28 5:56	文件夹
build-tools	2015/4/5 12:29	文件夹
docs	2015/4/5 12:33	文件夹
extras	2015/4/5 12:35	文件夹
platforms	2015/4/5 12:33	文件夹
platform-tools	2015/4/5 12:29	文件夹
samples	2015/4/5 12:35	文件夹
sources	2015/4/5 12:35	文件夹
system-images	2015/4/5 12:37	文件夹
temp	2015/4/7 10:36	文件夹
tools	2015/2/27 16:35	文件夹
AVD Manager.exe	2015/2/28 5:56	应用程序
SDK Manager.exe	2015/2/28 5:56	应用程序
SDK Readme.txt	2015/2/28 5:56	文本文档

图 1-10　Android 的 SDK 目录

Android SDK 目录中的内容如下。

（1）add-ons：里面保存一些附加的库，第三方公司为 Android 平台开发的附加功能系统，比如 GoogleMaps 等（一开始此包内容为空）。

（2）build-tools：里面保存了构建项目时用到的工具，当新建 Android 项目时，会用到这个包。如果没有此包，新建的项目会报错。其中还包括一些编译的工具。总之这个包不能缺少。

（3）docs：其中保存了 SDK 文档，包括对各种控件、类的官方说明，可以在里面找到所有的开发文档。

（4）extras：该文件夹中存放了 Google 提供的 USB 驱动、Intel 提供的硬件加速等附件工具包。

（5）platform-tools：该文件夹中存放了 Android 平台的相关工具，比如 adb. exe、sqlite3. exe。

（6）tools：里面存放了大量的 Android 开发、调试的工具和 platform-tools 有些重复，都是与开发有关的工具。

（7）platforms：platforms 是 SDK 中最重要的文件夹，其中可以有许多不同版本的 SDK。

（8）system-images：这里面存放的是创建 Android 虚拟机时的镜像文件。

（9）sources：里面存放的是源码。

5. 在 Eclipse 中配置 Android SDK

Android SDK 下载完成以后，启动 Eclipse，单击菜单栏中 Window > Preferences 命令，设置 Android SDK 目录，本书中将目录设为“D:\2017\Android\Android Software\android-sdk-windows-r24-updated”。完成设置后，可以看到此下载包仅包含一个 Android 5. 1. 1 版本，如图 1-11 所示。

若要增加版本，则运行图 1-10 中 Android SDK 目录中的 SDK Manager. exe 文件，如图 1-12 所示，选中没有安装的项，也就是 Status 标识为 Not installed 的项目，单击下面的安装按钮，即可安装。

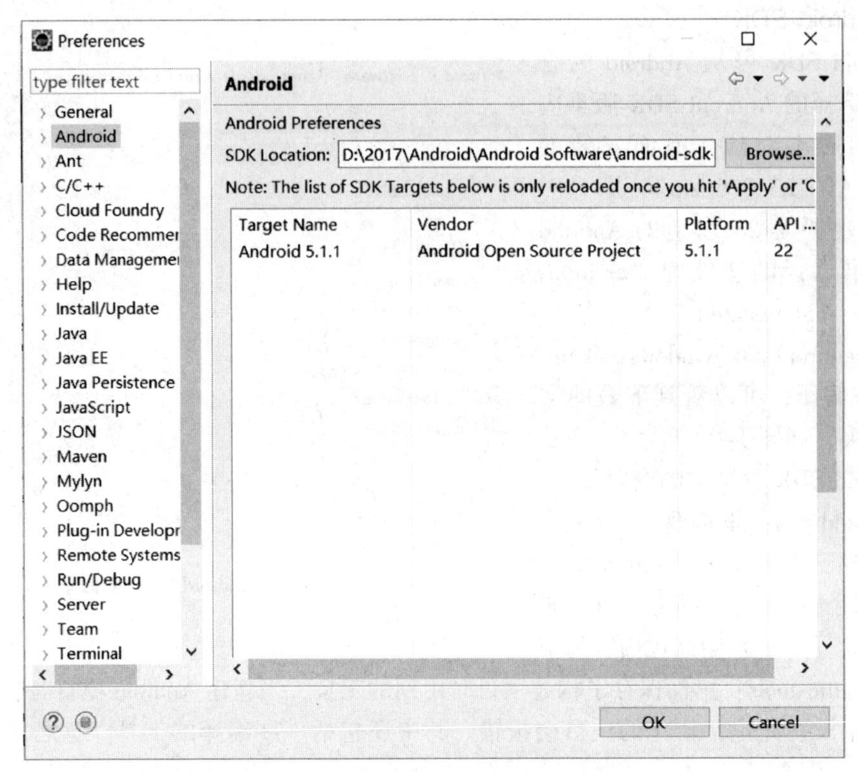

图 1-11　在 Eclipse 中配置 Android SDK

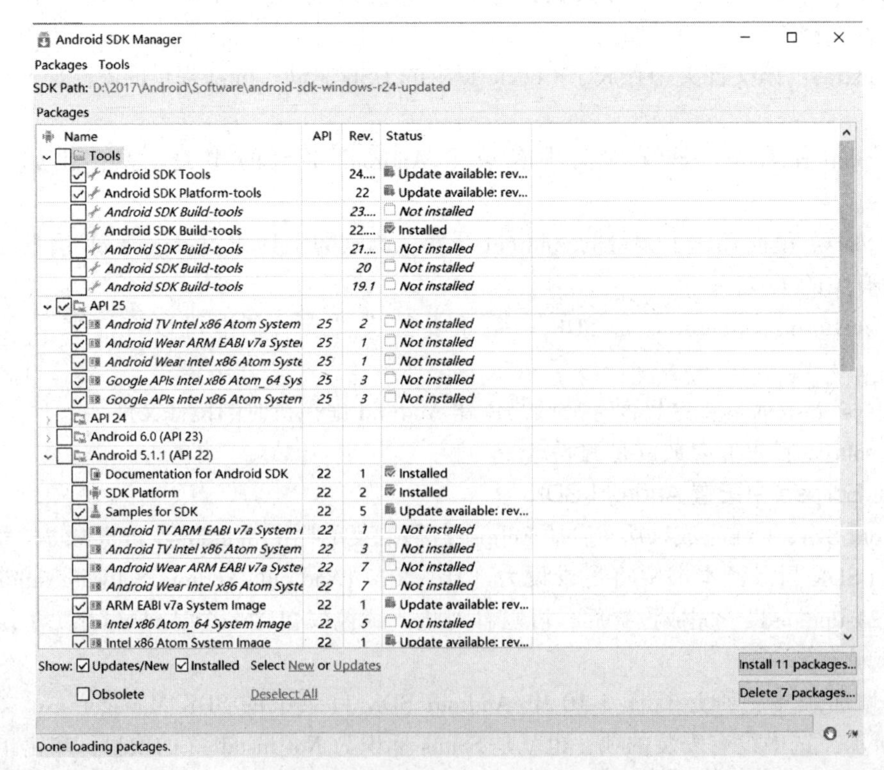

图 1-12　SDK Manager. exe 运行界面

6. 创建 Android 模拟器

Android SDK 自带一个移动模拟器，它是一个可以运行在计算机上的虚拟设备。该模拟器可以在不使用物理设备的情况下预览、开发和测试 Android 应用程序，一般开发程序先在模拟器运行，调试完成后下载到真机运行。

（1）建立 Android 虚拟机，在 Eclipse 主界面中单击菜单栏中的 Window > Android Virtual Device Manager 命令，如图 1-13 所示。

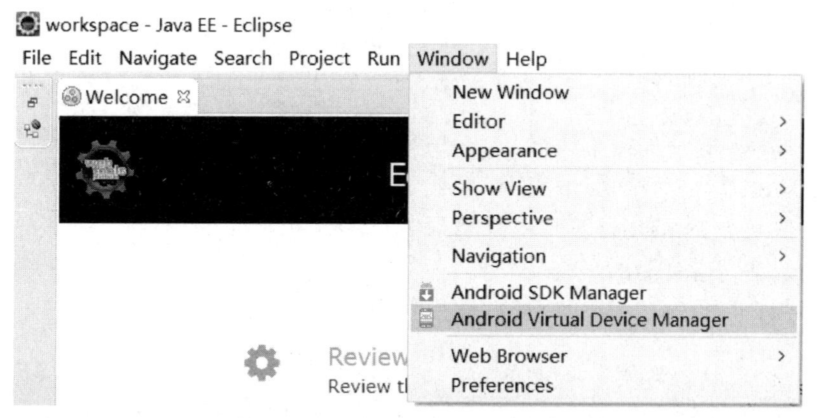

图 1-13 创建 Android 虚拟机

（2）弹出的窗口如图 1-14 所示。

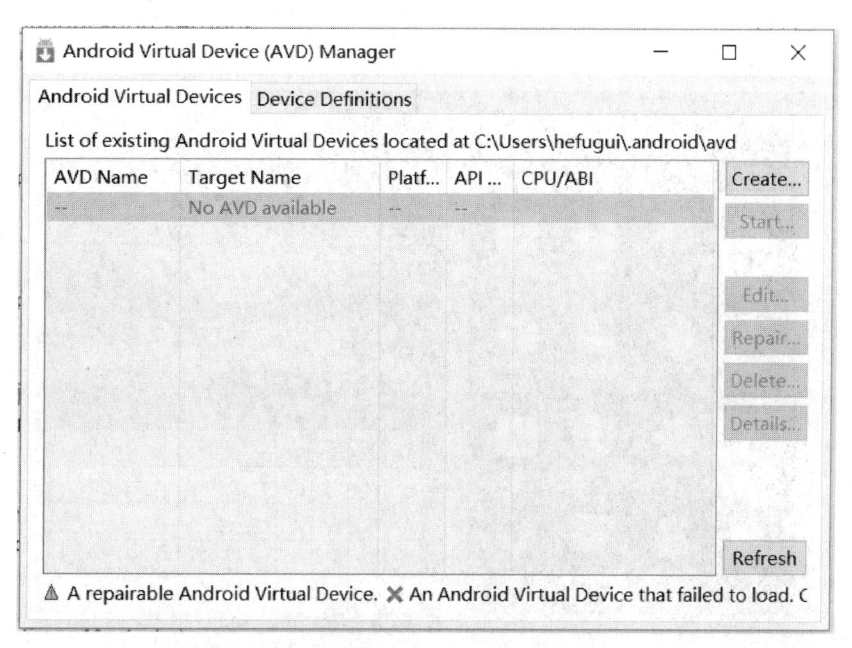

图 1-14 Android Virtual Device（AVD）Manager 窗口

（3）单击 Create 按钮，弹出虚拟机编辑窗口，进行配置，如图 1-15 所示。

（4）单击 OK 按钮，回到 Android Virtual Device（AVD）Manager 窗口，可以看到已经建立了一个虚拟机，如图 1-16 所示。

（5）单击 Start 按钮，启动虚拟机，如图 1-17 所示。

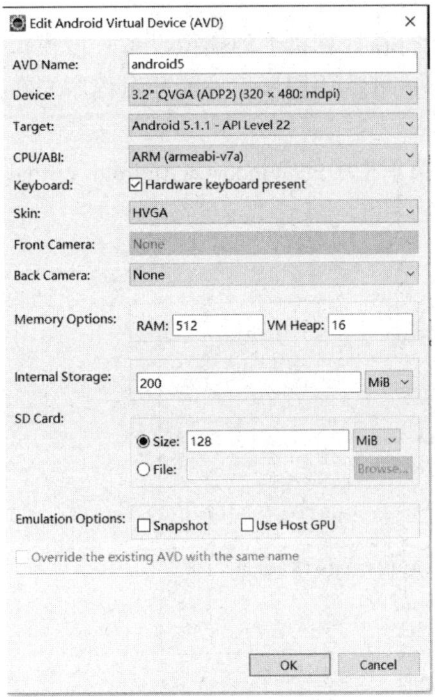

图 1-15　虚拟机编辑窗口

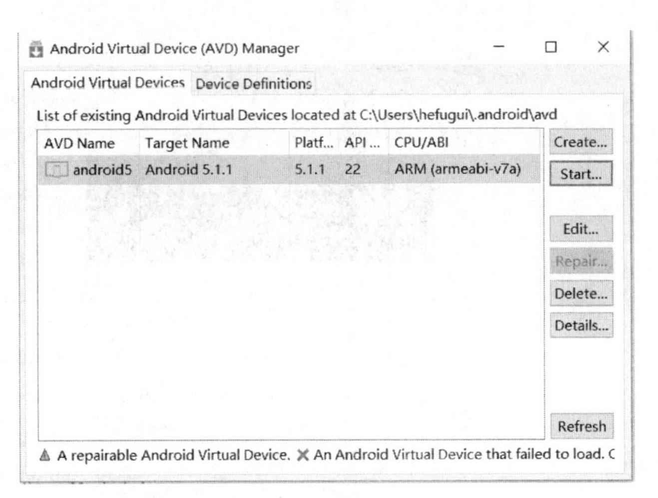

图 1-16　建立虚拟机后的 Android Virtual Device（AVD）Manager 窗口

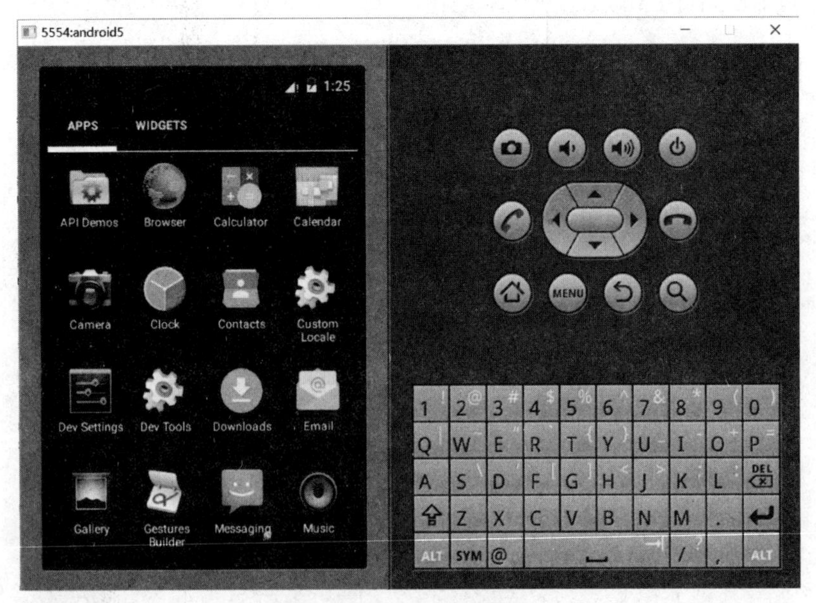

图 1-17　Android 虚拟机

1.2.2　项目结构

基于 Eclipse 的项目如图 1-18 所示，图的左侧是项目的目录，其含义如下。

➤ src：存放 Java 代码。

➢ gen：存放自动生成的文件，例如 R. java 表示 res 文件夹中对应资源的 id。

➢ assets：放置一些程序所需要的媒体文件。

➢ bin：工程的编译目录，存放编译时产生的一些临时文件和当前工程的 . apk 文件。

➢ libs：当前工程所依赖的 jar 包。

➢ res（resources）：资源文件。

■ drawable：存放程序所用的图片。

■ layout：存放 Android 的布局文件。

■ menu：存放 Android 的 OptionsMenu 菜单的布局。

■ values：应用程序所需要的数据，会在 R 文件中生成 id。

◆ strings. xml：存放 Android 字符串。

◆ dimens. xml：存放屏幕适配所用到的尺寸。

◆ style. xml：存放 Android 下显示的样式。

■ values-w820dp：表明这个目录下的资源所要求屏幕的最小宽度是 820dp。

■ values-v11：指定 3. 0 版本以上的手机显示的样式。

■ values-v14：指定 4. 0 版本以上的手机显示的样式。

➢ AndroidManifest. xml：Android 应用程序的配置文件，声明了 Android 里边的组件和相关配置信息以及添加的权限。

➢ proguard-project. txt：用于加密当前程序。

➢ project. properties：指定当前工程采用的开发工具包的版本。

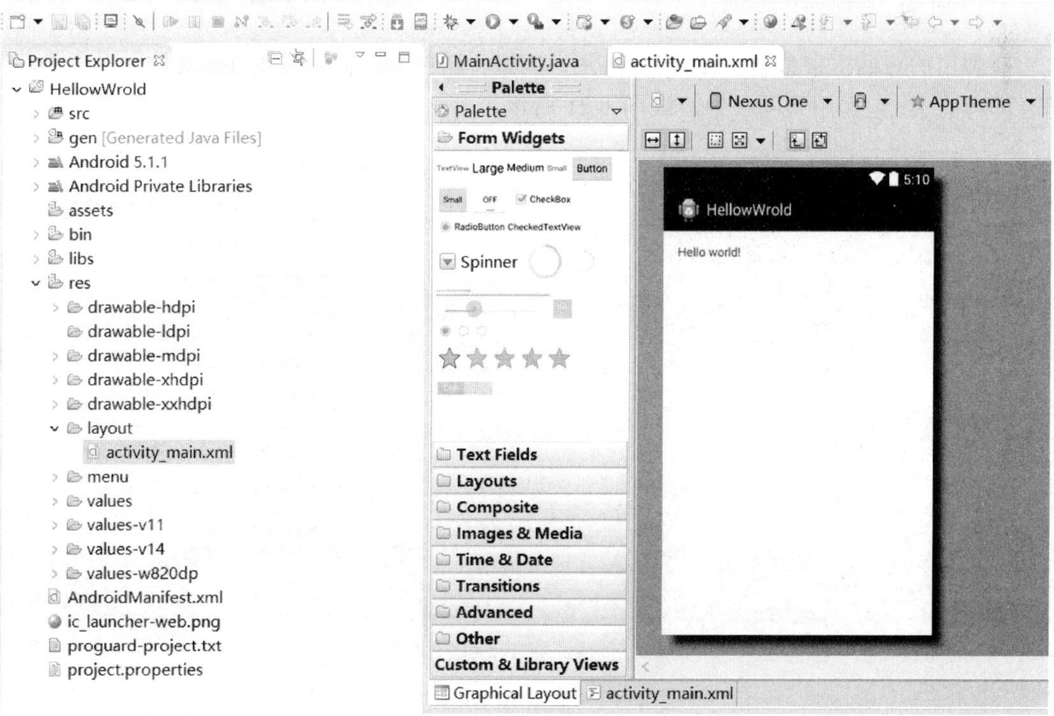

图 1-18　Eclipse 项目

1.2.3 使用第三方库

开发 Android 项目时会用到第三方提供的 Jar 包，通常情况下，首先下载第三方提供的 Jar 包到本地，然后复制到项目的 libs 目录中，如图 1-19 所示，但这样还不能使用，还需要引用库。选择项目，在右键菜单中选择 Build Path > Configure Build Path 命令，如图 1-20 所示。

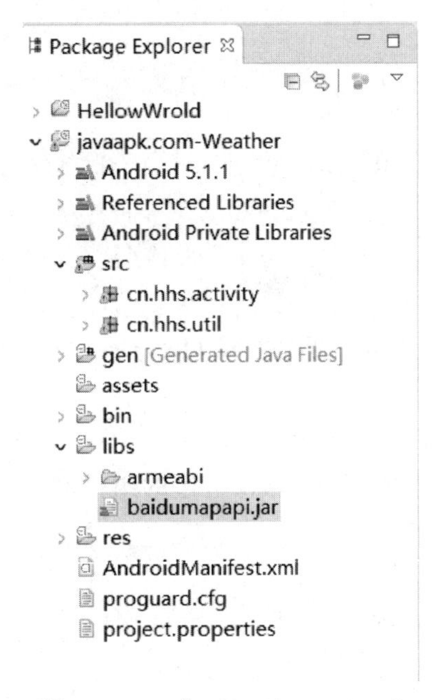

图 1-19　Jar 包“baidumapapi. jar”

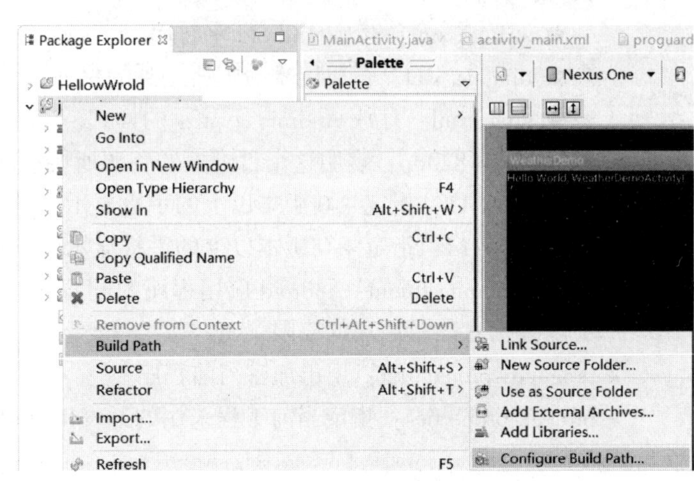

图 1-20　配置编译路径

在弹出的对话框中单击“Add JARS…”按钮，选择 libs 目录中的 Jar 文件，单击 OK 按钮，即可将 Jar 包引用到项目中，如图 1-21 所示。

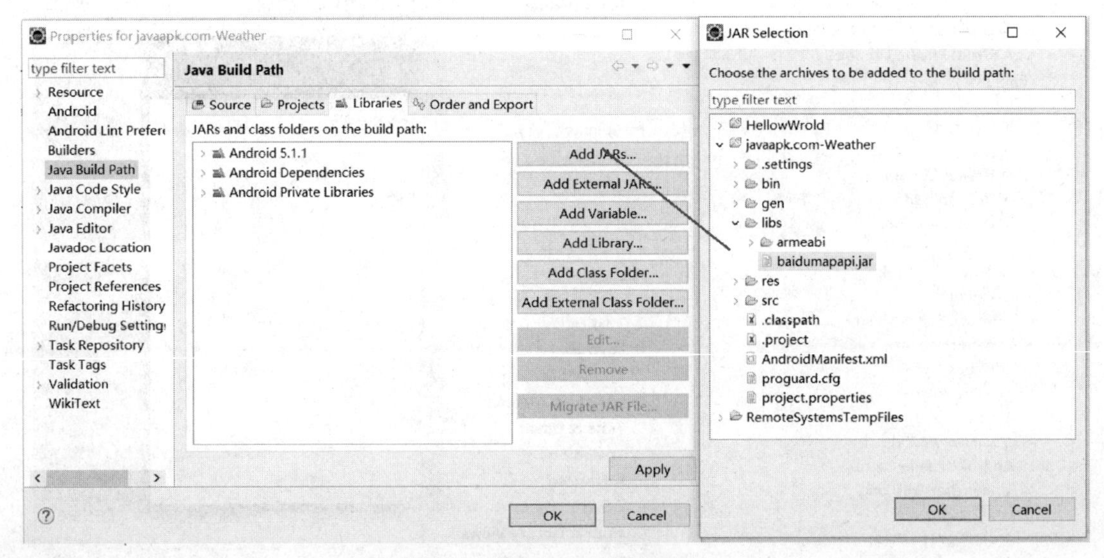

图 1-21　将 Jar 包引用到项目

1.2.4　项目的运行和调试

很多情况下，在 Eclipse 中调试 Android 应用程序，需要将程序的信息输出，这个功能是由 Android 的日志类 Log 实现的，Android 中的日志类 Log（android. util. log）提供了 5 个方法来供我们打印输出：Log. v( )、Log. d( )、Log. i( )、Log. w( ) 以及 Log. e( )，对应 VERBOSE、DEBUG、INFO、WARN、ERROR 这 5 个词语。

Log. v( )：这个方法用于打印那些最为琐碎的、意义最小的日志信息。对应的级别为 verbose，是 Android 日志里面级别最低的一种。

Log. d( )：这个方法用于打印一些调试信息，这些信息对于调试程序和分析问题应该有帮助。对应的级别是 debug，比 verbose 高一级。

Log. i( )：这个方法用于打印一些比较重要的数据，这些数据应该是您非常想了解的，可以帮助分析用户行为的那种数据。对应级别是 info，比 debug 高一级。

Log. w( )：这个方法用于打印一些警告信息，提示程序在这个地方可能存在潜在的风险，最好修复这些出现警告的地方。对应的级别为 warm，比 info 高一级。

Log. e( )：这个方法用于打印程序中的错误信息，比如程序中的 catch 语句的错误信息。对应的级别是 error，比 warn 高一级。

在 Eclipse 中调试 Android 应用程序的步骤如下。

（1）定位调试代码。为了演示调试过程，在前面的 HelloWorld 项目增加两行代码。

```
protected void onCreate(Bundle savedInstanceState)
{
    super.onCreate(savedInstanceState);
    setContentView(R.layout.activity_main);
    int i =6;
    Log.d (" MainActivity", " onCreate execute");
}
```

（2）设置断点。在设置断点的代码处，单击鼠标右键，选择 Toggle Breakpoint 命令，设置断点，如图 1-22 所示。

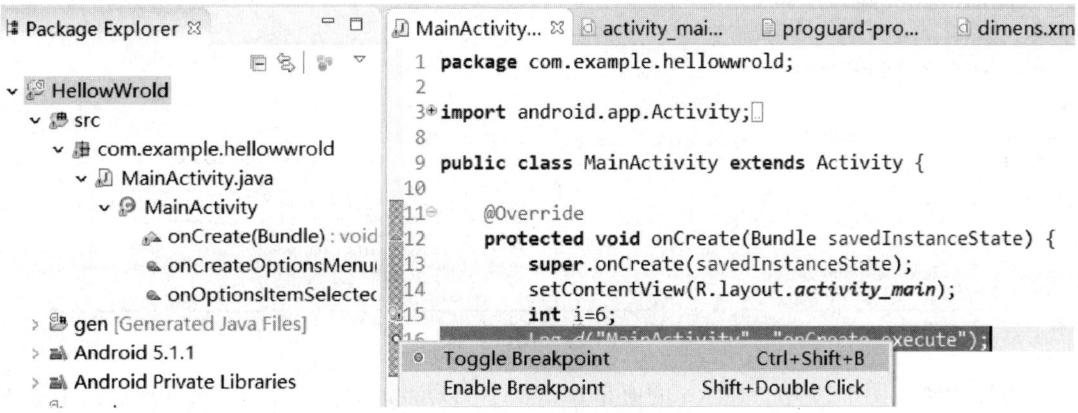

图 1-22　设置断点

（3）在 Eclipse 工具主界面的工具栏中，单击调试按钮，选择 HelloWorld 项目，启动调试，如图 1-23 所示。

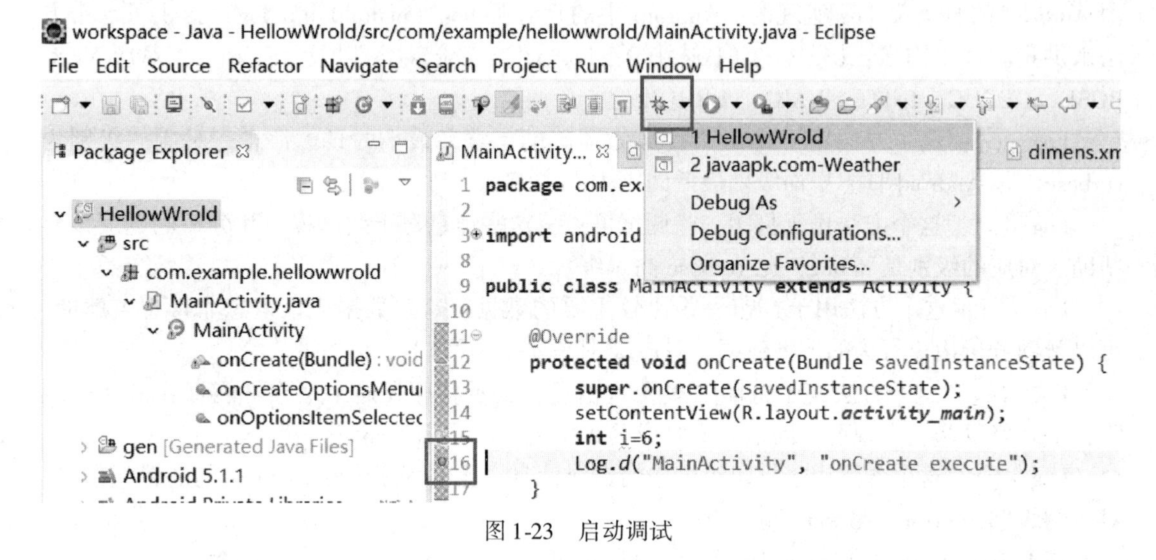

图 1-23　启动调试

（4）启动调试后，下载到真机，开始运行，运行到断点时停下来，如图 1-24 所示。

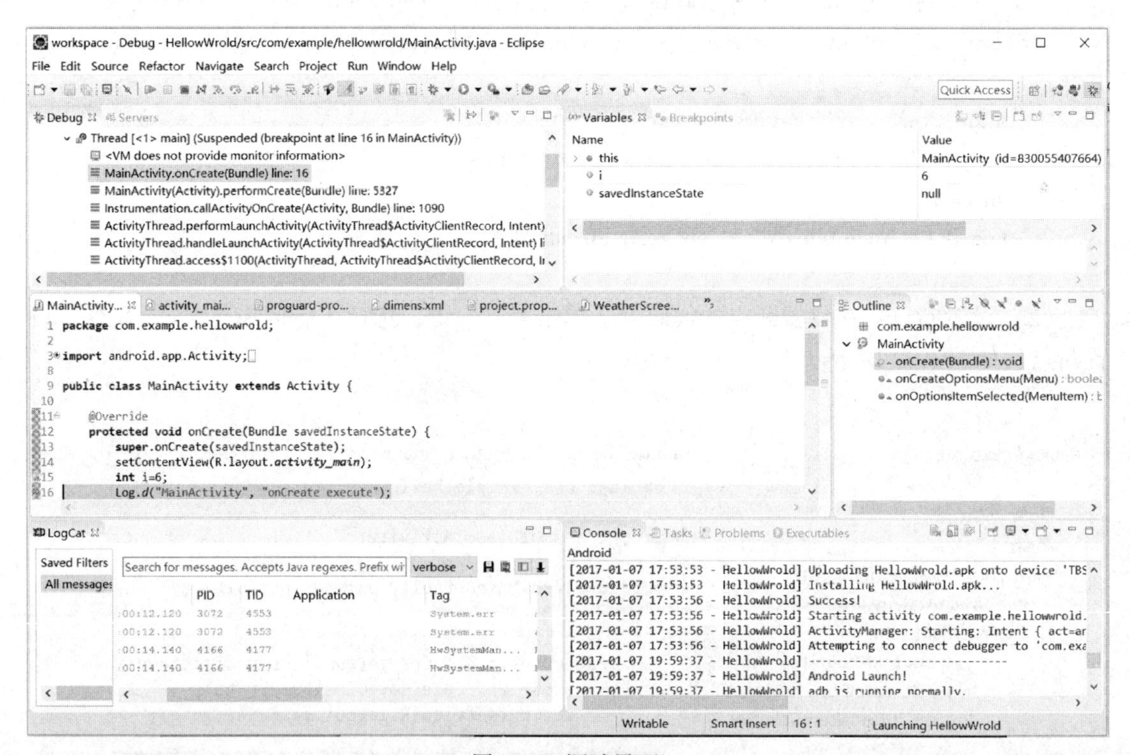

图 1-24　调试界面

可以看到在图 1-24 中，显示 i 的值是 6，然后可以继续运行，如果后面有断点，则继续调试。

1.3 基于 Android Studio 的开发环境

Android Studio 是 Google 推出的一个全新 Android 开发环境，提供了集成的 Android 开发工具用于开发和调试。Android Studio 能让开发者变得“更快、更具生产力”，是扩展开发平台 Eclipse 的替代平台。Android Studio 的开发源自集成开发环境 IntelliJ IDEA，按照 Google 的预想，这就意味着它是一套全功能开发环境。Google 打算将云消息以及其他服务整合到 Android Studio 中，它将成为一个开发中心，Android 开发者可以在这里开发新应用、更新旧应用。

Android Studio 在开发 Android 程序方面远比 Eclipse 强大和方便得多，不过 Android Studio 早期的版本不是非常稳定，而如今，其已经推出了 2.3 版本，稳定性完全不是问题，普及度也远超 Eclipse，所以未来 Android 开发会转向 Android Studio。

1.3.1 Android Studio 的特点

关注 Android Studio 的开发者越来越多，下面介绍 Android Studio 的优点。

（1）Google 开发

毫无疑问，这是它的最大优势，Android Studio 是 Google 推出，专门为 Android “量身定做”的，是 Google 大力支持的一款基于 IntelliJ IDEA 改造的 IDE，Google 的工程师团队肯定会不断完善这款软件，这个应该能说明为什么它是 Android 的未来。

（2）速度更快

Eclipse 的启动速度、响应速度、内存占用一直被诟病，相信大家应该深有体会，而且经常遇到卡死状态。Studio 不管在哪方面都全面领先 Eclipse。

（3）UI 更漂亮

I/O 上演示的那款黑色主题非常酷，采用极客范，Studio 自带的 Dracula 主题的炫酷黑界面实在是显得高大上，相比而言 Eclipse 的黑色主题差太多了。

（4）更加智能

提示补全对于开发来说意义重大，Studio 则更加智能，能够智能保存，从此再也不用常按 Ctrl + S 键了。熟悉 Studio 以后，效率会大大提升。

（5）整合了 Gradle 构建工具

Gradle 是一个新的构建工具，自 Studio 亮相之时就支持 Gradle，可以说 Gradle 集合了 Ant 和 Maven 的优点，不管是配置、编译、打包都非常便捷。

（6）强大的 UI 编辑器

Android Studio 的编辑器非常智能，除了吸收 Eclipse + ADT 的优点之外，还自带了多设备的实时预览，这对 Android 开发者来说简直是“神器”啊。

（7）内置终端

Studio 内置终端，这对于习惯命令行操作的人来说简直是福音，再也不用来回切换，一个 Studio 全部搞定。

（8）更完善的插件系统

Studio 支持各种插件，如 Git、Markdown、Gradle 等，想要什么插件，直接搜索下载即可。

（9）完美整合版本控制系统

安装的时候自带了如 GitHub、Git、SVN 等流行的版本控制系统。

1.3.2 搭建 Android Studio 应用开发环境

搭建 Android 应用开发环境需要三个工具：JDK、Android SDK 和 Android Studio，Android Studio 2.3 内部已经集成了 Java 执行环境，所以安装或不安装 JDK 都可以，安装 JDK 的方法与前面相同，Android Studio 安装完成后已经包含了 Android SDK 7.0，如需要其他版本，可以在 Android Studio 里下载，本节的末尾将介绍。

Android Studio 的下载页面里收集整理了 Android 开发所需的 Android SDK、开发中用到的工具、Android 开发教程、Android 设计规范、免费的设计素材等，网站中的 Android SDK Tools 界面如图 1-25 所示。

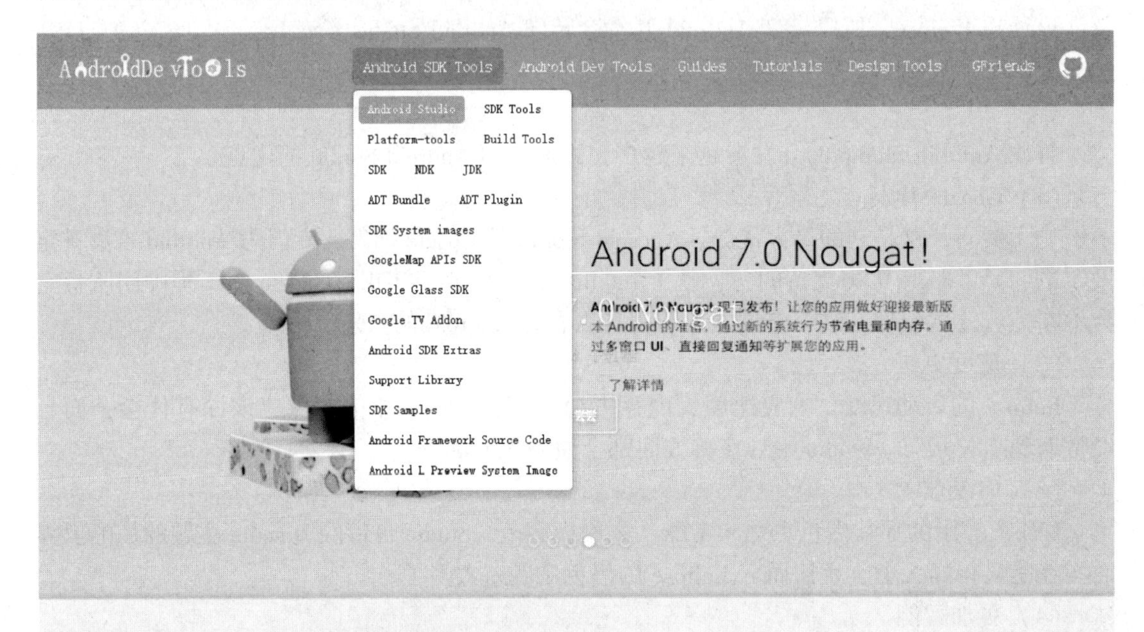

图 1-25　网站中的 Android SDK Tools 界面

网站中的 Android Dev Tools 菜单如图 1-26 所示。

从上面的网站可以下载 Android Studio 的各个版本，Android Studio 目前最新的稳定版本为 Android Studio 2.3，提供的平台有 Linux、Mac OS、Window 三种。

下面以 Windows 平台为例，说明环境搭建过程。

（1）执行下载的“android-studio-bundle-162.3871768-windows.exe”文件，即 Android Studio 2.3 安装文件，此文件中包含 Android SDK，如图 1-27 所示。

（2）选择安装组件，如图 1-28 所示。

图 1-26　网站中的 Android Dev Tools 菜单

图 1-27　开始安装

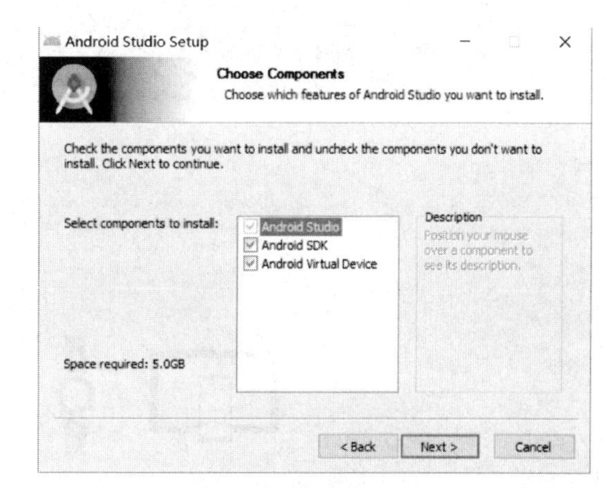

图 1-28　选择安装组件

（3）选择安装目录，如图 1-29 所示。完成安装，如图 1-30 所示。

（4）单击 Finish 按钮，启动 Android Studio，第一次启动时，会出现一些向导信息，以后启动将不再出现，如图 1-31 所示。在此可设置是否选择前面安装的版本的配置，本节选择不安装。

（5）单击 OK 按钮，进入 Android Studio 的配置界面，如图 1-32 所示。

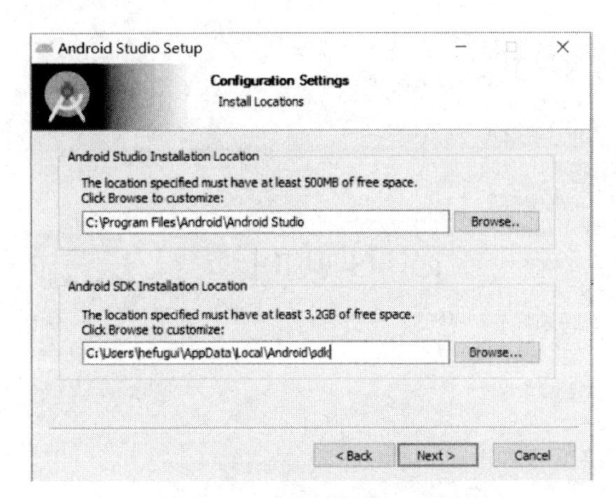

图 1-29　选择安装目录

图 1-30　安装完成

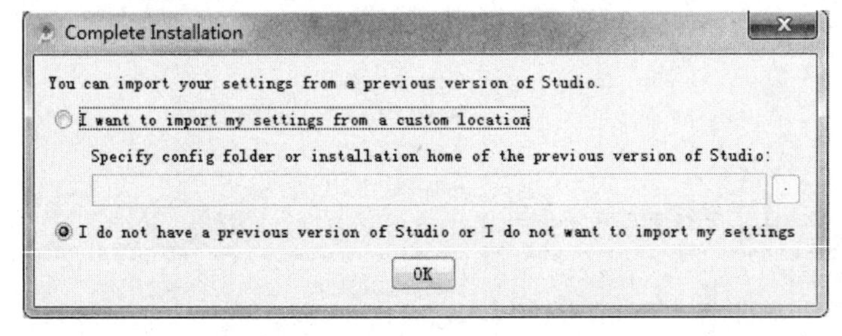

图 1-31　设置是否选择前面的配置

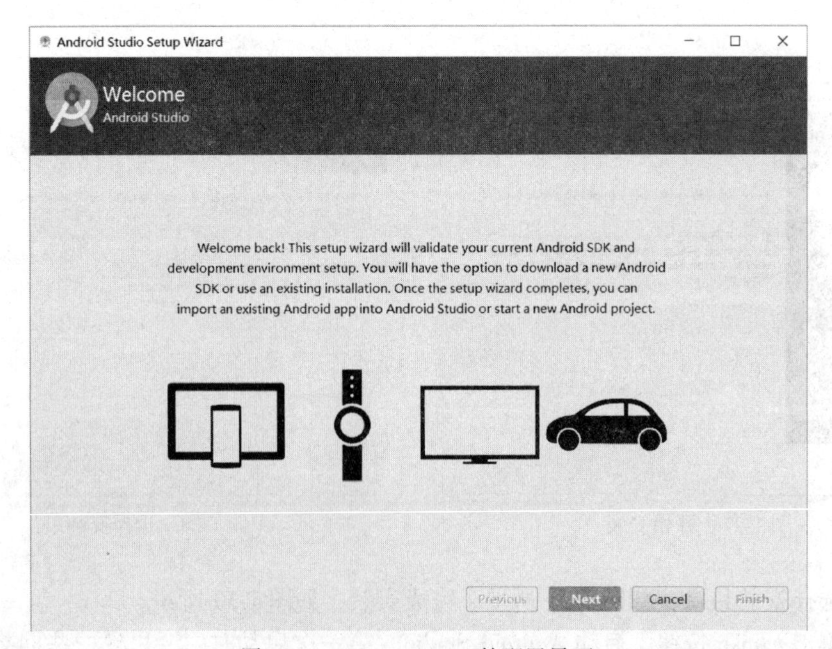

图 1-32　Android Studio 的配置界面

（6）单击 Next 按钮，开始进行具体的配置。在 Install Type 界面上有 Standard 和 Custom 两种选项，即标准和自定义类型，这里选中 Standard 选项，如图 1-33 所示。之后是验证设

置，如图 1-34 所示。

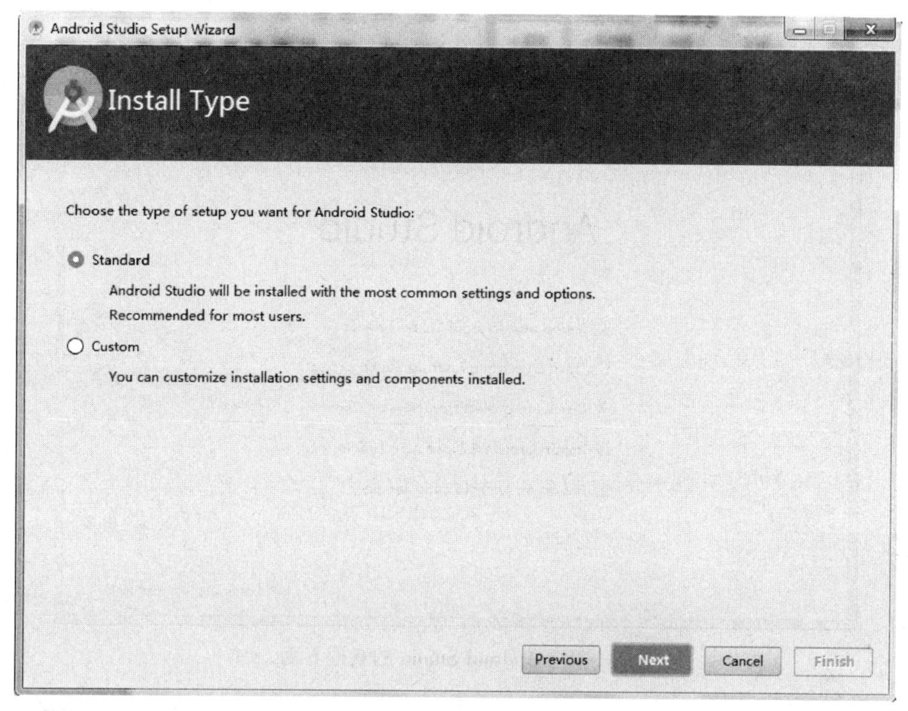

图 1-33　选中安装类型

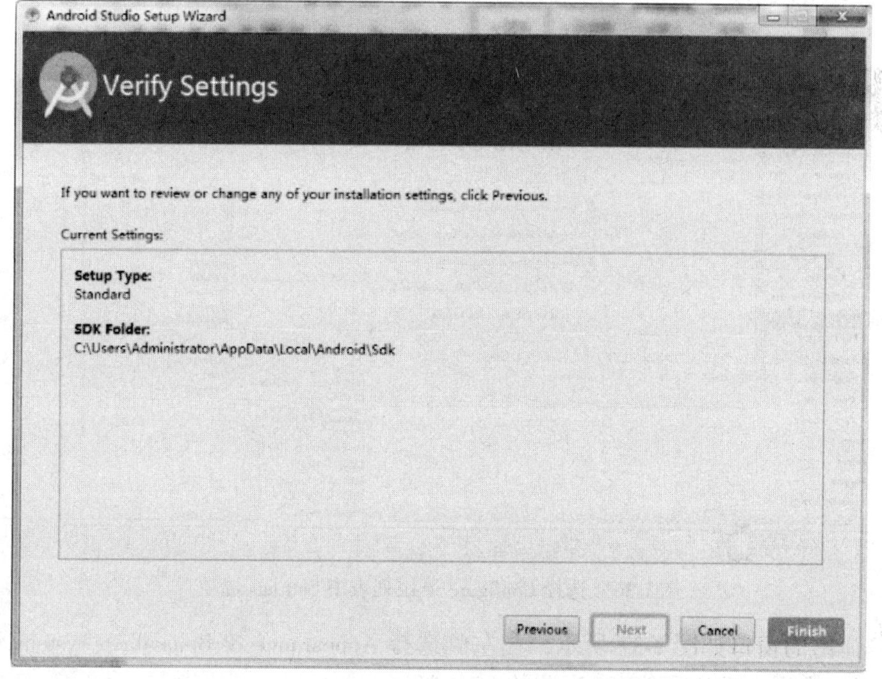

图 1-34　验证设置

（7）弹出 Android Studio 的欢迎界面，如图 1-35 所示，单击 Configure 按钮，打开下拉列表，选择 Settings 选项，如图 1-36 所示。

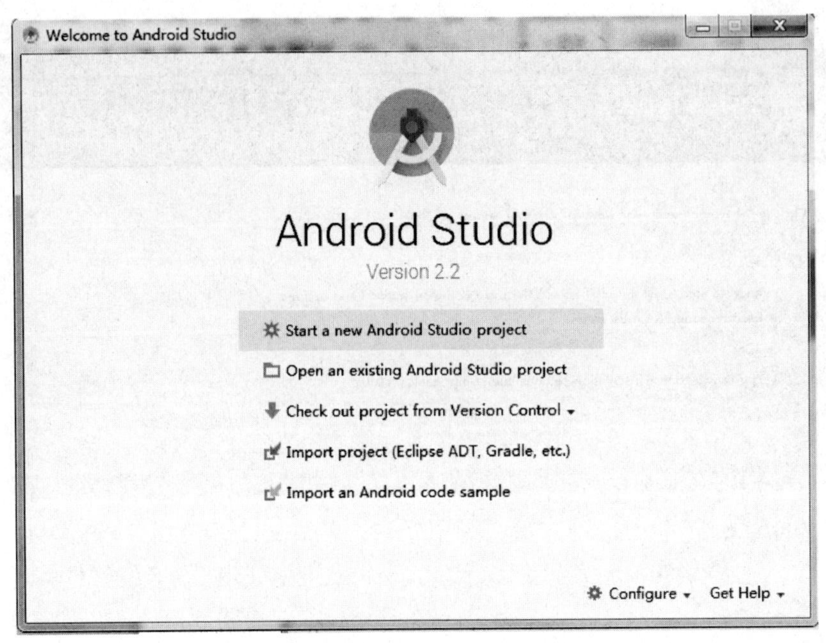

图 1-35　Android Studio 的欢迎界面

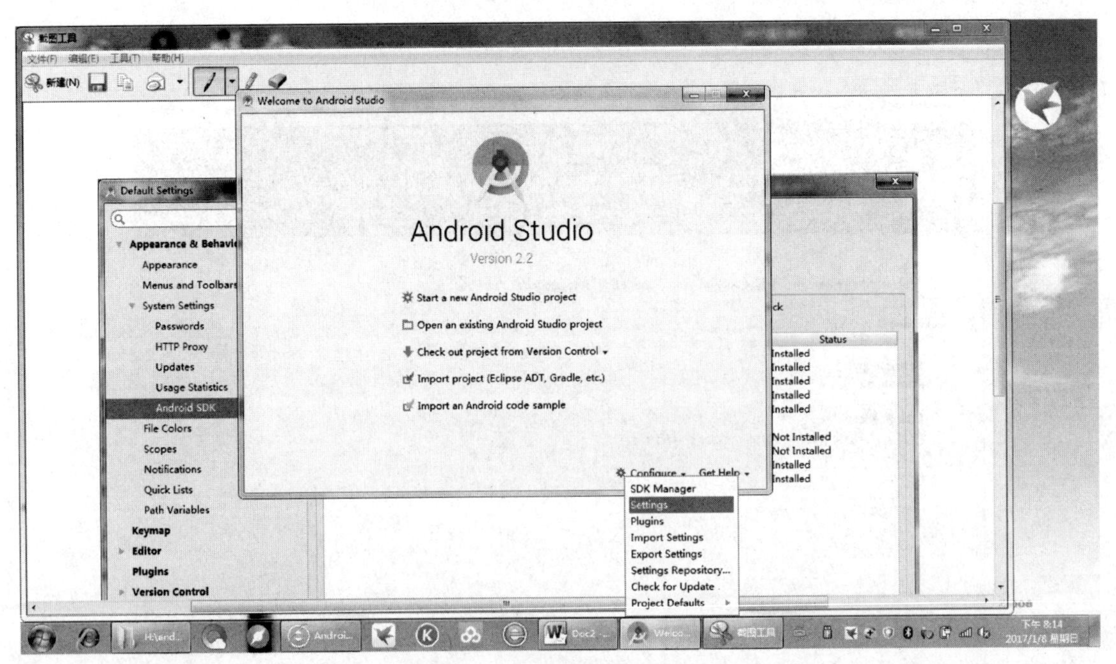

图 1-36　选择 Configure 下拉列表中 Settings 选项

（8）弹出的窗口如图 1-37 所示，在左侧选择 Appearance & Behavior > System Settings > Android SDK 选项，可以看到已经安装的 SDK 版本。

（9）单击 OK 按钮，开始创建新项目，此时会打开一个创建新项目的界面，如图 1-38 所示。

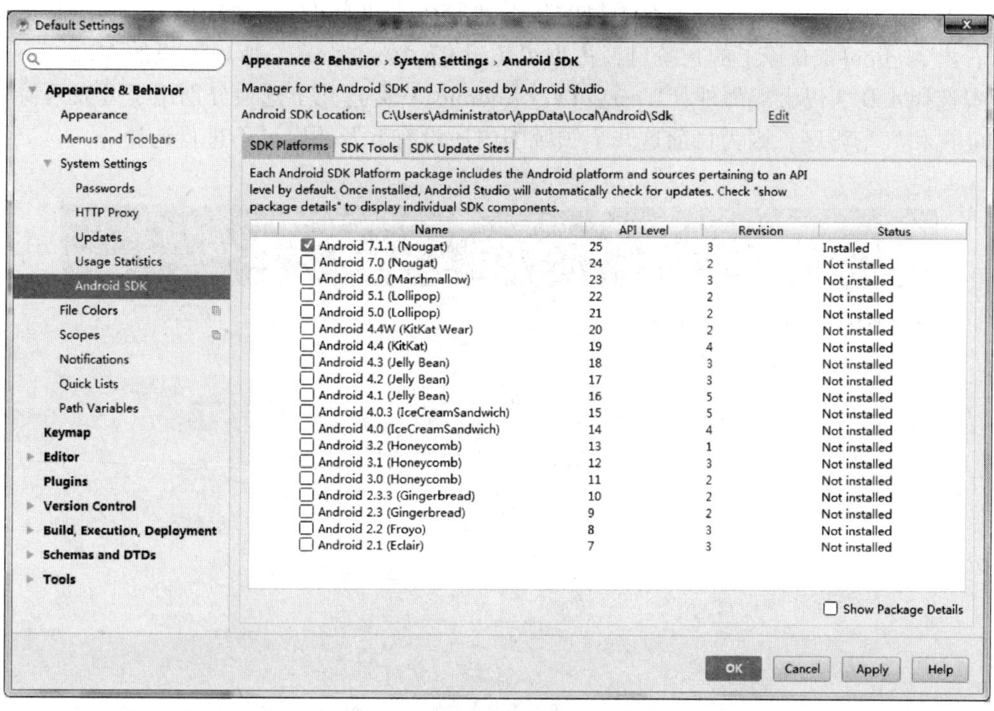

图 1-37　已经安装的 SDK 版本

Create New Project
New Project
Android Studio
Configure your new project
Application name: HelloWorld
Company Domain: administrator.example.com
Package name: com.example.administrator.helloworld
Edit
Include C++ Support
Project location: C:\Users\Administrator\AndroidStudioProjects\HelloWorld
Previous
Next
Cancel
Finish

图 1-38　创建新项目

（10）单击 Next 按钮，可以对项目的最低兼容版本进行设置，如图 1-39 所示。从图中可以看出 Android 4.0 以上的系统已经占据了超过 97.4% 的份额，所以这里的 Minimum SDK 的版本选择 4.0.3 以上。另外，Wear、TV、Android Auto 这几个选项分别用于开发可穿戴设备、电视和汽车程序，因为目前这几个领域国内还没有普及所以可不进行选择。

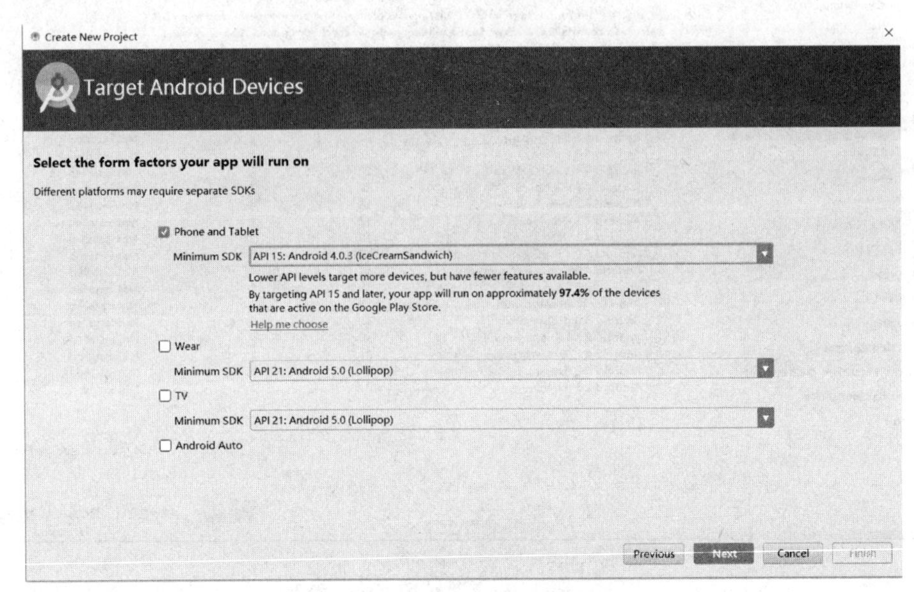

图 1-39　设置项目的最低兼容版本

（11）单击 Next 按钮，进入选择模板的界面，如图 1-40 所示，这里选择空模板。

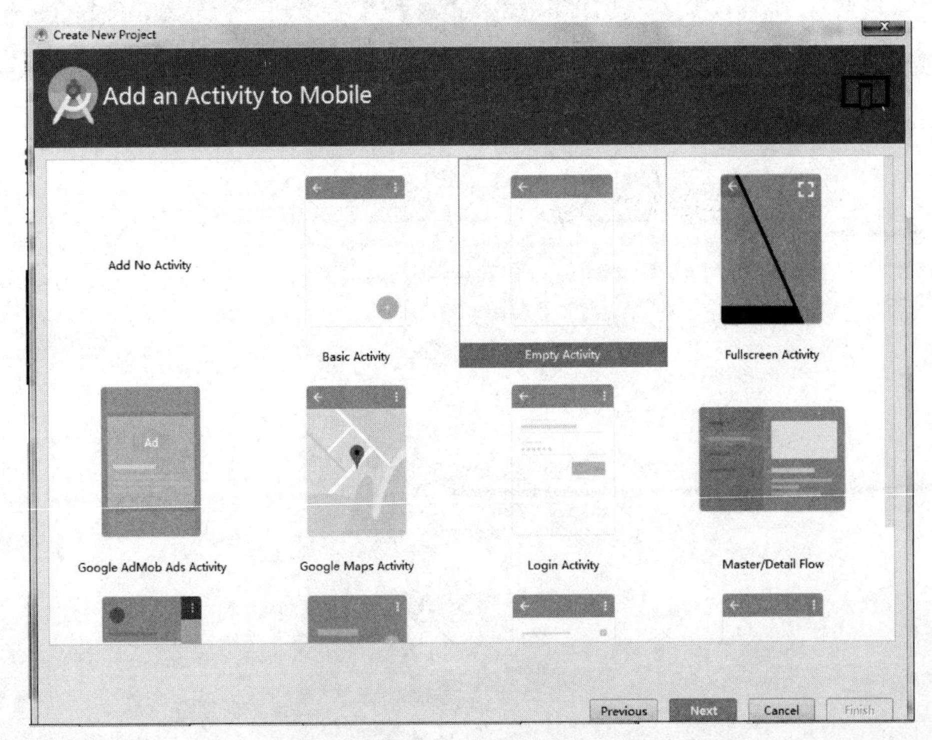

图 1-40　选择空模板

（12）单击 Next 按钮，填写活动的名称和布局名，如图 1-41 所示。

（13）单击 Finish 按钮，项目即创建成功，如图 1-42 所示。

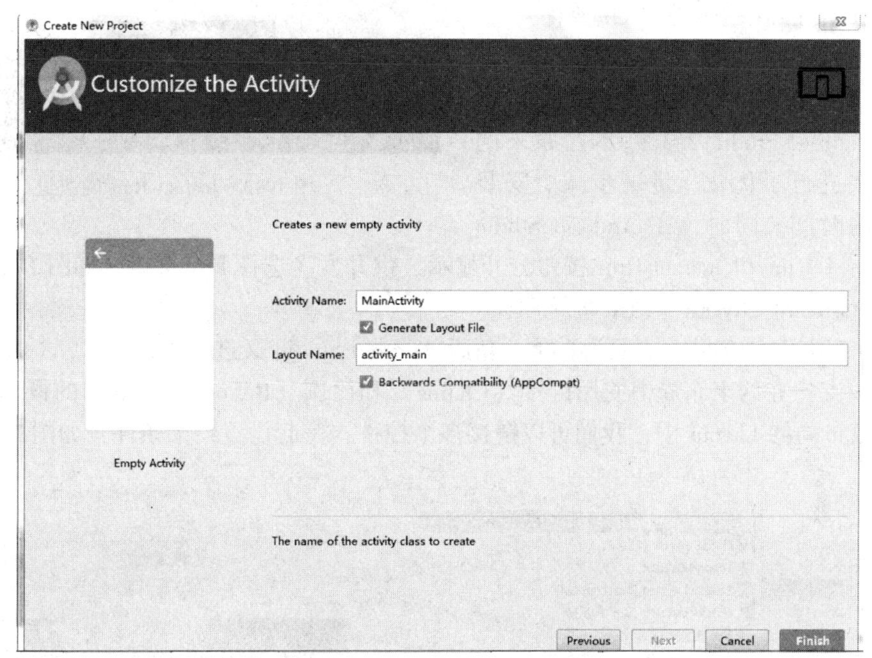

图 1-41　为活动和布局命名

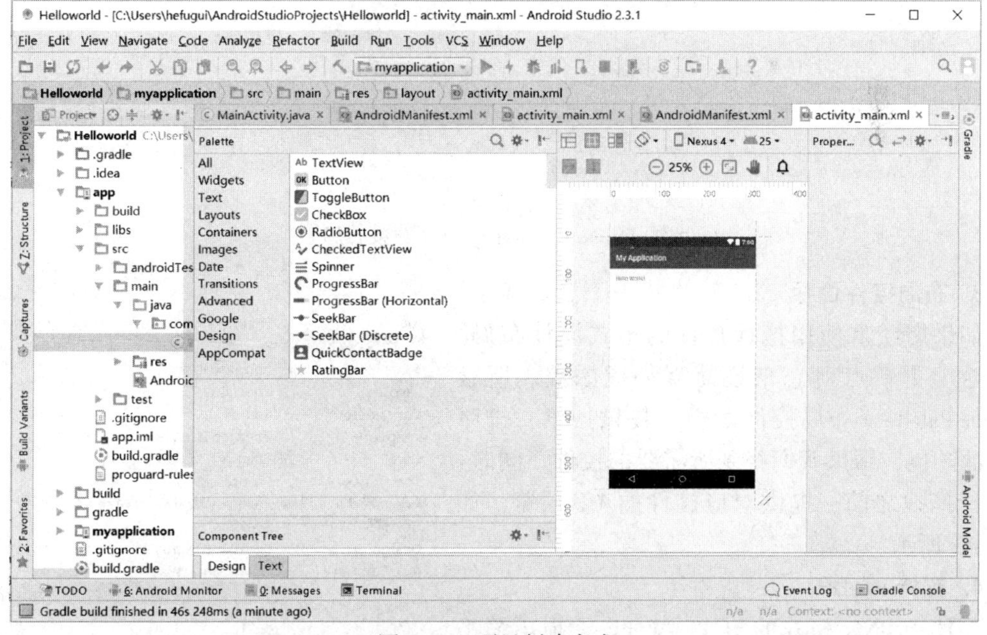

图 1-42　项目创建完成

1.3.3　Android Studio 2.3 的新特性

作为 Google 官方发布并维护的 IDE，被全球数以百万计的 Android 开发者钟爱并使用的开发工具，在 2017 年 3 月 2 日，Android Studio 2.3 正式版发布了，该版本包含了一些新特性，包括对 WebP 支持的更新、新布局 ConstraintLayout 等。

1. Instant Run 的改进和 UI 变化

Instant Run 基本能够解决中小型项目的编译缓慢问题。作为 Google 重点关注的一个功能，Android Studio 2.3 版本在原来的基础上再次进行了优化，进一步减少安装替换代码的时间。同时，在 Android Studio 的导航栏上将 Run 和 Instant Run 按钮分开显示，供开发者选择调试策略，如图 1-43 所示。

图 1-43　Instant Run 的改进

2. Constraint Layout（约束布局）

Google 官方对此布局方式尤为看重，新版 Android Studio 又进行了改进。2.3 版本的 Android Studio 支持在约束布局中使用链接（Chains）和比例（Ratios）。Chains 的概念大致是，在使用约束布局的 Layout 中，我们可以链接多个控件，一起设置约束条件，如图 1-44 所示。

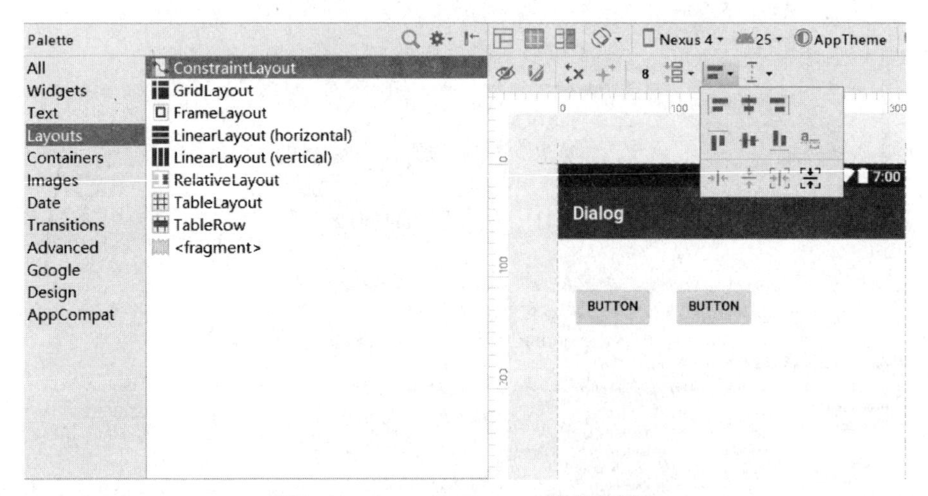

图 1-44　ConstraintLayout（约束布局）

3. 布局控件面板

如果您经常使用拖拽控件的形式设计布局，那么这个更新点对您来说简直如虎添翼。新版 Widget Palette（布局控件面板）提供搜索、排序和过滤功能，帮助我们找到所需要的控件。同时，在选择拖拽之前，提供对应控件的 UI 预览，如图 1-45 所示。

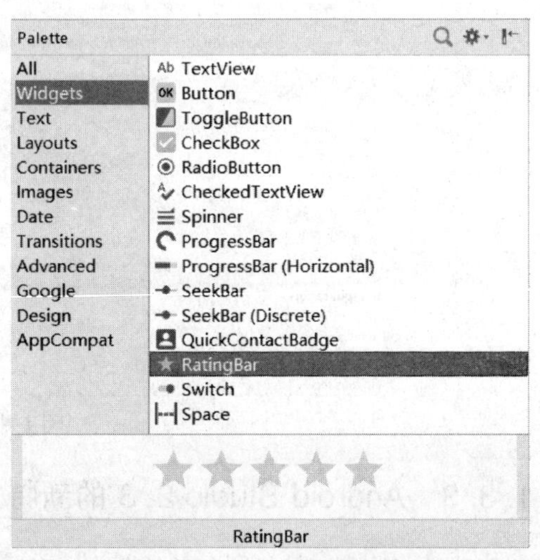

图 1-45　布局控件面板

4. 支持 WebP

相比于 PNG 格式的图片，WebP 无损压缩格式能够减少 25% 的文件大小。在 Android Studio 2.3 版本中，可以自由转换图片格式，如将 PNG 转为 WebP 格式，或者是将 WebP 转为 PNG 格式，同时您还可以通过控制质量调整文件大小。

5. Material Icon 库

新版的 Material Icon 矢量图标库支持搜索过滤功能，同时为每个 Icon 设置相应的 Label，以供搜索，这是一个非常人性化的改进。

6. Lint 基准线

Android Lint 是优化项目必不可少的一个工具，使用时可能会遇到这样的问题：执行 Lint 命令，该工具会自动遍历所有的目标文件，并将不符规范的问题分类列举出来，然后我们一一处理，但如果没有处理完，再次 Lint 时，将再次从头开始解决问题，新旧问题融合到一起。如果只想处理新的问题的话，将很难从中筛选出来。而基准线（BaseLine）的出现能解您燃眉之急。为每一次执行 Lint 设置一个 BaseLine，让你只想解决新问题的想法成为可能。

在 Android Studio 2.3 的主界面选中菜单栏中 Analyze > Inspect Code 命令，在弹出的窗口选中检查范围，如图 1-46 所示。

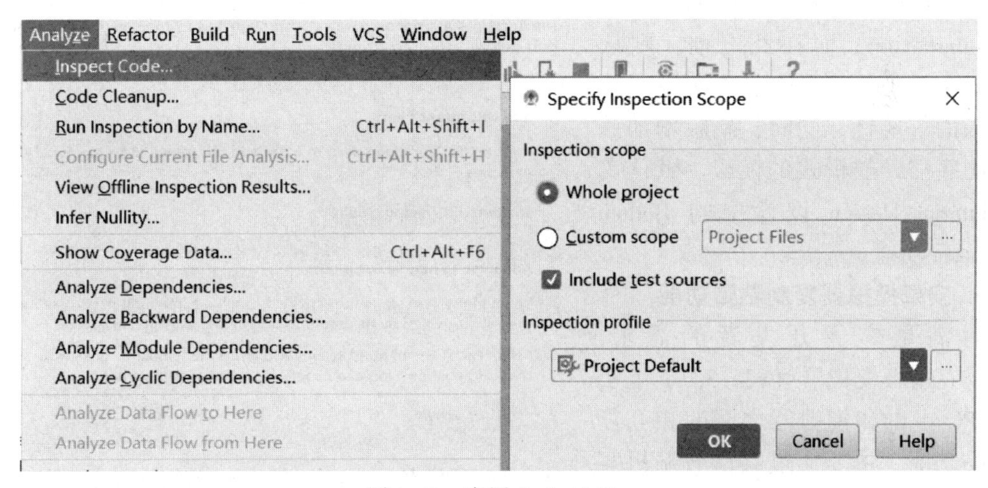

图 1-46　应用 Android Lint

在主界面的下方即显示 Android Lint 的检查结果，如图 1-47 所示。

图 1-47　Android Lint 窗口

7. App Links 助手

在 2015 年 I/O 大会上，Google 正式宣布 Android M 系统支持 App 链接，在 web url 到

native app 之间建立关联通道。比如，单击手机短信中的 url 链接和浏览器中的某个 url 就可以打开支持 App Links 的相应 App，这是一个非常赞的设计。要实现这个功能，需要在项目中添加相应的设置，修改 Manifest 文件等。新版本的开发工具提供了可视化的工具帮助我们进行这些设置。在 Android Studio 2.3 的主界面选中菜单栏中 Tools > App Links Assistant 命令，如图 1-48 所示。

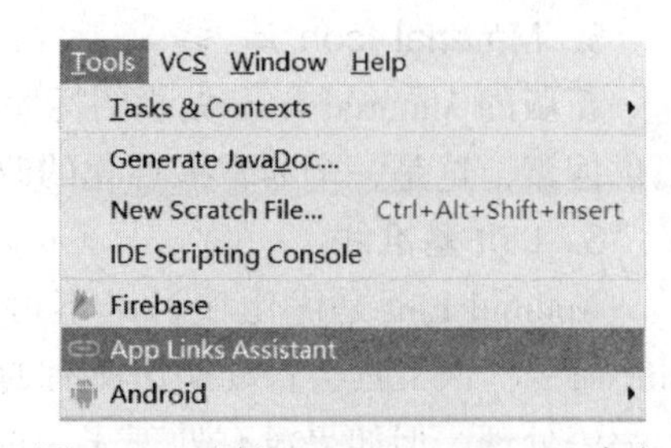

图 1-48　选择 App Links Assistant 命令

弹出的 App Links Assistant 窗口，如图 1-49 所示。

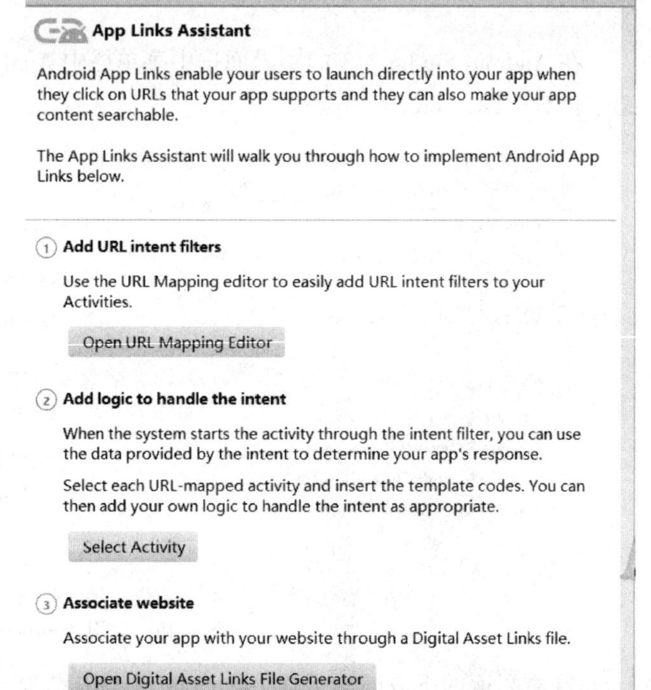

图 1-49　App Links Assistant 窗口

8. 模版更新

从 Android Studio 2.3 版本开始，新建项目时用到的所有模板默认使用 ConstraintLayout，而在此之前，默认使用 RelativeLayout。这也再次说明约束布局的重要性。同时，新版增加一个新的底部导航模式的模板，默认实现 Material Design 设计中的 Bottom Navigation 功能。

9. 安卓模拟器复制粘贴功能

为满足广大开发者的需求，Google 在新版模拟器（v25.3.1）上实现了 PC 主机和模拟器之间的相互复制粘贴功能，主要通过共享剪贴板实现。需要注意的是，Copy & Paste 功能仅在 x86 Google API Emulator、API Level 19（Android 4.4-Kitkat）和更高版本中起到作用。

1.3.4　安装 Android Studio 新插件

Android Studio 是一个功能全面的开发环境，装备了为各种设备——从智能手表到汽车——开发Android 应用程序所需要的所有功能。Android Studio 还提供了对第三方插件的支持，用好 Android Studio 插件，能大幅度减少我们的工作量。

Android Studio 2.3 已经预装了一些插件。主界面中选择菜单栏中 Files > Settings 命令，选中其中的插件（Plugins）选项，可以看到已经安装的插件，如图 1-50 所示。

JakeWharton 的 butterknife 帮我们有效解决了 findViewById 及各种 view 的监听事件泛滥的问题，极大简化了代码，如果配合使用 avast 的 android-butterknife-zelezny 插件，则可以一键注解所有 view，极大提高了编码效率。

下面以 butterknife 插件为例说明插件的使用过程。

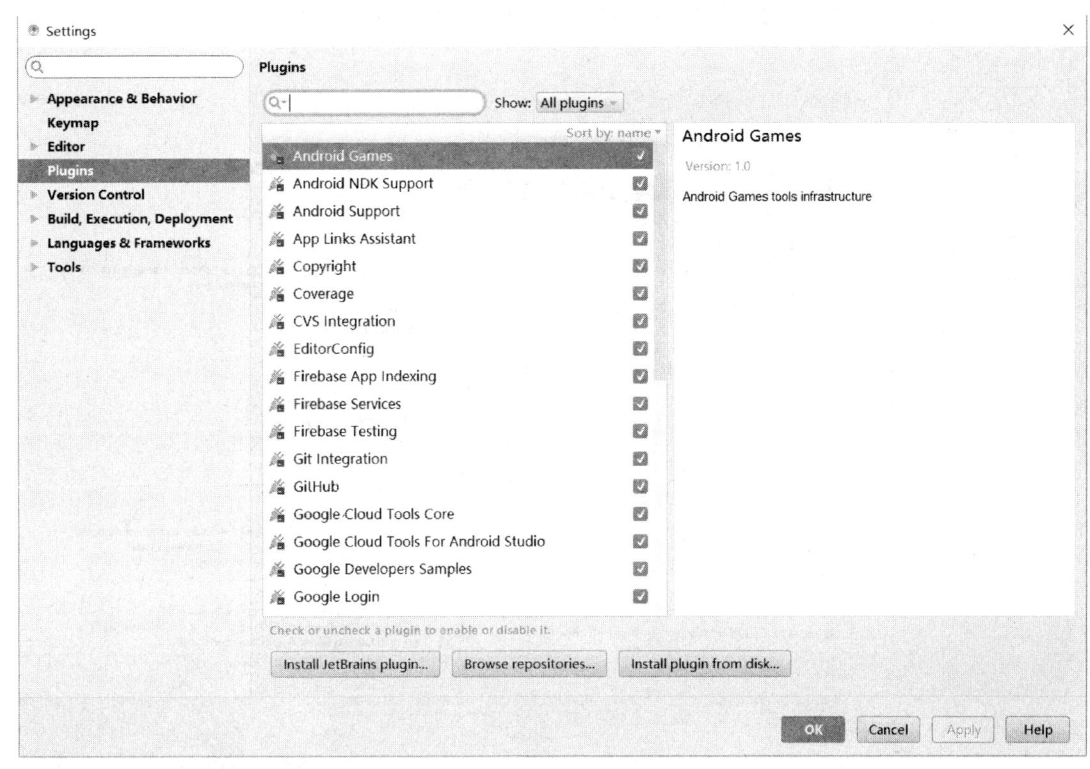

图 1-50　插件设置窗口

（1）下载 butterknife 插件，下载地址读者可行查询，这里不再赘述。然后在如图 1-50 所示的窗口中，单击下方 Install plugin from disk 按钮，弹出的窗口如图 1-51 所示，在其中选择下载的插件。

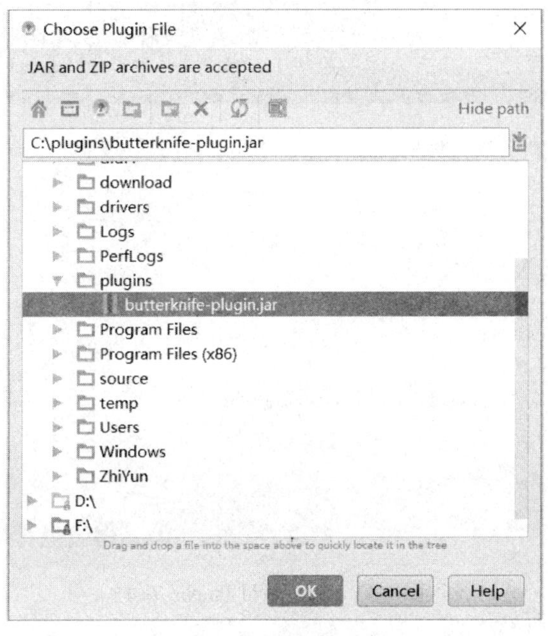

图 1-51　选择插件

（2）单击 OK 按钮，可以看到，插件安装完成，如图 1-52 所示。

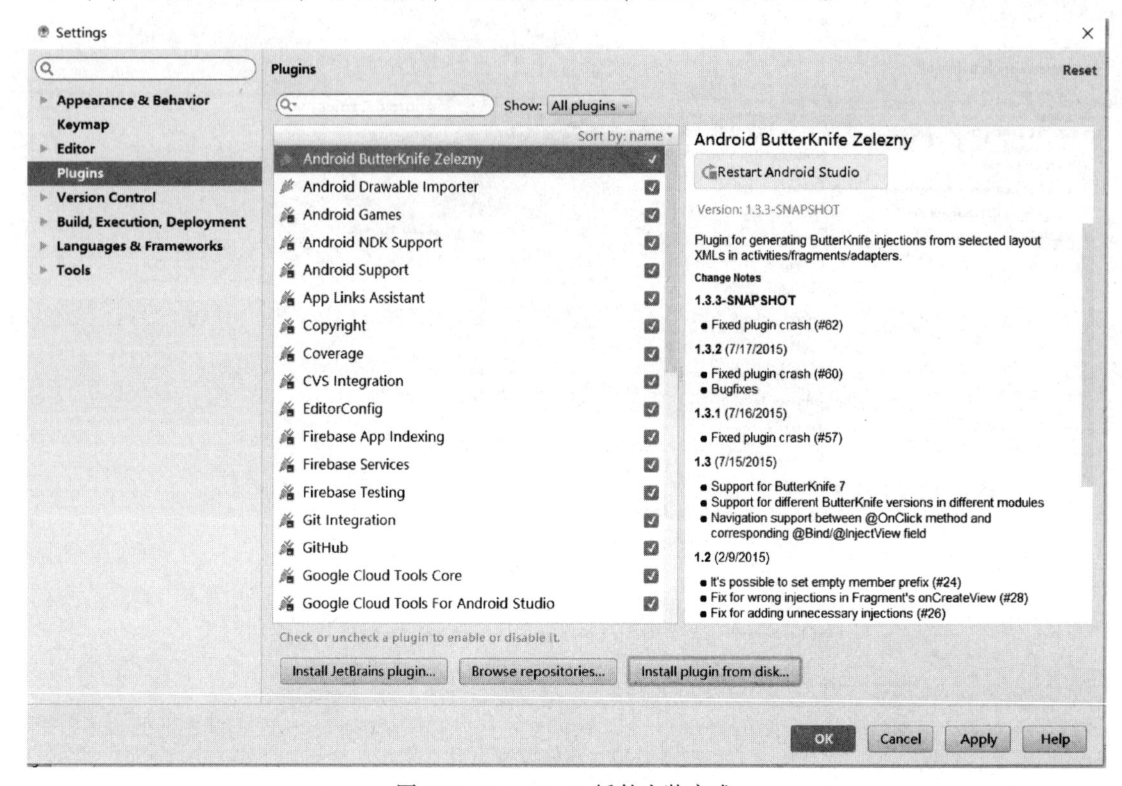

图 1-52　butterknife 插件安装完成

（3）重新启动，新建项目 Plugin_Test，如图 1-53 所示。

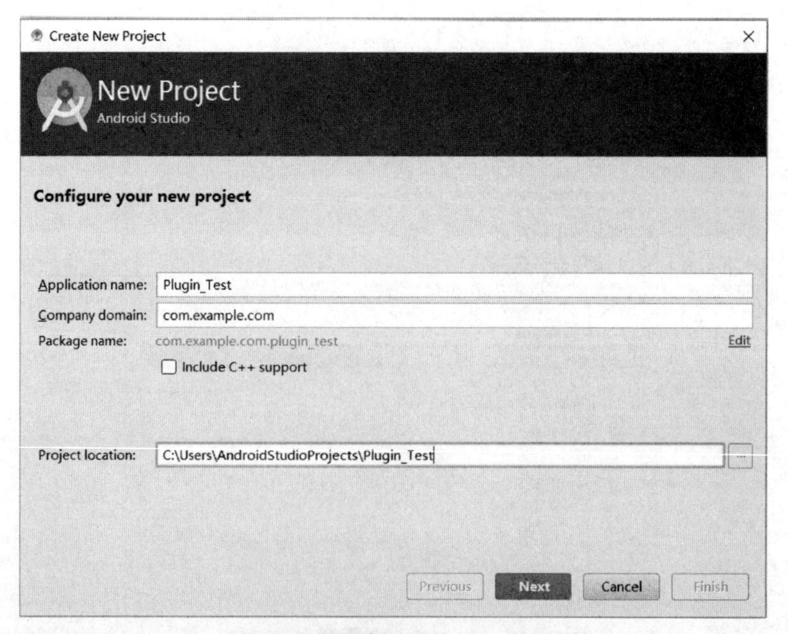

图 1-53　新建项目 Plugin_Test

（4）在项目外层的 build. gradle 文件，添加下面这行代码。

```
buildscript {
  repositories {
      jcenter()
  }
  dependencies {
    classpath 'com. android. tools. build:gradle:2. 3. 1'
    classpath 'com. neenbedankt. gradle. plugins:android-apt:1. 8' // 添加这行
  }
}
```

（5）在项目内层的 build. gradle 文件，添加下面三行代码。

```
apply plugin: 'com. android. application'
apply plugin: 'com. neenbedankt. android-apt' //添加这行
android {
  ……
}
dependencies {
  compile fileTree(dir: 'libs', include: ['* . jar'])
  compile 'com. android. support:appcompat-v7:25. 3. 1'
  compile 'com. android. support. constraint:constraint-layout:1. 0. 2'
  compile 'com. jakewharton:butterknife:8. 2. 1' //添加这行
  apt 'com. jakewharton:butterknife-compiler:8. 2. 1' //添加这行
}
```

（6）在项目的布局中增加两个 Button 控件，如图 1-54 所示。

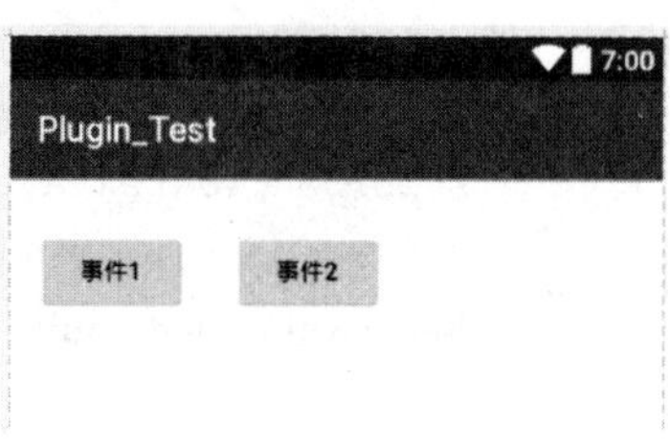

图 1-54　项目布局

（7）在项目的 Activity 文件中选择 R. lauout. activity _ main，单击右键，选择菜单中的 Generate 命令，在弹出的窗口中选择 Generate Butterknife Injections 选项，如图 1-55 所示。

```
public class MainActivity extends AppCompatActivity {

    @Override
    protected void onCreate(Bundle savedInstanceState) {
        super.onCreate(savedInstanceState);
        setContentView(R.layout.activity_main);
    }
}
```

Copy Reference　Ctrl+Alt+Shift+C
Paste　Ctrl+V
Paste from History...　Ctrl+Shift+V
Paste Simple　Ctrl+Alt+Shift+V
Column Selection Mode　Alt+Shift+Insert
Find Usages　Alt+F7
Find Sample Code　Alt+F8
Refactor
Folding
Analyze
Go To
Generate...　Alt+Insert

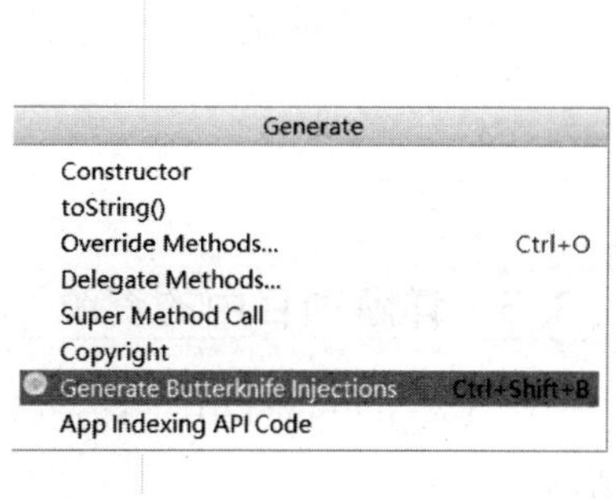

图 1-55　选择 Generate Butterknife Injections 选项

（8）弹出的窗口如图 1-56 所示，选择其中的控件 button 和 button2，单击 Confirm 按钮。

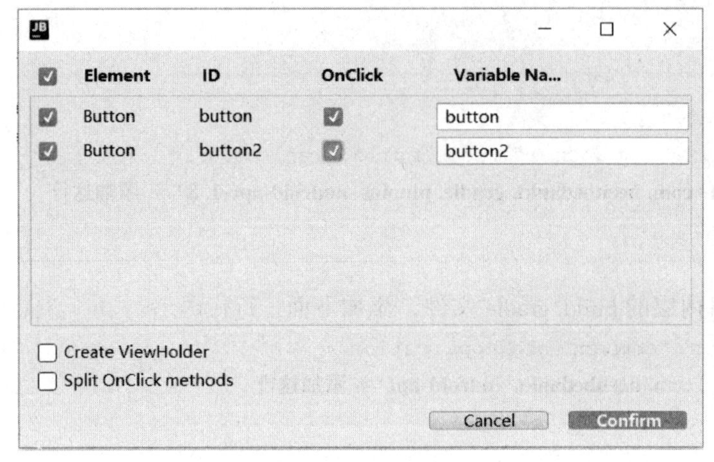

图 1-56　选择控件

（9）即会产生如下事件代码。

```
public class MainActivity extends AppCompatActivity {
  @BindView(R.id.button)
  Button button;
  @BindView(R.id.button2)
  Button button2;
  @Override
protected void onCreate(Bundle savedInstanceState) {
    super.onCreate(savedInstanceState);
    setContentView(R.layout.activity_main);
    ButterKnife.bind (this);
  }
  @OnClick ( {R.id.button, R.id.button2})
    public void onViewClicked (View view) {
      switch (view.getId()) {
          case R.id.button:
            break;
          case R.id.button2:
            break;
      }
  }
}
```

1.3.5　详解项目中的资源

在 Android Studio 中，首先展开前面的 Plugin_Test 项目，会看到如图 1-57 所示的目录结构。

任何一个新建的项目都会默认使用 Android 模式的项目结构，但这并不是真实的目录结

构，而是被 Android 转换过的，这种结构简洁明了，适合快速开发，模式结构可以切换，在 Android Studio 中，提供了以下几种项目结构类型，如图 1-58 所示。

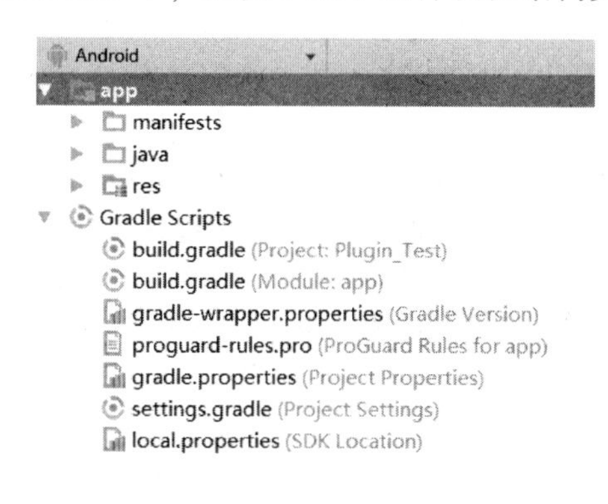

图 1-57　Android 模式的项目结构

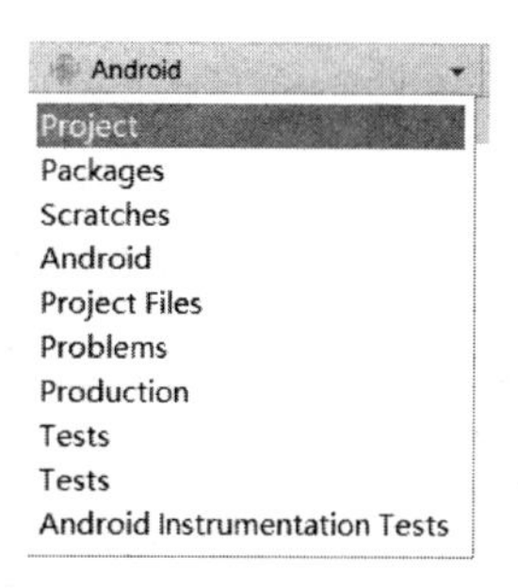

图 1-58　切换项目模式结构

现在将项目切换到 Project 模式，就是真实的项目结构了，如图 1-59 所示。

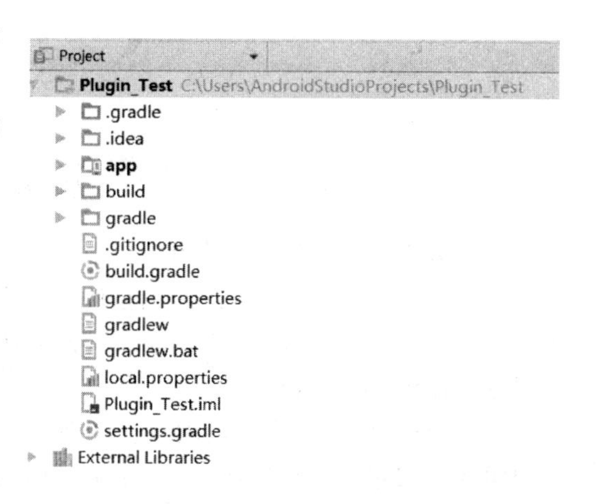

图 1-59　Project 模式项目结构

下面说明项目结构的内容，看完之后会感觉没有想象的复杂。

1. . gradle 和 . idea

该目录中放置的都是 Android Studio 自动生成的文件，一般不用管它。

2. app

项目中的代码、资源都在这个目录中，我们进行的开发工作都在这个目录中进行。

3. build

这个目录一般也不需要操作，包含一些编译自动生成的文件。

4. gradle

Gradle 是一个基于 Apache Ant 和 Apache Maven 概念的项目自动化建构工具。它使用一种基于 Groovy 的特定领域语言（DSL）来声明项目设置，抛弃了基于 XML 的各种繁琐配置。这个目录中包含 gradle wrapper 的配置文件，使用 gradle wrapper 的方式不需要提前将 gradle 下载好，而是会自动根据本地的缓存情况决定是否联网下载 gradle。Android Studio 默认不使用 gradle wrapper 的方式。如果需要打开，则在 Android Studio 主界面中单击 File > Settings 命令进行设置，如图 1-60 所示。

5. gitignore

这个文件用来指定将指定的文件或目录排除在版本控制之外，在 Git 部分将详细介绍。

6. build. gradle

该文件为这个项目全局的 gradle 构建脚本，通常这个文件中的内容是不需要修改的。

7. gradle. properties

这个文件是全局的 gradle 的配置文件，这里的配置将会影响到项目中所有 gradle 编译脚本。

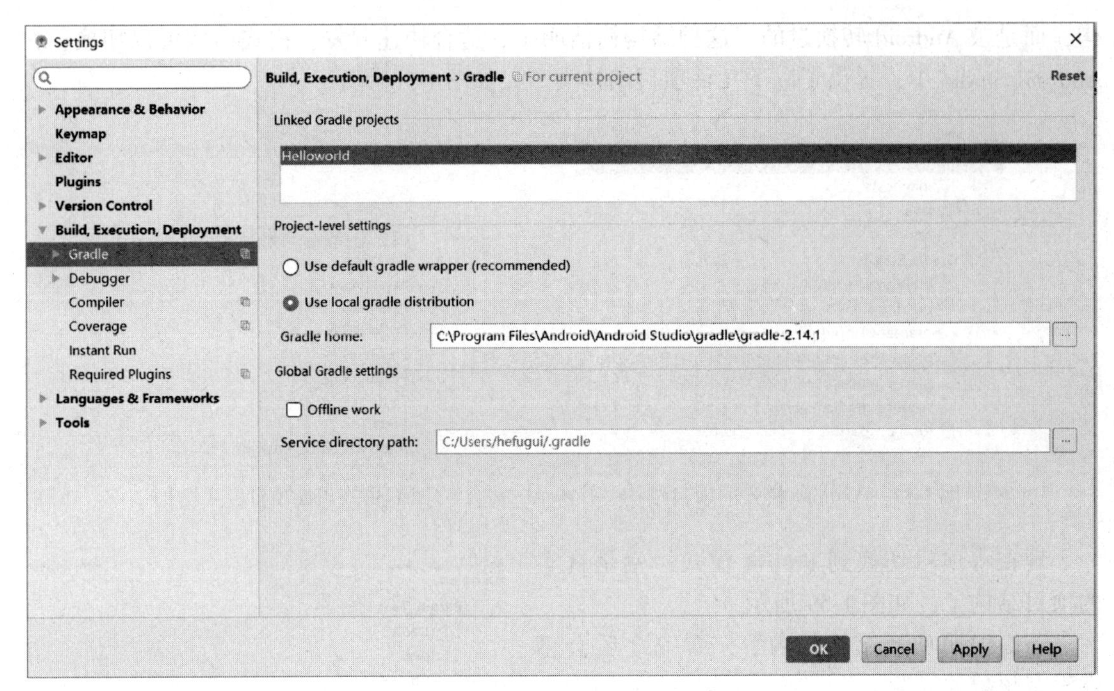

图 1-60　Gradle 设置

8. gradlew 和 gradlew. bat

这两个文件用来在命令行界面中执行 gradle 命令，其中 gradlew 是在 Linux 或 Mac 系统中使用的，gradlew. bat 是在 Windows 系统中使用的。

9. Plugin_Test. iml

Iml 文件是所有 IntelliJ IDEA 项目都会自动生成的一个文件（Android Studio 是基于 IntelliJ IDEA 开发的），用于标识是一个 IntelliJ IDEA 项目，不需要修改。

10. local. properties

这个文件用于指定本机中的 Android SDK 路径，通常内容是自动生成的，除非本机的 Android SDK 位置发生了变化。

11. settings. gradle

这个文件用于指定项目中所有引入的模块。由于 HelloWorld 项目中只有 app 模块，因此该文件中只引入 app 一个模块。通常是自动完成的。

现在整个目录介绍完了，您会发现，除了 app 目录外，大多数文件和目录是自动生成的，一般不需要修改。app 目录中的内容是我们介绍的重点，如图 1-61 所示。

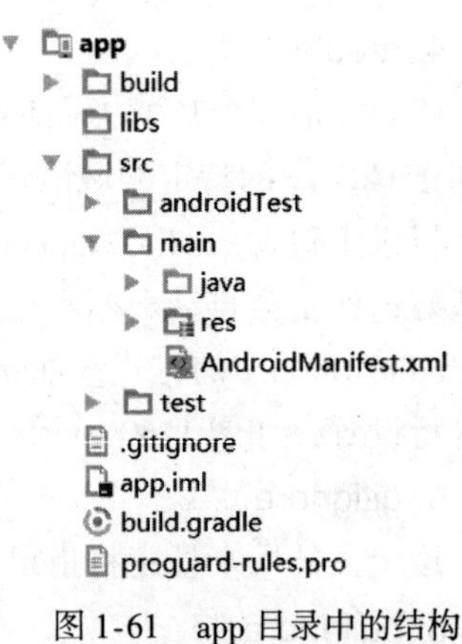

图 1-61　app 目录中的结构

下面分析这些内容。

1. build

这个目录和外层的 build 目录类似，主要包含一些在编译自动生成的文件，一般不需要关心。

2. libs

如果项目中使用到了第三方 jar 包，就需要把这些 jar 包都放在 libs 目录下，放在这个目

录下的 jar 包会被自动添加到构建路径中。

3. androidTest

此处是用来编写 AndroidTest 测试用例的，可以对项目进行一些自动化测试。

4. java

存放 java 源代码。

5. res

存放项目的资源，和 Eclipse 的 res 目录内容相同。

6. AndroidManifest. xml

Android 应用程序的配置文件，声明了 Android 里边的组件和相关配置信息、添加的权限。和 Eclipse 的 AndroidManifest. xml 基本相同。

7. test

用来编写 Unit Test 测试用例的，是对项目进行自动化测试的另一种方式。

8. . gitignore

这个文件用来指定将 app 模块内指定的文件或目录排除在版本控制之外，作用和外层的 . gitignore 类似。

9. app. iml

IntelliJ IDEA 项目自动生成的文件，不需要修改。

10. build. gradle

这是 app 模块的 gradle 构建脚本，这个文件中会指定很多项目构建相关配置。

11. proguard-rules. pro

这个文件用于指定代码的混淆规则，当代码开发完成后打包成安装包文件，如果不希望代码被别人破解，通常会将代码进行混淆。

1. 3. 6　详解 build. gradle 文件

不同于 Eclipse，Android Studio 采用 Gradle 来构建项目，Eclipse 开发是通过 ADT 进行项目编译、打包的，Android Studio 中把 ADT 这块彻底抛弃了，引入了 gradle 这个自动化构建工具。Gradle 是一个非常先进的项目构建工具。

项目中有两个 build. gradle 文件，一个在外层目录中，一个在 app 目录中，如图 1-62 所示，这两个文件对构建 Android Studio 项目起到了非常关键的作用。

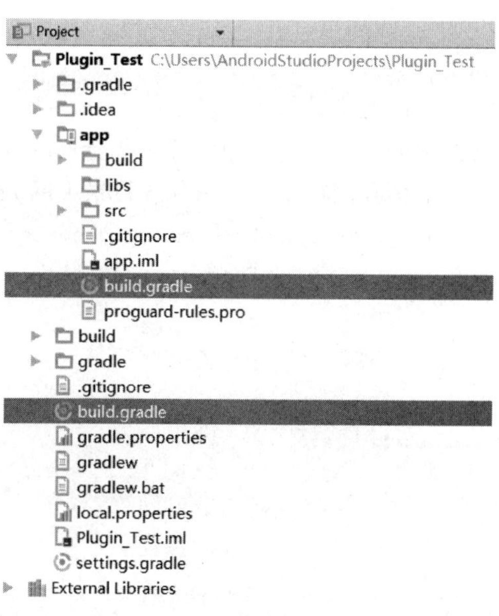

图 1-62　项目的两个 build. gradle 文件

下面是最外层的 build. gradle 文件。

```
buildscript {
  repositories {
       jcenter()
  }
  dependencies {
```

```
        classpath 'com.android.tools.build:gradle:2.3.1'
        classpath 'com.neenbedankt.gradle.plugins:android-apt:1.8' //添加这行
    }
}
allprojects {
    repositories {
        jcenter()
    }
}
task clean(type: Delete) {
    delete rootProject.buildDir
}
```

这些代码是自动生成的，虽然语法结构看上去可能有点难以理解，但是如果忽略语法结构，只看最关键的部分，其实还是很好懂的。

首先，两处 repositories 的闭包中都声明了 jcenter()这行配置，jcenter 是什么意思呢？其实它是一个代码托管仓库，很多 Android 开源项目都会选择将代码托管到 jcenter 上，声明了这行配置之后，我们就可以在项目中轻松引用任何 jcenter 上的开源项目了。

jcenter 用来查找和分享常用 Apache Maven 包，可通过 Maven、Gradle、Ivy 和 SBT 等工具使用。

接下来，dependencies 闭包中使用 classpath 声明了一个 Gradle 插件。为什么要声明这个插件呢？因为 Gradle 并不是专门为构建 Android 项目而开发的，Java、C++等很多种项目都可以使用 Gradle 来构建。因此如果我们要想使用它来构建 Android 项目，则需要声明 com.android.tools.build：gradle：2.3.1 这个插件。其中，最后面的部分是插件的版本号。

task clean 声明了一个任务，任务名称为 clean（也可以改为其他名称），任务类型是 Delete（也可以是 Copy），就是每当修改 settings.gradle 文件后单击同步，就会删除 rootProject.buildDir 下的文件（实际上看到的效果是清除了 External Libraries 里的包，然后又添加了一次）。

这样就将最外层目录下的 build.gradle 文件分析完了，通常情况下您并不需要修改这个文件中的内容，除非想添加一些全局的项目构建配置，例如前面介绍的插件项目。

下面我们再来看一下 app 目录下的 build.gradle 文件，代码如下：

```
apply plugin: 'com.android.application'
apply plugin: 'com.neenbedankt.android-apt' //添加这行
android {
    compileSdkVersion 25
    buildToolsVersion "25.0.2"
    defaultConfig {
        applicationId "com.example.com.plugin_test"
        minSdkVersion 21
        targetSdkVersion 25
        versionCode 1
        versionName "1.0"
```

```
            testInstrumentationRunner "android.support.test.runner.AndroidJUnitRunner"
        }
        buildTypes {
            release {
                minifyEnabled false
                proguardFiles getDefaultProguardFile('proguard-android.txt'), 'pro-
                guard-rules.pro'
            }
        }
    }

    dependencies {
        compile fileTree(dir:'libs', include:['*.jar'])
        androidTestCompile('com.android.support.test.espresso:espresso-core:2.2.2', {
            exclude group:'com.android.support', module:'support-annotations'
        })
        compile 'com.android.support:appcompat-v7:25.3.1'
        compile 'com.android.support.constraint:constraint-layout:1.0.2'
        testCompile 'junit:junit:4.12'
        compile 'com.jakewharton:butterknife:8.2.1' //添加这行
        apt 'com.jakewharton:butterknife-compiler:8.2.1' //添加这行
    }
```

这个文件中的内容相对复杂一些，下面一行行地进行分析。首先，第一行应用了两个插件，一般有三种值可选：com.android.application 表示这是一个应用程序模块，com.android.library 表示这是一个库模块，com.neenbedankt.android-apt 表示这是一个外部插件。应用程序模块和库模块的最大区别在于，一个可以直接运行，一个只能作为代码库依附于别的应用程序模块来运行。

接下来是一个大的 Android 闭包，在这个闭包中我们可以配置项目构建的各种属性。其中，compileSdkVersion 用于指定项目的编译版本，这里指定成 25，表示使用 Android 7.0 系统的 SDK 编译。buildToolsVersion 用于指定项目构建工具的版本，目前最新的版本是 25.0.2，如果有更新的版本，Android Studio 会进行提示。

这里在 Android 闭包中又嵌套了一个 defaultConfig 闭包，defaultConfig 闭包中可以对项目的更多细节进行配置。其中，applicationId 用于指定项目的包名，前面我们在创建项目的时候其实已经指定过包名了，如果想对其进行修改，那么在这里修改即可。minSdkVersion 用于指定项目最低兼容的 Android 系统版本，这里指定成 21，表示最低兼容到 Android 5.0 系统。targetSdkVersion 指定的值表示在该目标版本上已经做过了充分的测试，系统将会为应用程序启用一些最新的功能和特性。比如说 Android 7.0 系统中引入了运行时权限这个功能，如果将 targetSdkVersion 指定成 24 或者更高，那么系统会为程序启用运行时权限功能，而如果将 targetSdkVersion 指定成 25，那么就说明程序最高只在 Android 7.1.1 系统上做过充分的测试，Android 8.0 系统中引入的新功能自然不会启用。剩下的两个属性都比较简单，versionCode 用于指定项目的版本号，versionName 用于指定项目的版本名，这两个属性在生成安装文件的时候非常重要。

testInstrumentationRunner 这一行表明要使用 AndroidJUnitRunner 进行单元测试。

分析完了 defaultConfig 闭包，接下来看一下 buildTypes 闭包。buildTypes 闭包中用于指定生成安装文件的相关配置，通常只会有两个子闭包，一个是 debug，一个是 release。debug 闭包用于指定生成测试版安装文件的配置，release 闭包用于指定生成正式版安装文件的配置。另外，debug 闭包是可以忽略不写的，因此我们看到上面的代码中只有一个 release 闭包。下面来看一下 release 闭包中的具体内容：minifyEnabled 用于指定是否对项目的代码进行混淆，true 表示混淆，false 表示不混淆。proguardFiles 用于指定混淆时使用的规则文件，这里指定了两个文件，第一个 proguard-android. txt 是在 Android SDK 目录下的，里面是所有项目通用的混淆规则，第二个 proguard-rules. pro 是在当前项目的根目录下的，里面可以编写当前项目特有的混淆规则。需要注意的是，通过 Android Studio 直接运行项目，生成的都是测试版安装文件。

这样整个 Android 闭包中的内容就分析完了，接下来还剩下一个 dependencies 闭包。这个闭包的功能非常强大，它可以指定当前项目所有的依赖关系。通常 Android Studio 项目一共有 3 种依赖方式：本地依赖、库依赖和远程依赖。本地依赖可以对本地的 Jar 包或目录添加依赖关系，库依赖可以对项目中的库模块添加依赖关系，远程依赖则可以对 jcenter 库上的开源项目添加依赖关系。观察一下 dependencies 闭包中的配置，第一行的 compile fileTree 就是一个本地依赖声明，它表示将 libs 目录下所有 . jar 后缀的文件都添加到项目的构建路径当中。而第二行的 compile 则是远程依赖声明，com. android. support：appcompat-v7：25. 3. 1 就是一个标准的远程依赖库格式，其中 com. android. support 是域名部分，用于和其他公司的库做区分；appcompat-v7 是组名称，用于和同一个公司中不同的库做区分；25. 3. 1 是版本号，用于和同一个库不同的版本做区分。加上这句声明后，Gradle 在构建项目时会首先检查本地是否已经有这个库的缓存，如果没有的话则会去自动联网下载，然后再添加到项目的构建路径当中。这里没有用到库依赖声明，它的基本格式是 compile project 后面加上要依赖的库名称，比如说有一个库模块的名字叫 helper，那么添加这个库的依赖关系只需要加入 compile project（'：helper'）这句声明即可。Apt 表示是一个 Gradle 插件，协助 Android Studio 处理 annotation processors，它有两个目的：（1）允许配置只在编译时作为注解处理器的依赖，而不添加到最后的 APK 或 library；（2）设置源路径，使注解处理器生成的代码能被 Android Studio 正确地引用。另外剩下的一句 testCompile，它用于声明测试用例库，我们暂时用不到，先忽略即可。

1. 3. 7 项目运行

Android 应用程序的运行需要一个载体，可以是真实的手机，也可以是 Android 模拟器，这里先使用模拟器来运行程序。

1. 建立模拟器

（1）在 Android Studio 的主界面中，单击工具栏的 AVD Manager 工具按钮，如图 1-63 所示。

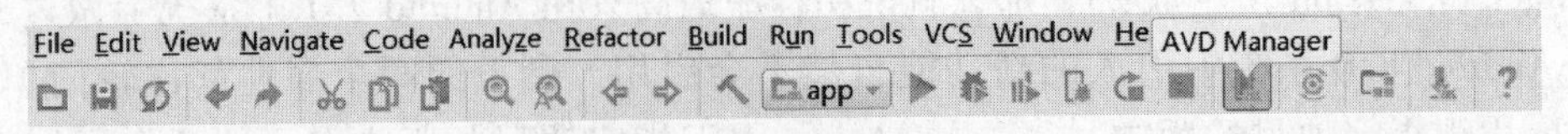

图 1-63 启动模拟器

或者选择主界面菜单栏中 Tools > Android > AVD Manager 命令，如图 1-64 所示。

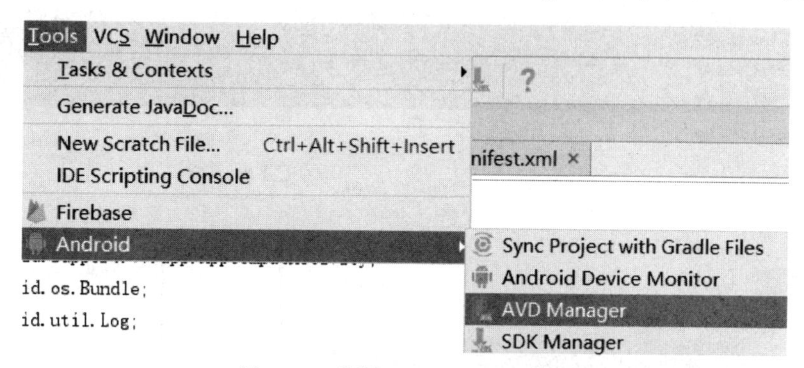

图 1-64　选择 AVD Manager 命令

（2）弹出的窗口如图 1-65 所示，单击 Create Virtual Device 按钮。

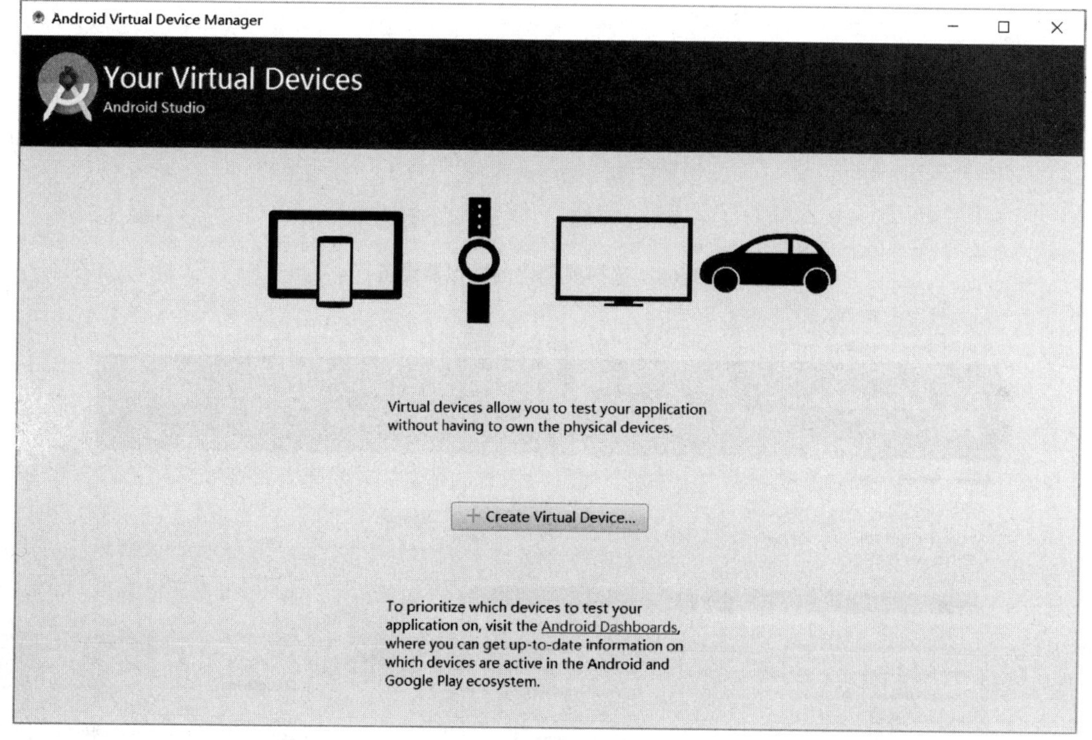

图 1-65　创建模拟器

（3）弹出的窗口如图 1-66 所示，这里有很多可供我们选择的设备，包括 TV、Wear（穿戴设备）、Phone、Tablet（平板），这里选择 Nexus 5X。

（4）单击 Next 按钮，进入 System Image 选择界面，在此选择操作系统版本，这里选择 Nougat，Android 新系统：Android Nougat（牛轧糖），即 Android 7.0，如图 1-67 所示。

（5）单击 Next 按钮，弹出的窗口如图 1-68 所示，在这里可对模拟器进行配置，保持默认设置。

（6）单击 Finish 按钮，弹出的窗口如图 1-69 所示。

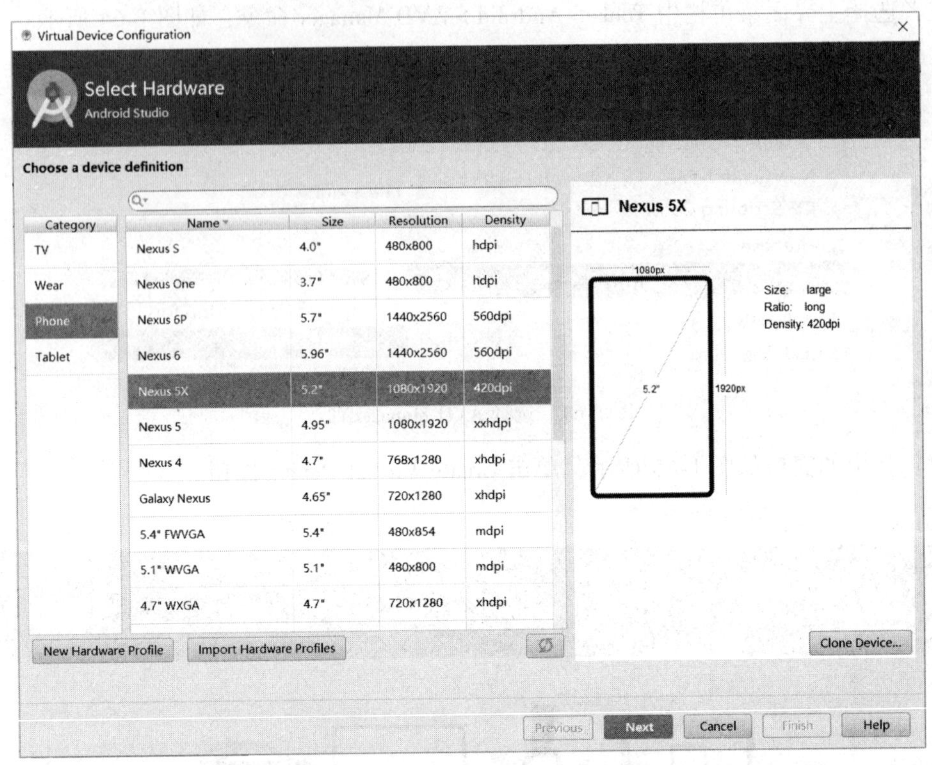

图 1-66　选择要创建的模拟器设备

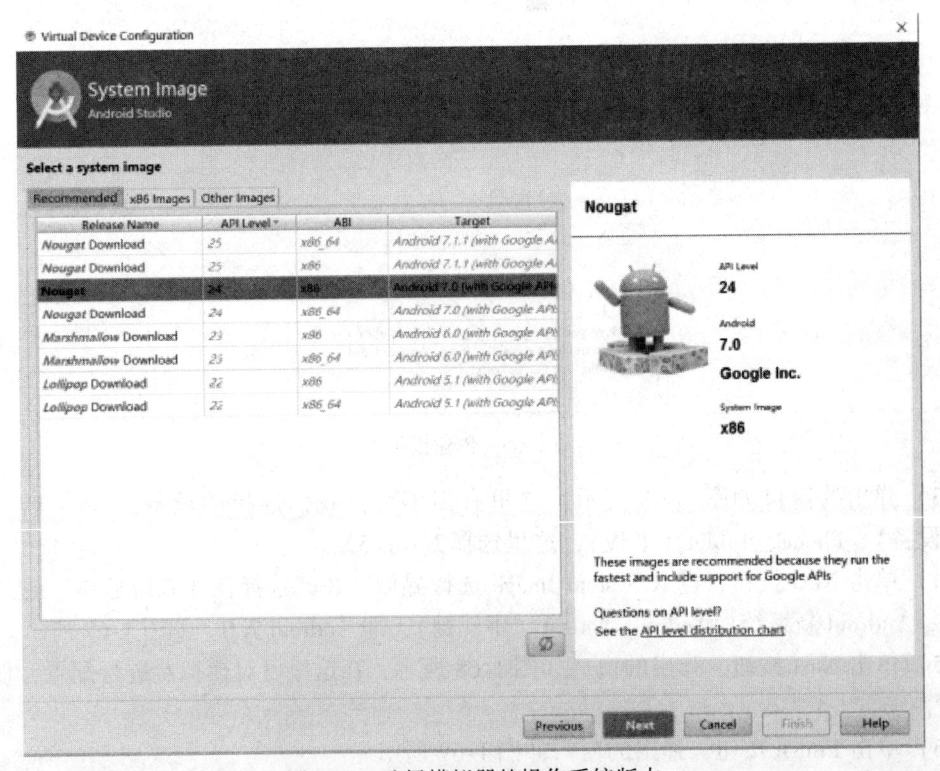

图 1-67　选择模拟器的操作系统版本

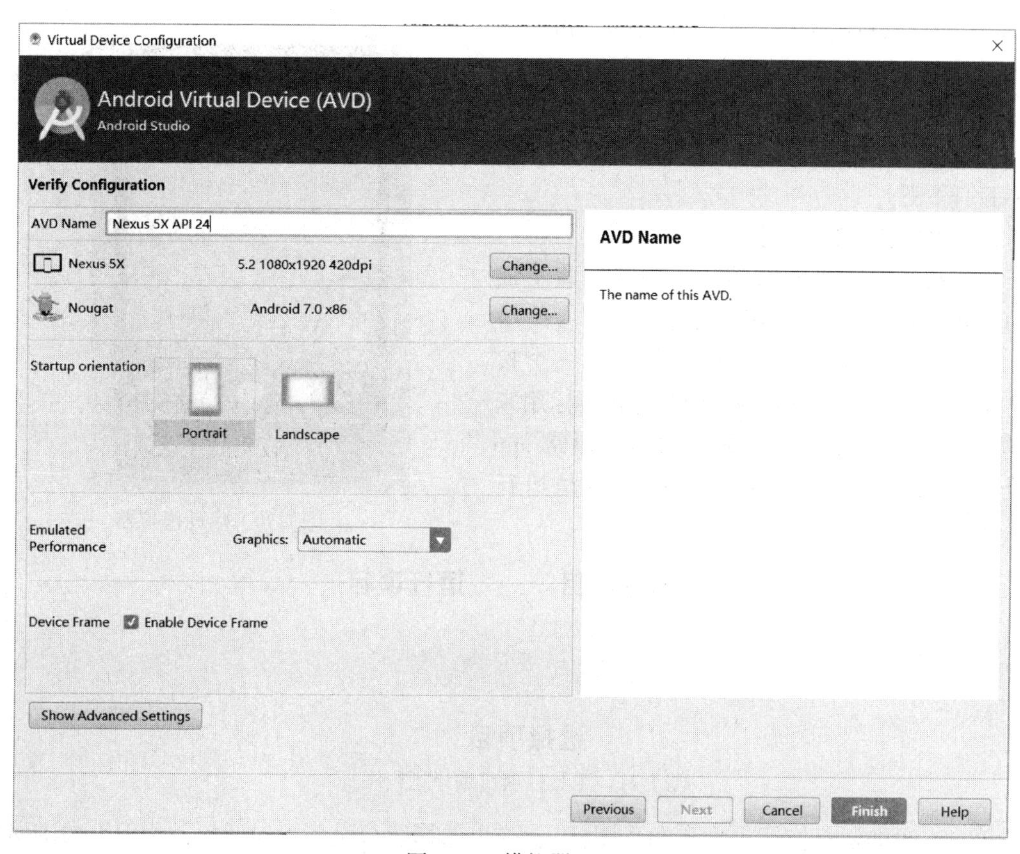

图 1-68　模拟器配置

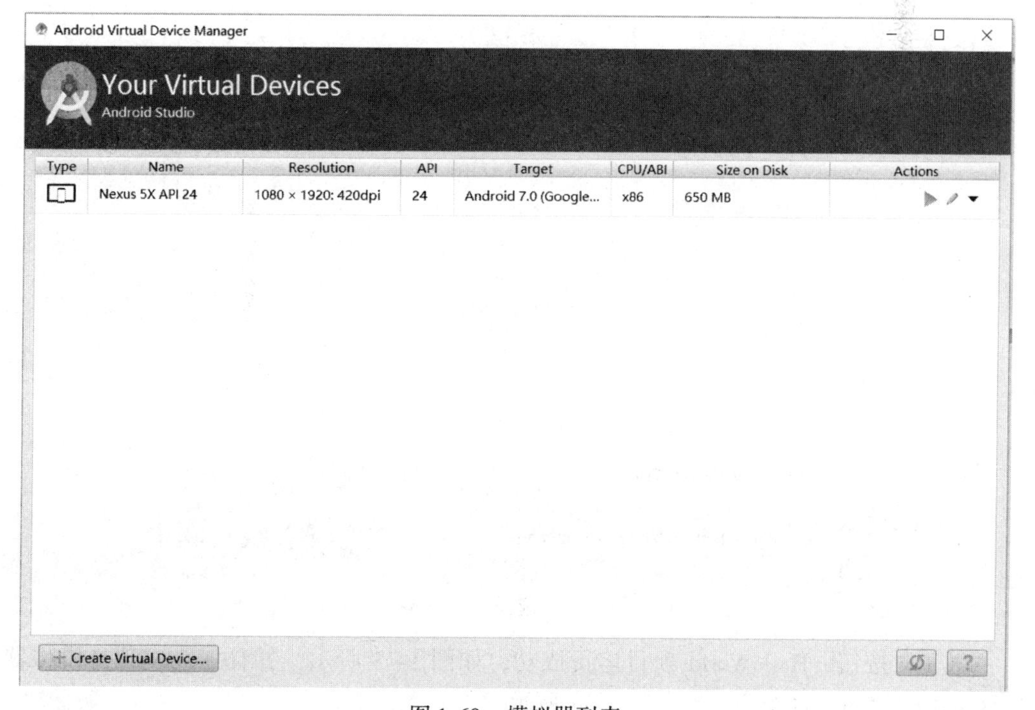

图 1-69　模拟器列表

(7) 单击右边 Actions 栏最左边的三角形按钮，即可启动模拟器，如图 1-70 所示。

出现很清新的 Android 界面，看上去还不错吧，Android 模拟器对手机的模仿度非常高，赶紧体验一下吧。

图 1-70　启动模拟器

2. 下载项目到模拟器

现在已经启动了模拟器，下面我们将上面的 HelloWorld 项目运行到模拟器上。

在 Android Studio 的工具栏中部有三个按钮，如图 1-71 所示，其中左边的锤子图标用来编译项目，中间按钮用于选择项目，通常 app 就是当前的项目，右边的三角形图标是运行项目。

图 1-71　工具栏项目编译运行图标

单击右边的运行按钮（即三角形图标），弹出选择运行设备的对话框，如图 1-72 所示。

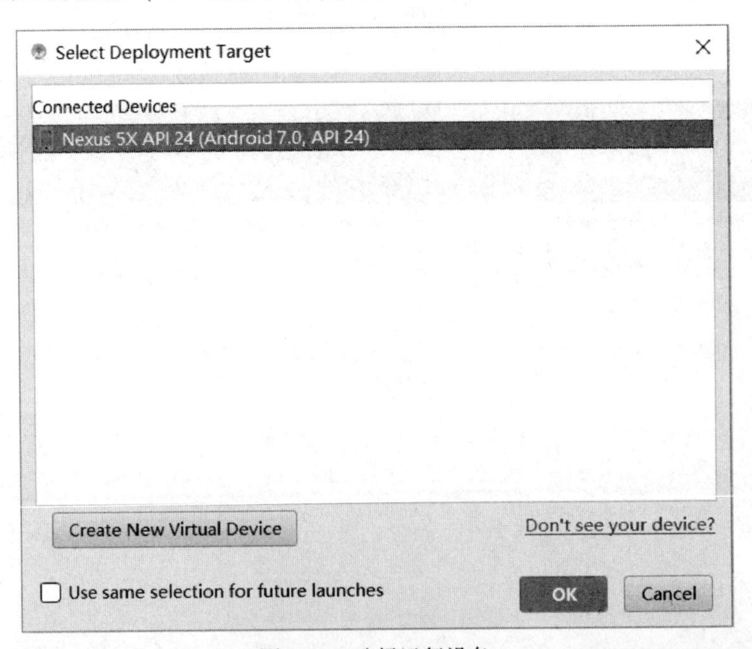

图 1-72　选择运行设备

单击 OK 按钮，HelloWorld 项目运行成功，如图 1-73 所示，并且您会发现，模拟器已经安装了 HelloWorld 这个应用，如图 1-74 所示。

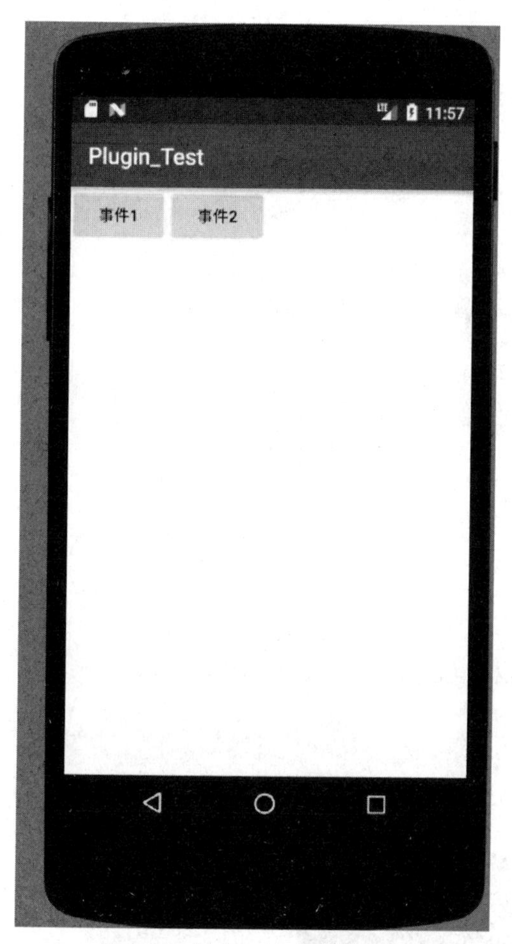

图 1-73　运行 HelloWorld 项目

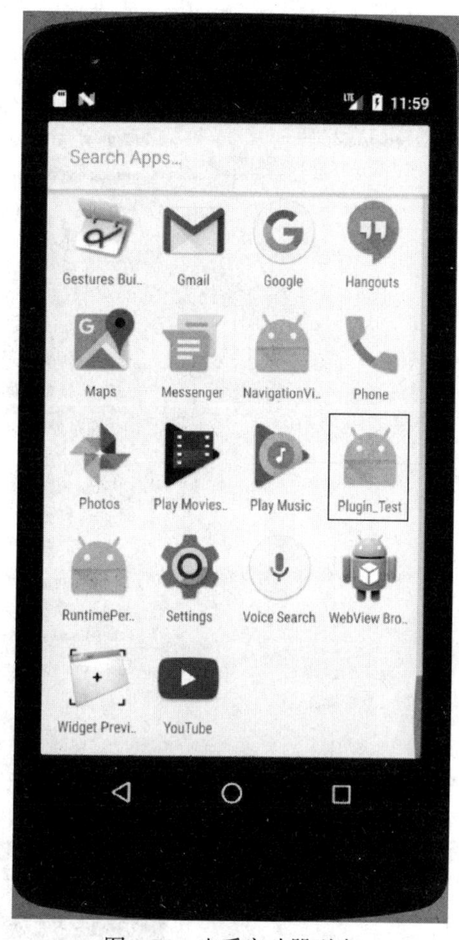

图 1-74　查看启动器列表

3. 下载到真机

Android Studio 开发的应用程序最终要下载到真机运行，首先按前面的介绍，将 PC 和移动终端通过 USB 线连接起来，然后下载安装 Universal Adb Driver，或者 360 手机助手、豌豆荚、应用宝、91 助手等。

（1）单击 Android Studio 主界面的工具栏中 app 下拉按钮，选择 Edit Configurations 选项，如图 1-75 所示。

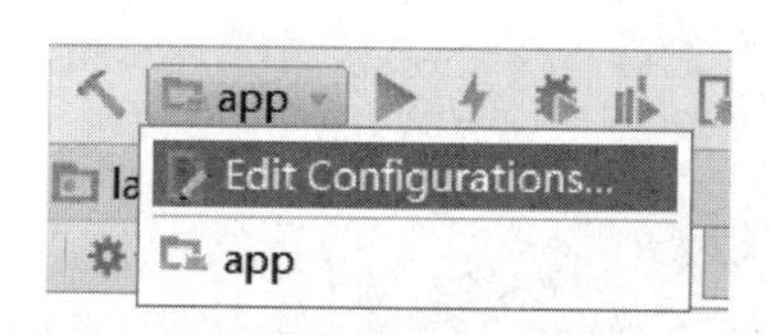

图 1-75　选择 Edit Configurations 选项

（2）弹出的窗口如图 1-76 所示，选中左边的 app 选项，在右侧选择 Target 下拉列表中的 USB Device 选项。

（3）单击图 1-71 中 app 按钮后面的三角形运行按钮，Plugin_Test 应用项目即下载到 USB 连接的手机中运行了，如图 1-77 所示。

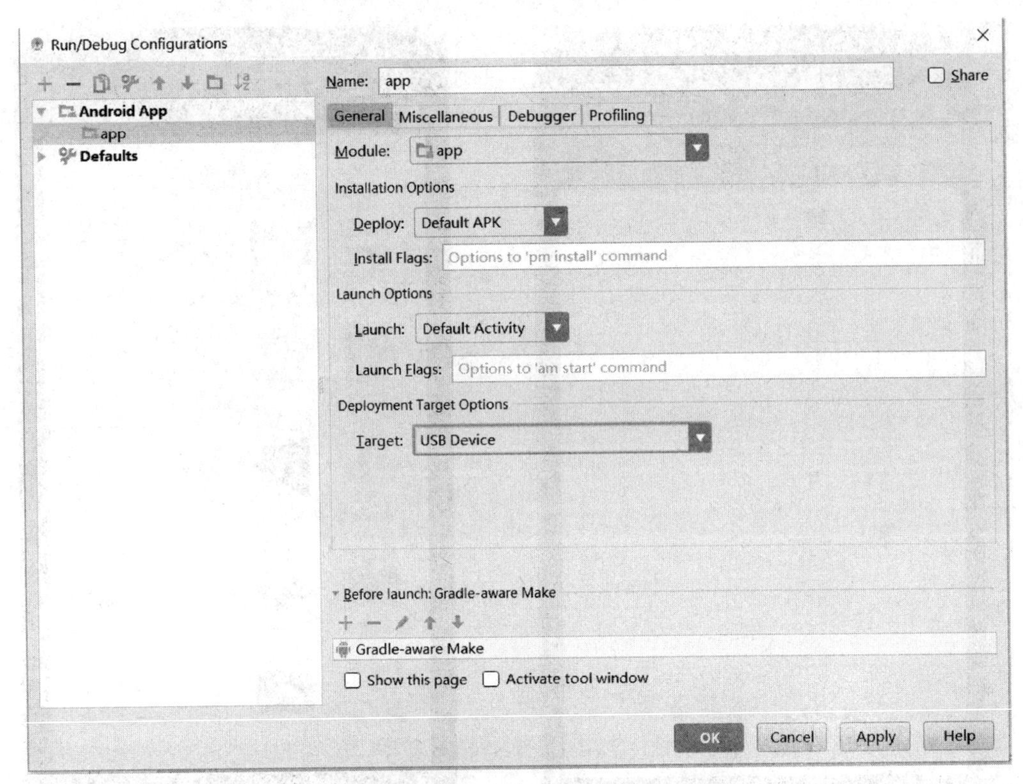

图 1-76　运行/调试配置

图 1-77　Plugin_Test 应用项目真机运行

4. 第 3 方模拟器 Genymotion

Genymotion 是一套完整的工具，它提供了 Android 虚拟环境，支持 Windows、Linux 和 Mac OS 等操作系统。

Genymotion 安卓模拟器不是普通的模拟器，严格来说，Genymotion 是虚拟机，被网传定义为模拟器，Genymotion 虚拟机能够带来更好的 Android 模拟体验，目前具备以下特性。

- 支持 OpenGL 加速，提供较好的 3D 性能体验。
- 可以从 Google Play 安装应用。
- 支持全屏并改善了使用感受。
- 全控制。
- 可同时启动多个模拟器。
- 支持传感器管理，如电池状态、GPS、Accelerator 加速器。
- 支持 Shell 控制模拟器。
- 完全兼容 ADB，可以从主机控制模拟器。
- 管理设备。
- 易安装。
- 兼容 Microsoft Windows 32/64 bits、Mac OSX 10. 5 + 和 Linux 32/64 bits。
- 可以配置模拟器参数，如屏幕分辨率、内存大小、CPU 数量。
- 轻松下载、部署最新的 Genymotion 虚拟设备。
- 从 Eclipse 启动虚拟设备。
- 测试应用。

1. 3. 8　导入 Eclipse 项目

Android Studio 可以导入从网上下载的 Android Studio 项目和 Eclipse 项目，这两种项目的导入都是通过菜单操作完成的，步骤如下。

（1）在 Android Studio 主界面中单击菜单栏中 File > New > Import Project 命令，如图 1-78 所示。

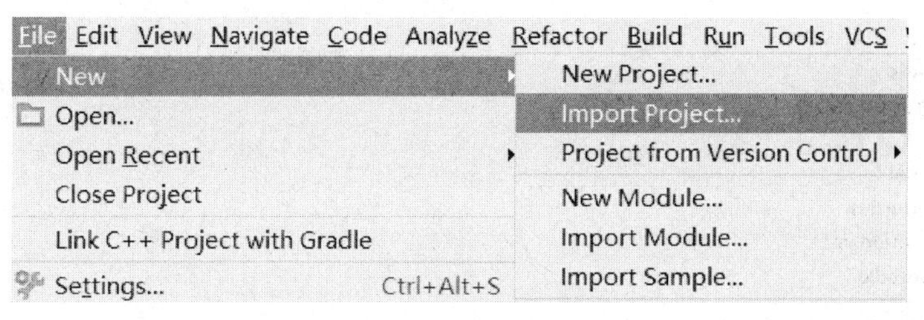

图 1-78　导入外部项目

（2）弹出的窗口如图 1-79 所示，可以选择 Eclipse 项目和 Android Studio 项目。

（3）如果打开的是 Android Studio 项目，则选择文件“build. gradle”，或者选择项目目录，如图 1-80 所示，单击 OK 按钮即可，如图 1-81 所示。

（4）如果打开的是 Eclipse 项目，则选择项目目录，单击 OK 按钮，弹出的窗口如图 1-82 所示，选择导出目录。单击 Next 按钮，进入转换窗口，如图 1-83 所示。

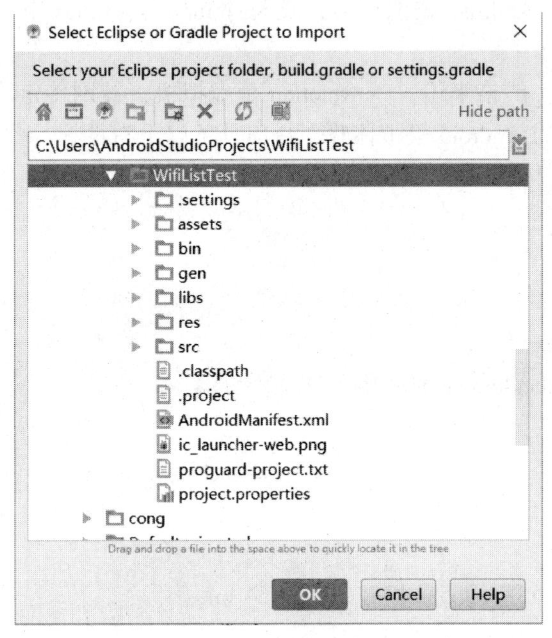

图 1-79　选择项目

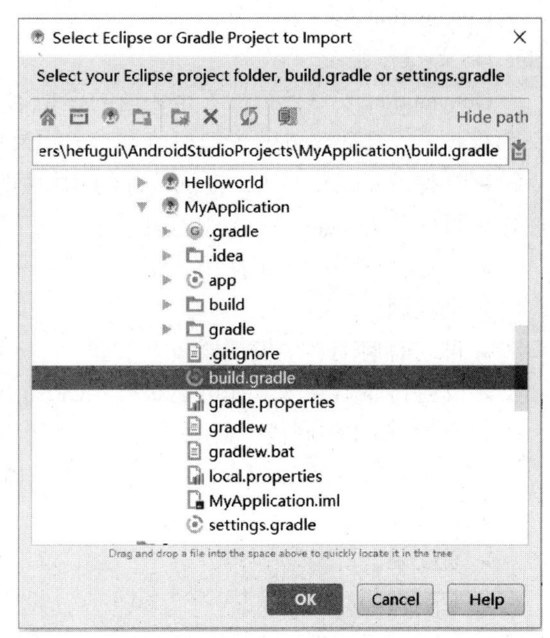

图 1-80　打开 Android Studio 项目

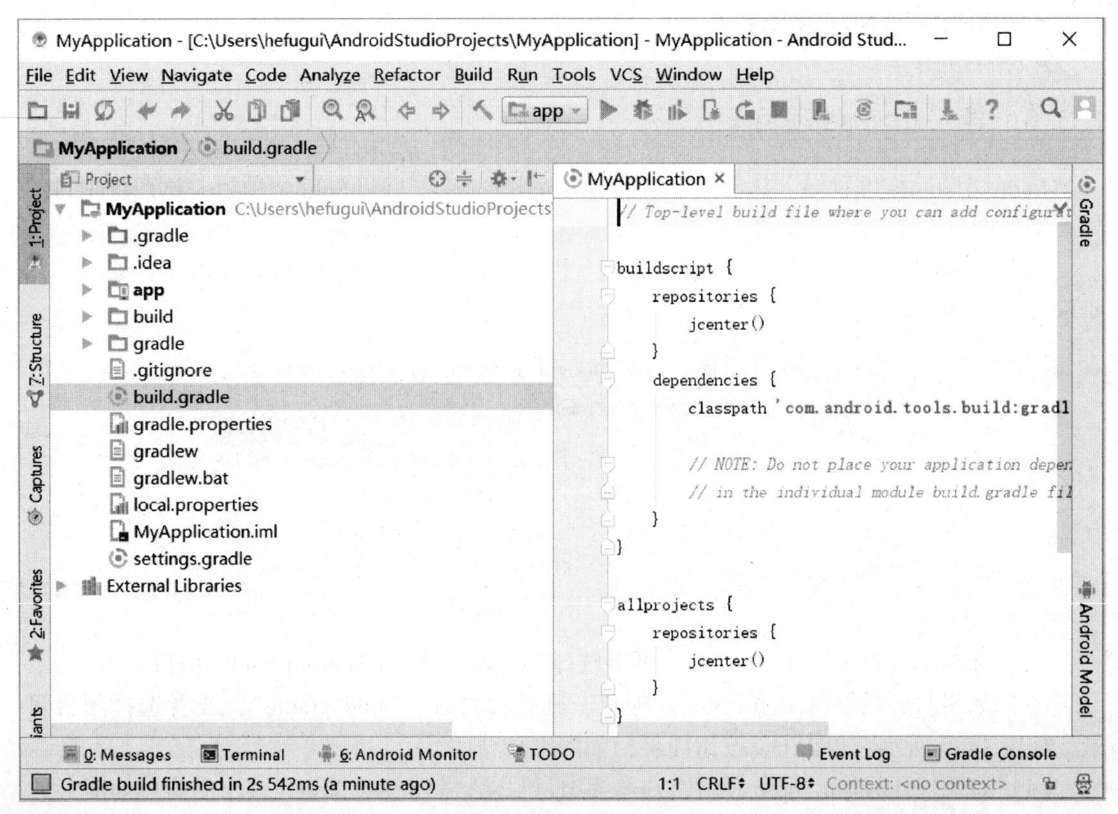

图 1-81　打开的项目

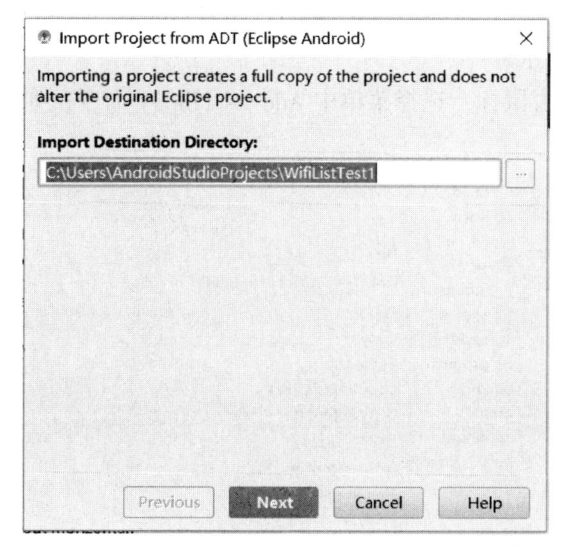

图 1-82　选择导出目录

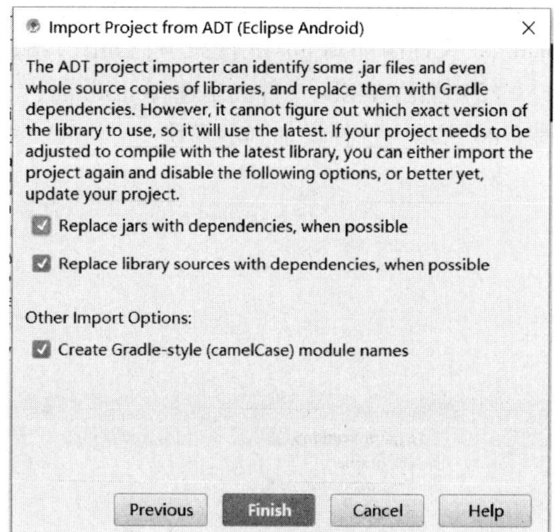

图 1-83　转换窗口

（5）单击 Finish 按钮，完成导入过程，如图 1-84 所示。

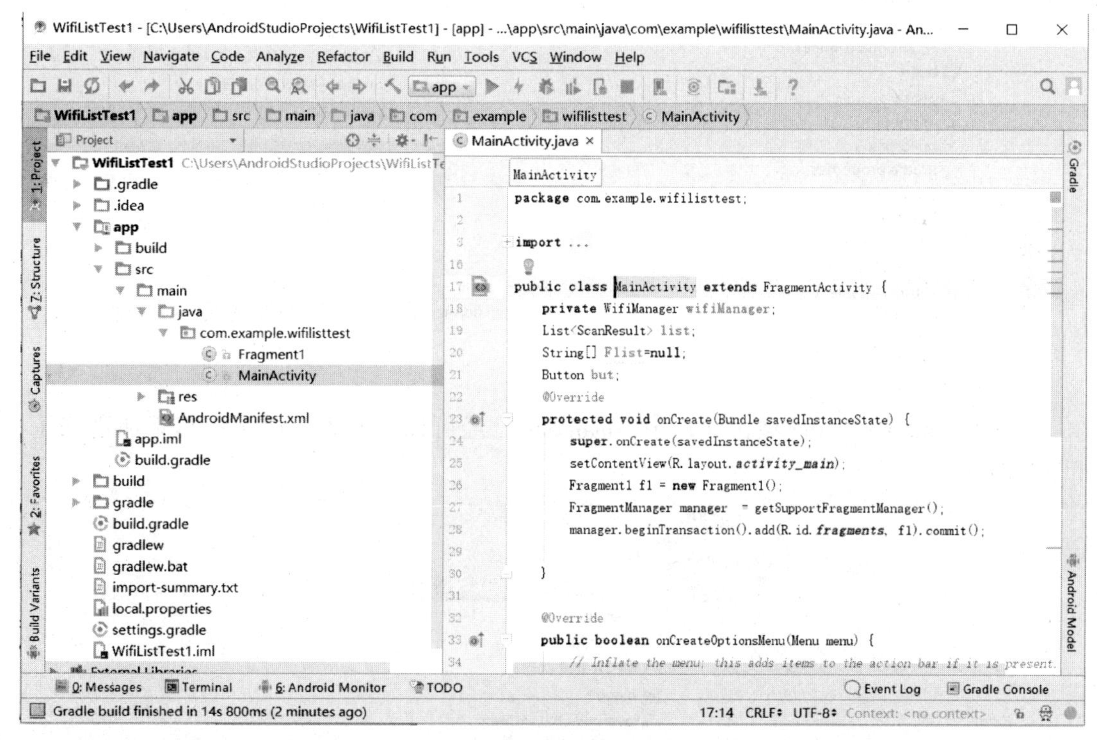

图 1-84　导入的项目

1.3.9　导入 JAR 文件

Android Studio 添加外部库的步骤如下。

（1）选择 Android Studio 项目，添加一个第三方的 jar 文件：mail. jar，添加 mail jar 包，

用 Java 实现邮件发送。在项目中选中或添加 libs 文件夹，直接通过 Copy/Paste 把下载的 mail. jar 文件添加到 libs 文件夹中，如图 1-85 所示。

（2）然后在 libs 文件夹中选择 mail. jar，右击鼠标，选择菜单中 Add As Library 命令，如图 1-86 所示，选择增加到的模块。

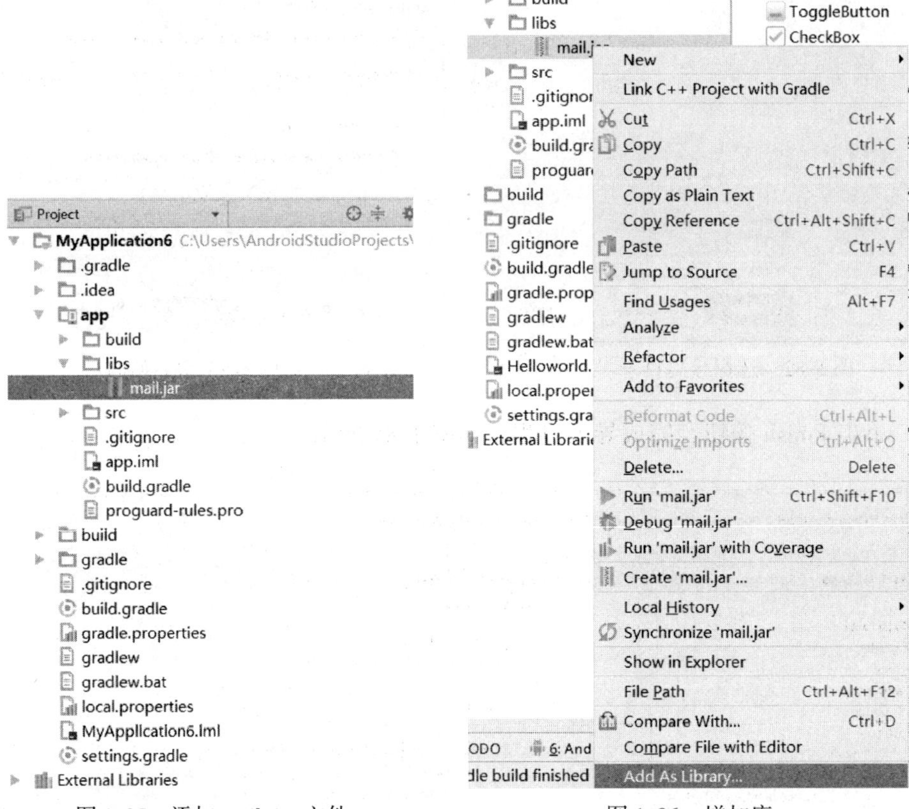

图 1-85　添加 mail. jar 文件　　　　图 1-86　增加库

（3）在 Android Studio 项目面板中，mail. jar 已经可以展开了，说明将其添加进来了，如图 1-87 所示。

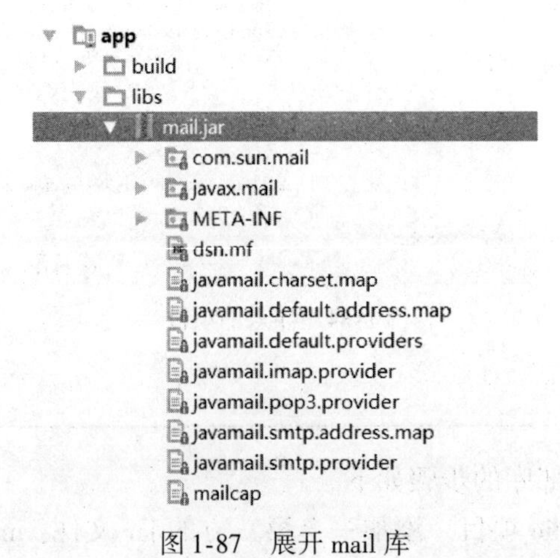

图 1-87　展开 mail 库

1.3.10　调试

在项目的开发过程中和程序的运行过程中，会出现各种各样需要解决的问题，这时候需要调试来排查和定位问题的原因，然后解决问题。

Android Studio 允许在模拟器和真机上调试应用程序，我们可以根据需要设置不同类型的断点，也可以查看内存中数据的变化，还可以在运行时添加日志或计算表达式。

调试程序的一般步骤如下。

（1）添加断点。

（2）运行调试。

（3）执行到断点。

（4）显示调试器窗口。

（5）查看调试信息。

（6）使用步进调试工具分析代码。

（7）使用控制调试工具管理断点和程序运行。

1.3.10.1　调试应用程序

下面演示在 Android Studio 中调试 Android 应用程序的过程。

（1）定位调试代码。为了演示调试过程，在 onCreate( ) 函数中增加三行代码。

```
protected void onCreate(Bundle savedInstanceState) {
    super.onCreate(savedInstanceState);
    setContentView(R.layout.activity_main);
    int i=6;
    Log.d("MainActivity", "onCreate execute");
    Log.d("MainActivity", "over");
}
```

（2）设置断点，选定要设置断点的代码行，在行号的区域后面单击鼠标左键即可，如图 1-88 所示，此处设置 3 个断点。

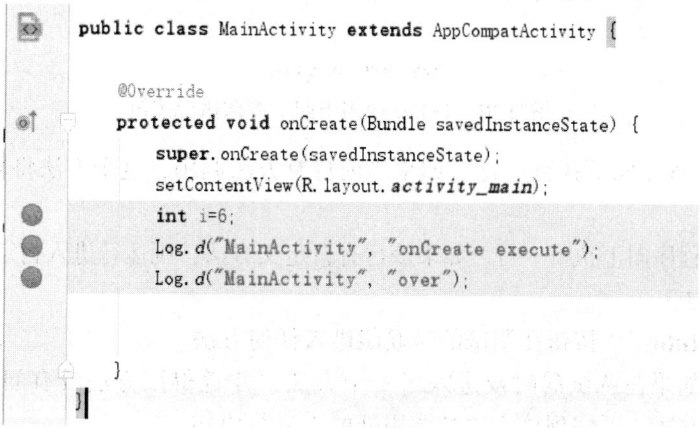

图 1-88　设置断点

（3）开启调试会话，单击黑色箭头指向的按钮，开始调试，如图 1-89 所示。

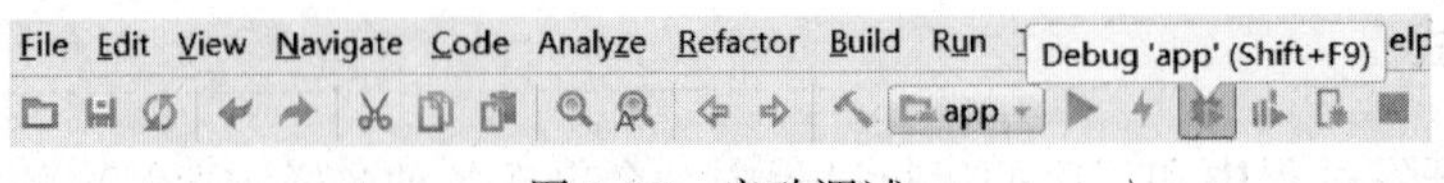

图 1-89　启动调试

（4）IDE 下方出现 Debug 视图，在其中显示了当前调试程序停留的代码行，如图 1-90 所示。

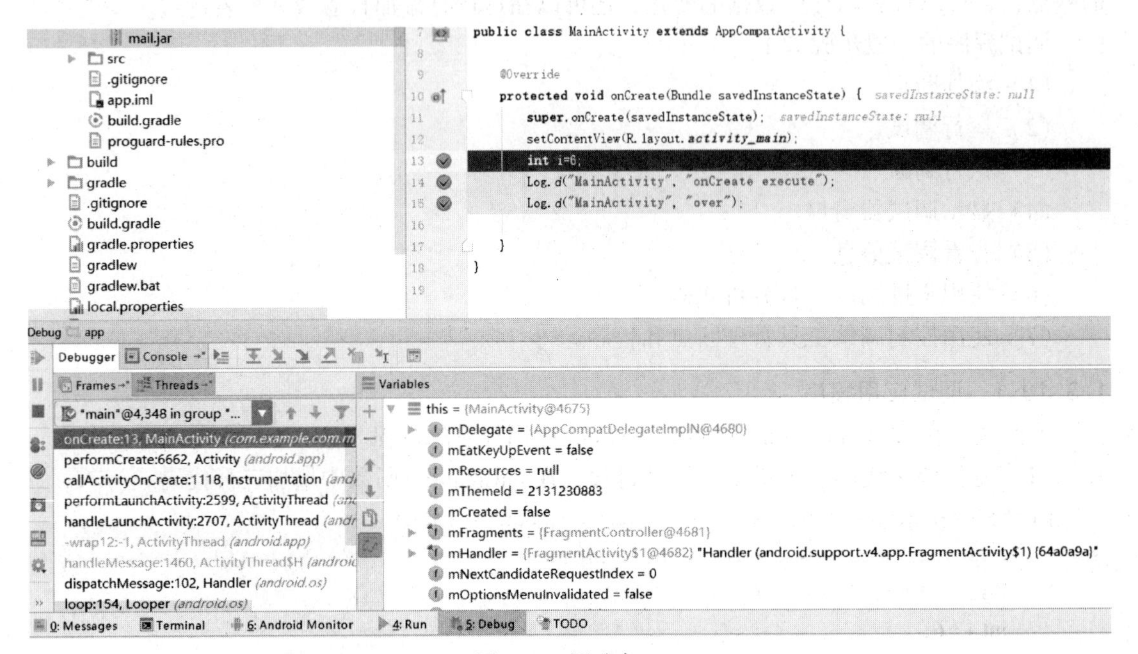

图 1-90　调试窗口

（5）在 Debug 视图中有四个调试执行按钮，如图 1-91 所示。

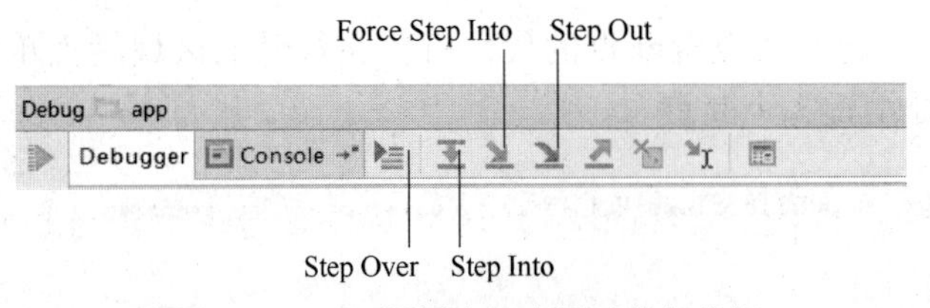

图 1-91　Debug 视图中四个调试执行按钮

Step Over：程序向下执行一行（如果当前行有方法调用，这个方法将被执行完毕返回，然后到下一行）。

Step Into：程序向下执行一行。如果该行有自定义方法，则运行进入自定义方法（不会进入官方类库的方法）。

Force Step Into：该按钮在调试的时候能进入任何方法。

Step Out：如果在调试的时候进入了一个方法，并觉得该方法没有问题，则可以使用 Step Out 跳出该方法，返回到该方法被调用处的下一行语句。

（6）在 Debug 视图中，单击 Step Over 按钮，程序执行下一行，如图 1-92 所示。

（7）停止调试。单击工具栏中红色方块按钮，即停止调试，如图 1-93 所示。

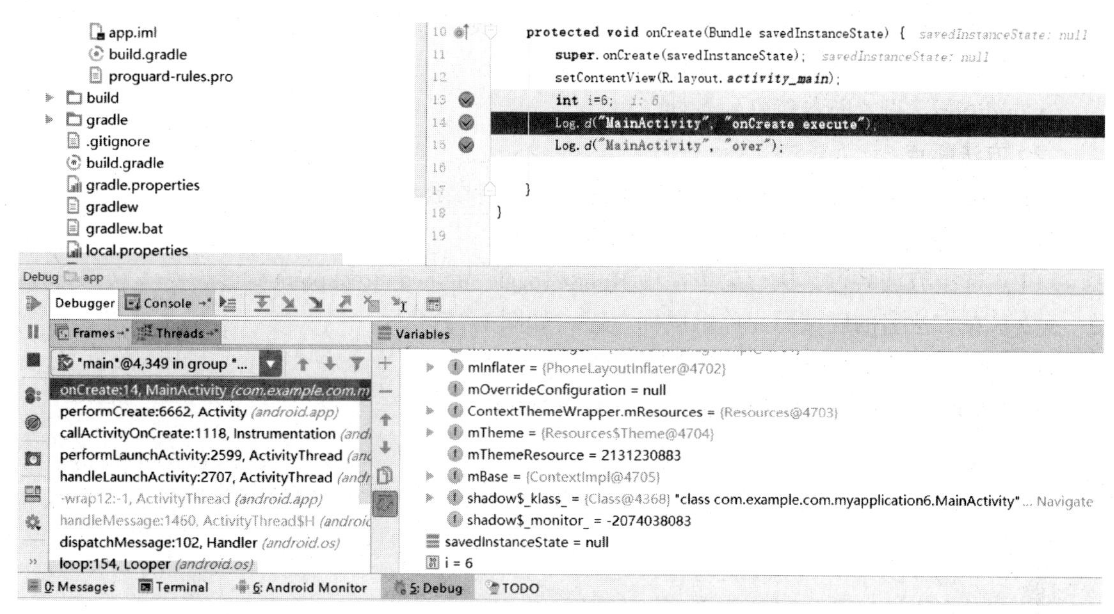

图 1-92　执行到下一个断点

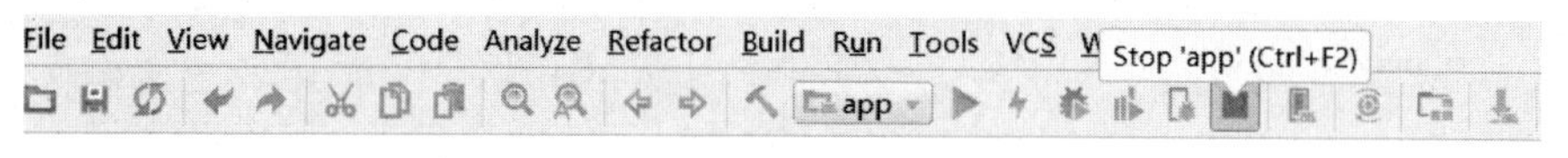

图 1-93　停止调试

1.3.10.2　断点

断点会暂停应用程序的执行，线程被挂起，然后可以通过调试器查看信息。Android Studio 中的断点类型有很多，每一种断点都有它适用的场合和特殊的作用。

1. 行断点

行断点是最常用的断点类型，用于对代码中特定的行进行调试。设置断点的方法如下。

（1）选中代码行，选择菜单栏中 Run > Toggle Line Breakpoint 命令。

（2）在行号的区域后面单击鼠标左键。

取消断点的方法和设置断点相同。

右击行断点，弹出属性对话框，在此可进行断点属性设置，如图 1-94 所示。

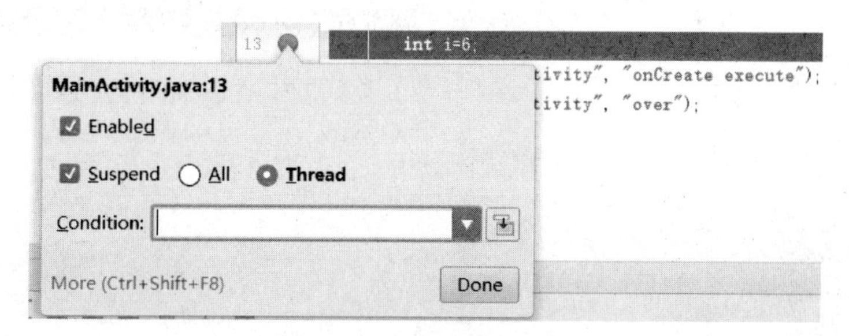

图 1-94　断点属性对话框

对话框中的选项含义如下。

Enabled：断点禁用和启用。

Suspend：勾选 All 选项，执行到断点时，所有线程都会被挂起；勾选 Thread 选项，执行到当前断点，只有当前断点所在的线程挂起。

Condition：设置断点暂停的条件。

2. 方法断点

方法断点主要用来检查方法的输入和输出，包括参数和返回值。

设置方法断点的方法如下。

（1）选中方法名行，执行菜单栏中 Run > Toggle Method Breakpoint 命令。

（2）在方法名的区域后面单击鼠标左键，如图 1-95 所示。

取消断点的方法与设置断点的方法相同。

右击行断点，弹出属性对话框，如图 1-96 所示。

```
10     protected void onCreate(Bundle savedInstanceState) {
11         super.onCreate(savedInstanceState);
12         setContentView(R.layout.activity_main);
13         int i=6;
14         getSum(7,3);
15         Log.d("MainActivity", "onCreate execute");
16         Log.d("MainActivity", "over");
17
18     }
19     private int getSum(int a, int b)
20     {
21         return a+b;
22     }
```

图 1-95　设置方法断点

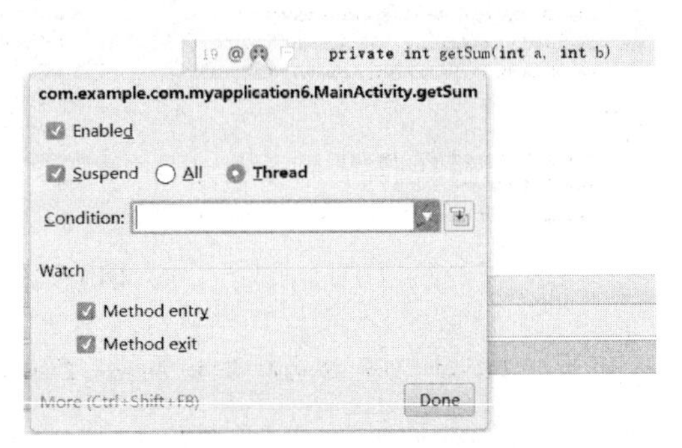

图 1-96　断点属性对话框

与行断点相比，方法断点的属性多了一个 Watch，用来监视方法的进入（Method entry）和退出（Method exit）。

单击调试运行按钮，开始进入调试，运行到方法头停下来，如图 1-97 所示。

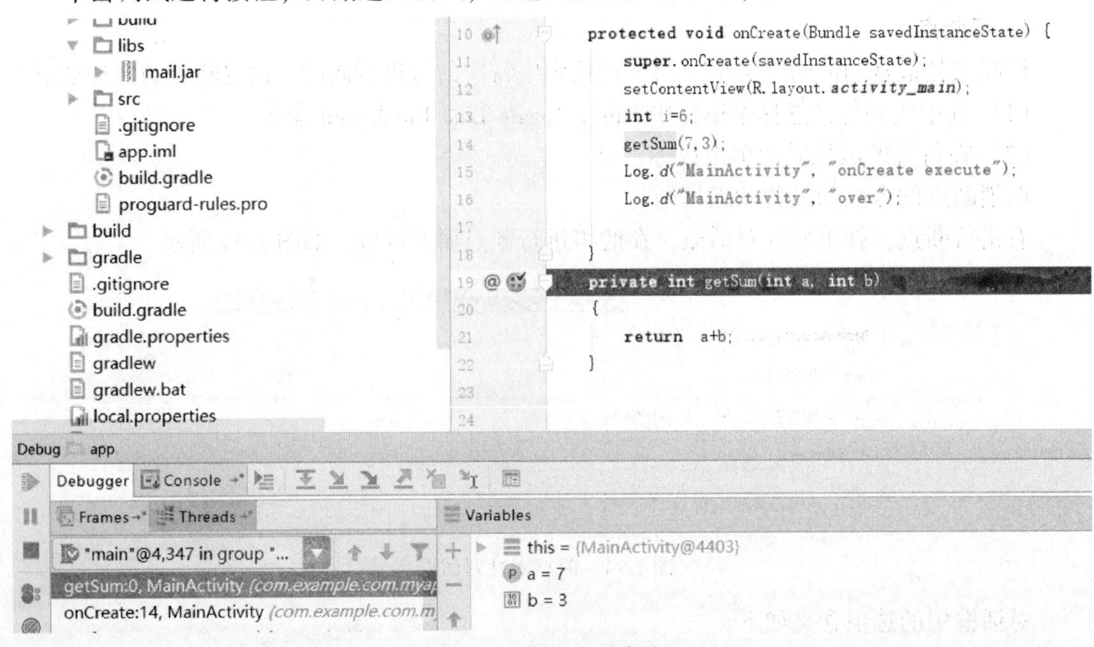

图 1-97　调试方法断点

3. 临时断点

若想某个断点只被触发一次后即自动删除，可以使用临时断点。

设置临时断点的方法：选中行，执行菜单栏中 Run > Toggle Temporary Breakpoint 命令。

右击行断点，弹出属性对话框，如图 1-98 所示。

如果想把临时断点变为普通断点，在属性中取消勾选 Remove once hit 复选框即可。

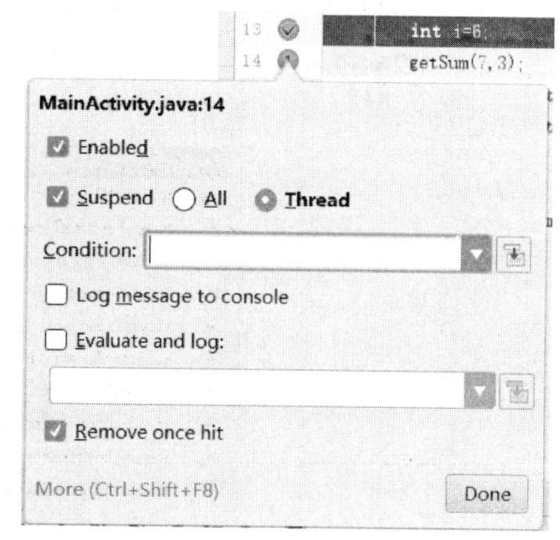

图 1-98 临时断点属性对话框

4. 异常断点

异常断点会在某个异常发生时触发断点，这样我们就可以在第一时间得到异常信息，便于排查问题。

设置异常断点的方法：执行菜单栏中 Run > View Breakpoints 命令，弹出的窗口如图 1-99 所示，选择左上角的 + 号，选择第 3 个选项 Java Exception Breakpoints。

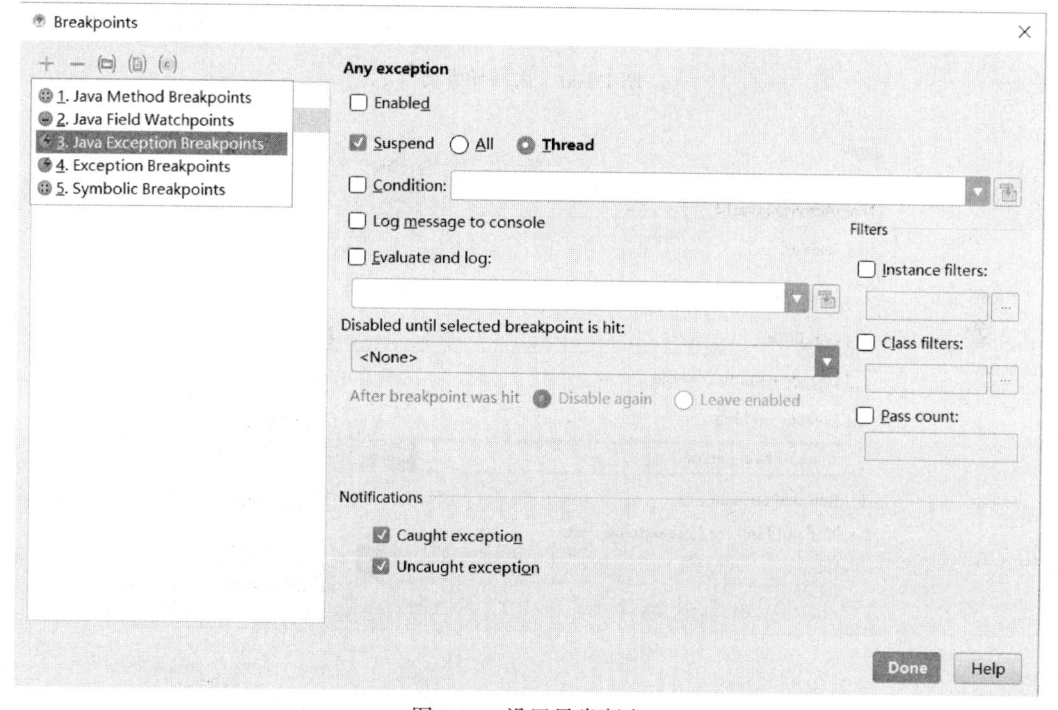

图 1-99 设置异常断点

弹出的窗口如图 1-100 所示，在其中选择异常类。

如果要取消断点，则选中左上角的-号。

5. 日志断点

在调试的时候，若想临时多加一些日志，但又不想重新构建应用程序，则可以使用日志断点。

断点设置方法：右击断点，取消勾选 Suspend 复选框，在展开的面板中勾选 Evaluate and log 复选框，输入日志表达式，如图 1-101 所示。

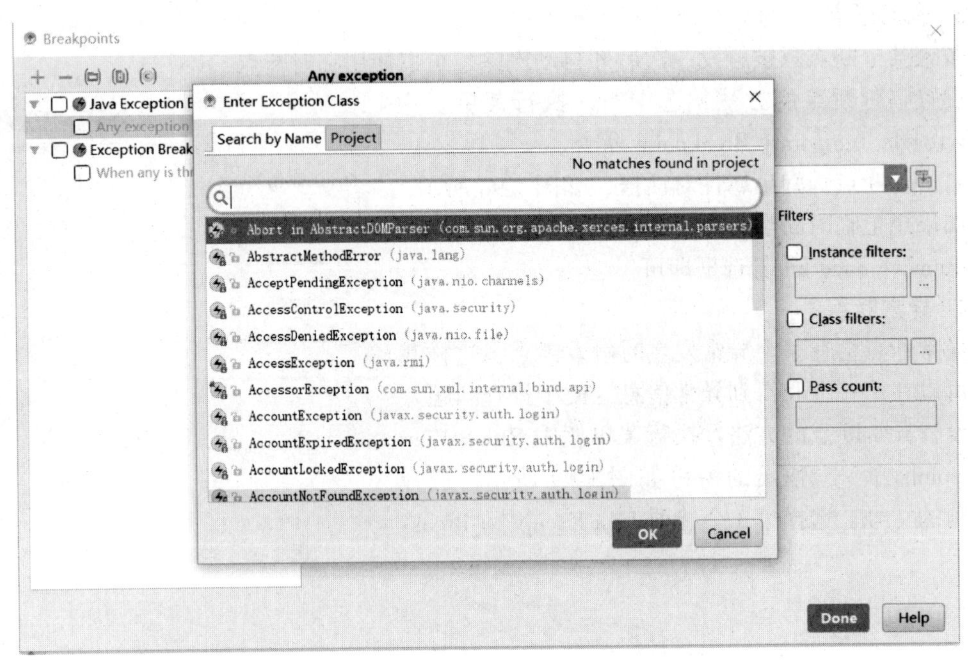

图 1-100　选择异常类

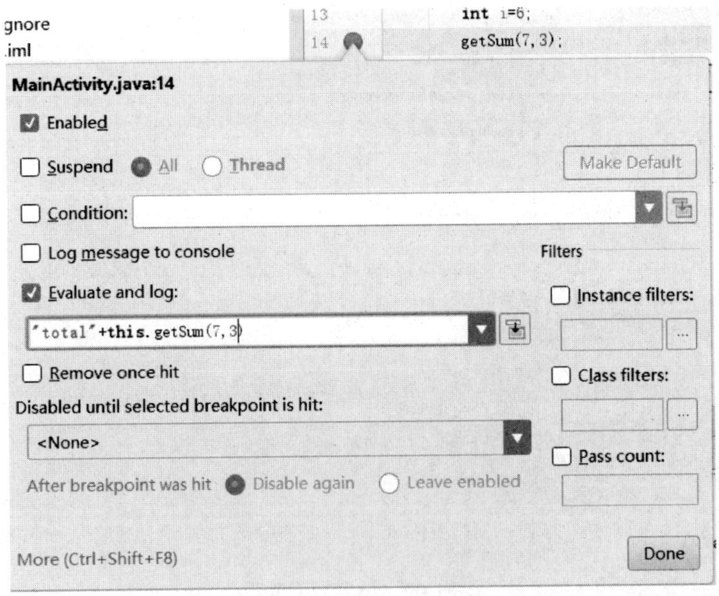

图 1-101　设置日志断点

1.3.10.3　帧调试窗口

帧调试窗口显示了当前断点所在的线程以及执行到该断点所调用过的方法。

1. 堆栈帧

堆栈帧是用来存储数据和部分过程结果的数据结构，同时也被用来处理动态链接（Dynamic Linking）、方法返回值和异常分派（Dispatch Exception）。堆栈帧随着方法调用而创建，随着方法结束而销毁——无论方法是正常完成还是异常完成（抛出了在方法内未被捕获的异常)，都算作方法结束，每一个堆栈帧都有自己的局部变量表、操作数栈和指向当前方法所属的类的运行时常量池的引用。

2. 当前堆栈帧

一个线程在执行过程中，执行到断点处暂停时，如果只有当前正在执行的那个方法的堆栈帧是活动的，这个堆栈帧就叫当前堆栈帧。

帧调试窗口如图 1-102 所示。

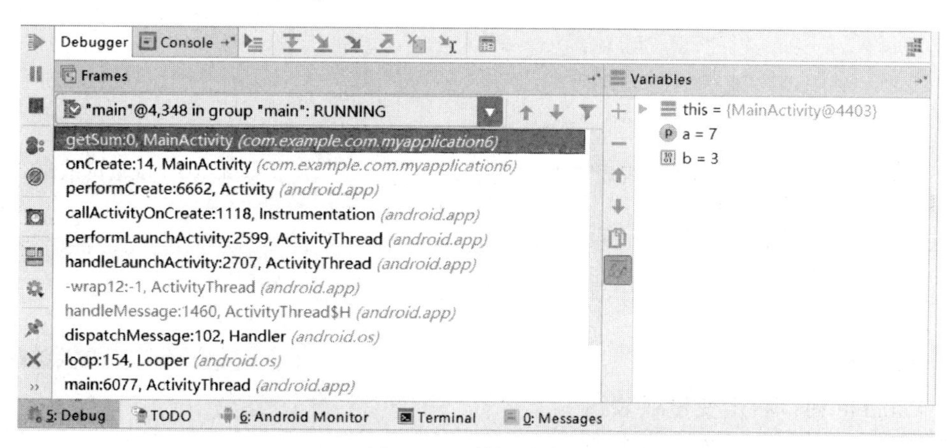

图 1-102　帧调试窗口

1.3.10.4　监视窗口

监视窗口用来计算当前堆栈帧范围内的变量或表达式，在调试过程中通过监视窗口来监视变量或表达式的值。

1. 监视变量

在 Variable 窗口右击变量，选择 Add to Watches 命令，右边的监视窗口即会出现变量，如图 1-103 所示。

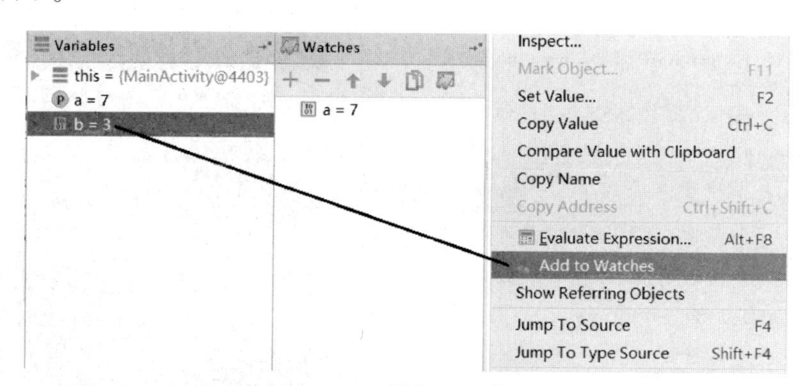

图 1-103　增加监视变量

2. 监视表达式

单击 Watches 窗口左上角 + 按钮，输入表达式，如图 1-104 所示。

1.3.10.5　变量调试窗口

可以在变量窗口中检查应用程序中对象的值。当选择堆栈帧的时候，变量调试窗口就会显示范围内的所有数据，如图 1-105 所示。

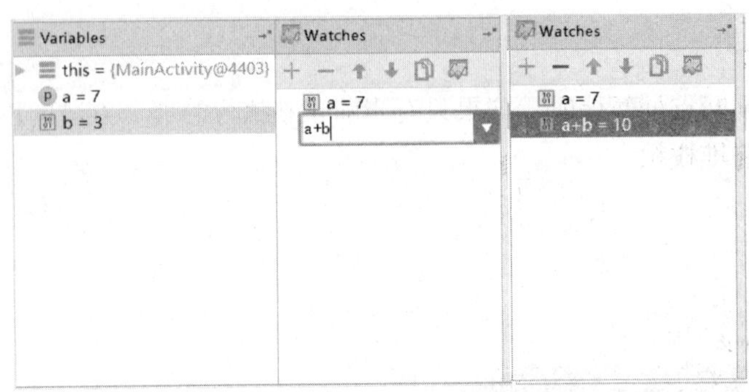

图 1-104　输入监视表达式

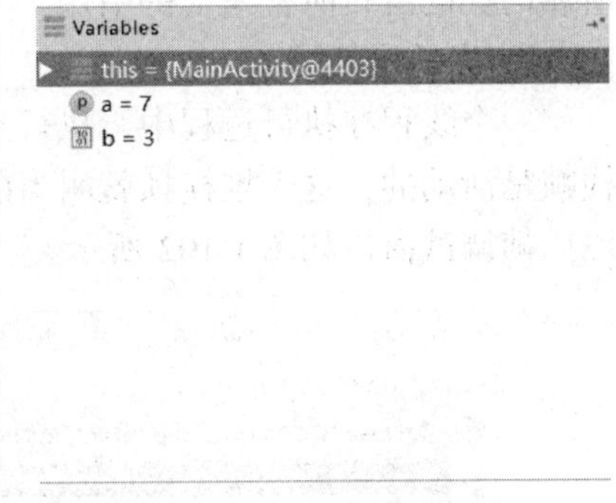

图 1-105　变量调试窗口

在这里可以设置对象的标签、检查对象、计算表达式、添加变量到监视窗口等。

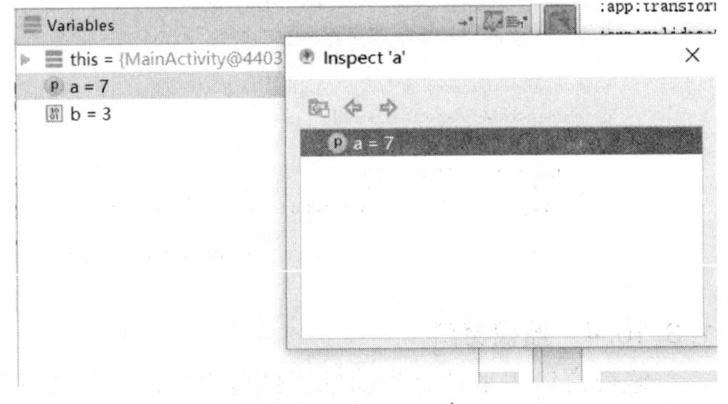

图 1-106　Inspect 窗口

1. Inspect

在 Variable 窗口右击变量或表达式，选择 Inspect 命令，弹出一个非模式检查窗口，如图 1-106 所示。这个窗口可以脱离主窗口。

2. Mark Object

在 Variable 窗口右击变量或表达式，选择 Mark Object 命令，弹出窗口，输入标签，也可以选中颜色，如图 1-107 所示，为对象添加标签，看起来更加直观。

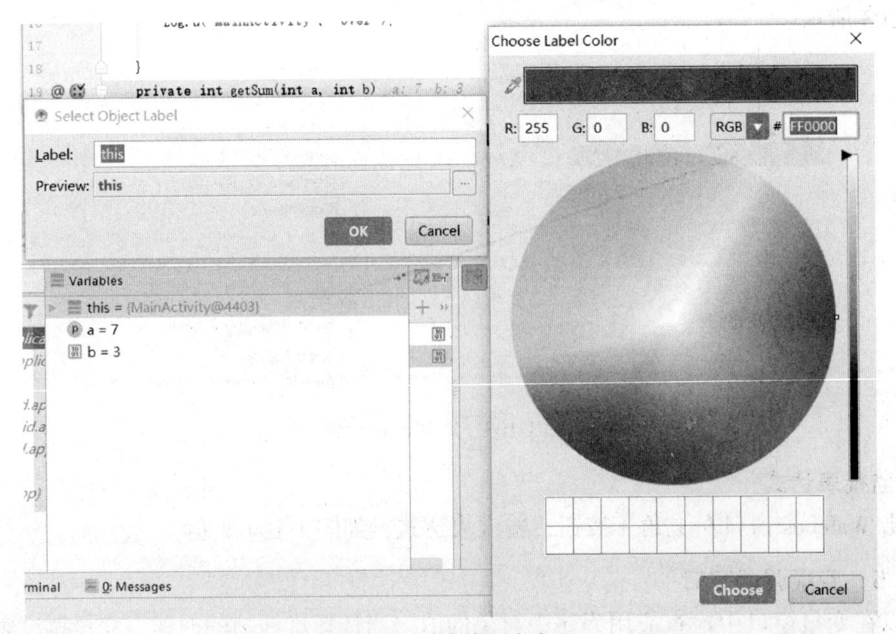

图 1-107　为对象添加标签

1.3.10.6　调试控制工具

调试控制工具用来管理调试程序的运行，提供了以下常用功能。

➢ 暂停、恢复程序执行。

➢ 终止进程。

➢ 查看、禁止断点。

➢ 获取线程堆栈。

Debugger 窗口的调试控制工具按钮及对应的功能如图 1-108 所示。

图 1-108　调试控制工具

1.4　两种开发环境的比较和应用程序转化

Eclipse 是老牌的开发工具，相信早期开发 Android 程序的“码农”都使用过这个软件，添加 ADT 插件之后就能开发 Android 程序。直到遇到 Google 亲自操刀的 Android Studio 这匹黑马，曾经的王者也只能俯首称臣了！Android Studio 是由 Google 亲自研制的用来开发 Android 项目的工具，它的强大也是理所应当的。

Android Studio 是一项全新的基于 IntelliJ IDEA 的 Android 开发环境。Android Studio 提供了集成的 Android 开发工具用于开发和调试，主要有以下特点。

➢ 基于 Gradle 的构建支持。

➢ Android 特定重构和快速修复。

➢ 提示工具更好地对程序性能、可用性、版本兼容和其他问题进行控制捕捉。

➢ 支持 ProGuard 和应用签名功能。

➢ 自带布局编辑器，可以拖放 UI 组件，并在多个屏幕配置上预览布局等。

Android Studio 与 Eclipse 的比较情况如下。

（1）Android Studio 构建程序界面更方便

Android Studio 从一面世就打着所见即所得的旗号，以迅雷不及掩耳之势占领了 Android 项目开发工具的市场。在 Eclipse 中构建 App 的界面，不仅效果和真机上差别太大，而且速度也不快。而 Android Studio 的界面显示非常清晰，而且修改起来很方便。

（2）Android Studio 打印信息更详细

Android Studio 打印的信息可谓是应有尽有，几乎所有在项目中遇到的问题，包括编写、设计、开发、打包、构建等的错误信息都可以在控制台上打印出来，便于问题的准确发现和定位。反观 Eclipse 中的打印信息则少得多，除了 LogCat 之外就是控制台，有时布局文件中多了逗号都发现不了。

（3）Android Studio 编辑历史更详细

Android Studio 对工作台上修改代码、修改布局文件或者删除文件等，记录得非常细致，每一个操作都有记录，每一个操作都能够撤销。而 Eclipse 中删除文件后，之前的编辑记录会被清空，恢复文件会非常麻烦，之前那么多的操作如何回滚是一个问题。

（4）Android Studio 智能识别更强大

智能识别是 Android Studio 中的一大亮点，只要输入 Fp，之后会自动推送含有 FP 或者 fp、甚至是%F（f）%P（p）的选项，中间不管隔着多少个字符，或者大小写不同，系统都能够识别出来并向您推送。

（5）Android Studio 的资源文件可以在代码中预览

使用 Android Studio 进行开发时，资源文件的内容可以在代码中实时预览，不仅布局文件、图片文件能够预览，甚至在 colors. xml 文件中定义的颜色，都能在代码编辑器中看到，这对于 Eclipse 来说是不可思议的。

（6）Android Studio 提供了超过 10 种视图

Android Studio 开发界面中为我们提供了超过 10 种视图，每种视图显示的内容和重点，以及最后呈现出来的代码结构都不一样，非常强大和方便。比如您倾向于显示各个项目的内容，则可在左侧选择 Project 后，在上方切换到 Project 或者 Project files，各个项目的信息就会单独显示。

（7）Eclipse 中的项目体积比较小

在 Eclipse 中所有的文件都是必需的，没有多余的配置文件，所以项目的体积很小。一个项目几十万行的代码，顶多 30M。但是在 Android Studio 中就不一样了，项目中包含各种配置文件，这些文件包含了工具自身的历史文件，还有 gradle 的构建文件，一个项目超过 90M 是常见的事。

（8）Eclipse 中的配置文件无须更新

创建好一个项目后到项目上线，可能无须更新任何 Eclipse 的文件，这个时间的跨度有可能是一年，而 Android Studio 更新 gradle 文件是比较频繁的。

总体来说，Android Studio 比 Eclipse 更强大，通过 Android Studio 进行 Android 项目开发是不可颠覆的趋势和潮流，毕竟 Eclipse 可以做的东西很多，不够专注，而 Android Studio 只面向手机开发，在开发 Android 项目方面的优势肯定是很明显的。但 Android Studio 的低版本也有缺点，特别是在使用 gradle 文件方面，用户体验有待提高，经过几次更新之后，Android Studio 已经成为了非常强大的 IDE 开发环境。安卓产品经理 Jamal Eason 在声明中写到“Google 将会全力专注于 Android Studio 编译工具的开发和技术支持，中止为 Eclipse 提供官方支持。包括中止对 Eclipse ADT 插件以及 Android Ant 编译系统的支持。”Android 开发者是时候正式与 Eclipse 说再见了。

假如您以前使用 Eclipse 进行开发，现在想迁移到 Android Studio 上。首先需要导出工程，导出的目的是生成 Gradle 文件。然后将导出的工程导入到 Android Studio 即可。

1. 从 Eclipse 导出

（1）更新 Eclipse 的 ADT 插件（ADT 的版本必须大于等于 22.0）。

（2）在 Eclipse 中，选择 File > Export 命令。

（3）在弹出的对话框中，单击 Android 并选择 Generate Gradle build files。

（4）选择要导出的工程后单击 Finish 按钮。

所选择导出的工程依旧在原来的路径下，只是多了一个为 Android Studio 准备的 build. gradle 文件。

2. 导入到 Android Studio 中

（1）在 Android Studio 中，关闭当前的工程。页面会跳到欢迎页面。

（2）选择 Import Project 命令。

（3）定位到想要导入的工程所在的目录，选择 build. gradle 文件。

（4）在弹出的对话框中，不进行任何更改直接单击 OK 按钮。

这时，工程即被导入到 Android Studio 中了。

注意：即使工程没有生成 build. gradle 文件，也可以导入到 Android Studio 中。Android Studio 也可以使用 Ant 来进行编译工程。为了更好地使用其他的功能（如 build variants），我们强烈建议您使用 ADT 插件生成一个 gradle 文件或者在 Android Studio 中直接写 gradle 文件。

1.5 本章小结

通过本章的学习，读者对 Android 有了基本的认识，成功地将两种开发环境搭建起来了，了解了环境的构成和意义，并能够实现 Eclipse 项目到 Android Studio 项目的转化。现在已经完成了开发前的准备，搭建了环境，创建了第一个简单的 Android 应用程序，了解了项目结构，从而开启了 Android 应用程序的实践之旅。

第二章 Android开发基础知识

在开发 Android 项目前，首先要了解项目开发的总体流程，本章介绍 Android 软件项目开发流程的相关内容。

2.1 总体流程

一个完整的软件开发流程包括策划、交互、视觉、软件、测试、维护和运营这七个环节，这七个环节并不是孤立的。它们是开发一款成功产品的前提，但每一项也都可以形成一个学科，是一个独立的岗位，随着敏捷开发的流行，以及体验为王时代的到来，现代软件开发更多的是注重效率和敏捷，而不是循规蹈矩地遵循这些开发流程，比如软件开发的岗位不再仅仅是个技术岗位，需要去参与前期的设计和评审，可以在视觉和交互方面提出自己的见解，在开发的过程中需要自测程序尽快解决现有问题，运营和维护的过程中也需要软件的帮助。Android 项目开发的总体流程如图 2-1 所示。

2.2 各阶段描述

下面详细介绍各阶段流程的主要工作。

1. 项目需求分析阶段描述

在此阶段，项目只是一些抽象的想法，需要对想法进行讨论、研究，并对可行性进行评估，将想法一步步拆分，最后分解成一个个明确的需求功能点。

输出：《项目产品需求规格说明书》。

2. 项目设计阶段

（1）原型设计：产品经理根据已明确的需求，对 App 功能进行规划，对页面及布局进行设计，并设计各个页面的跳转逻辑，最终输出 App 各个页面的原型设计图。

（2）UI 设计：UI 设计师根据产品的原型页面设计进行 UI 界面的配色设计，最终输出各个 App 页面的高保真设计效果图。UI 效果图基本跟最终看到的 App 页面效果一样。

输出：《产品概要设计说明书》。

3. 项目实施阶段

App 开发：App 开发人员拿到 UI 设计图后，根据各个 UI 界面效果图进行功能和界面的开发。

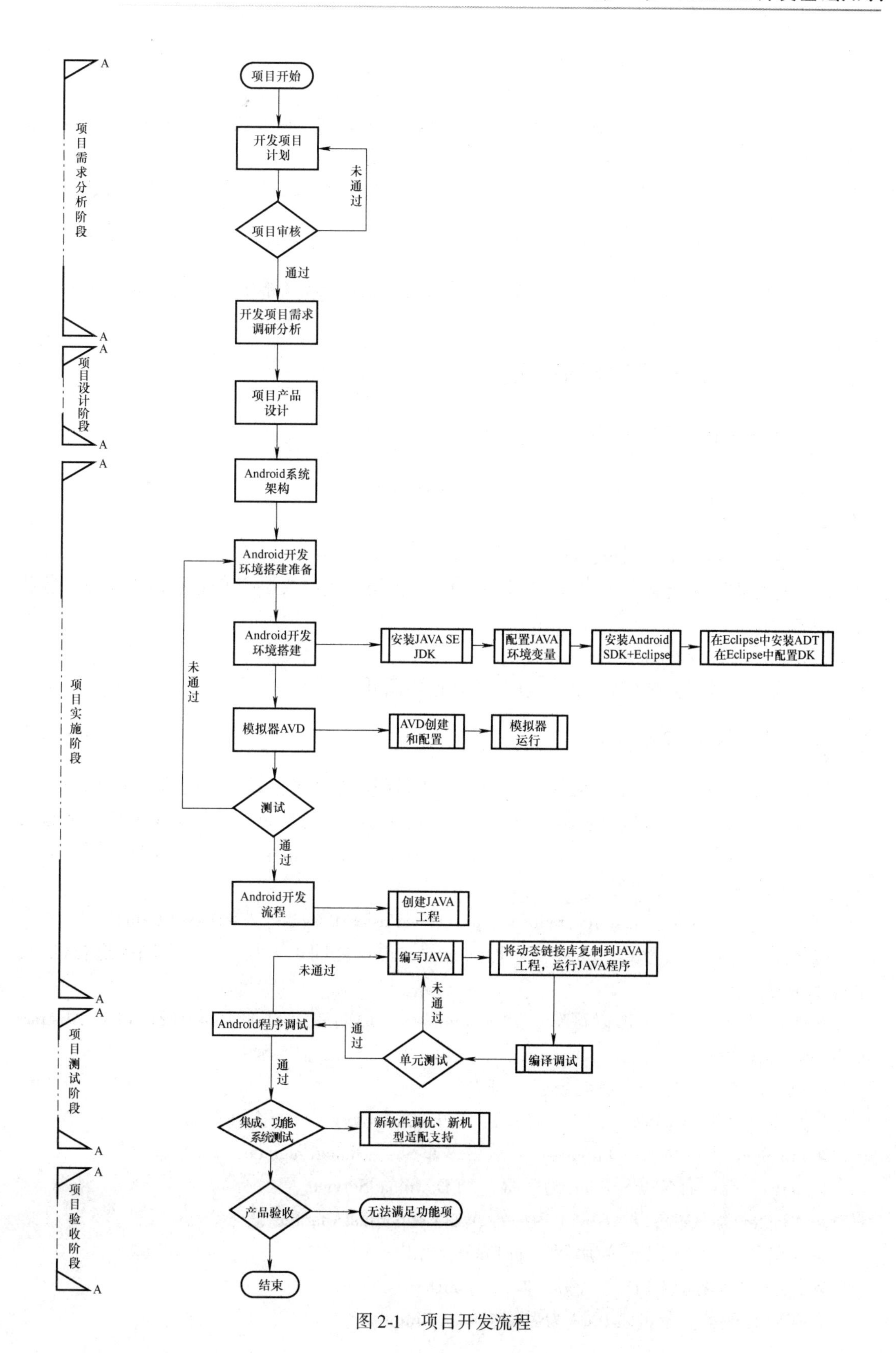
项目需求分析阶段
项目设计阶段
项目实施阶段
项目测试阶段
项目验收阶段
A
项目开始
开发项目计划
未通过
项目审核
通过
开发项目需求调研分析
项目产品设计
Android系统架构
Android开发环境搭建准备
Android开发环境搭建
安装JAVA SE JDK
配置JAVA环境变量
安装Android SDK+Eclipse
在Eclipse中安装ADT
在Eclipse中配置DK
模拟器AVD
AVD创建和配置
模拟器运行
测试
Android开发流程
创建JAVA工程
编写JAVA
将动态链接库复制到JAVA工程，运行JAVA程序
Android程序调试
单元测试
编译调试
集成、功能、系统测试
新软件调优、新机型适配支持
产品验收
无法满足功能项
结束

图 2-1　项目开发流程

输出：《产品详细设计说明书》。

4. 项目测试阶段

App 功能开发完成后，测试人员会对整个 App 进行测试，从而发现程序中的一些问题，一般来说，开发人员需要同步调试测试人员发现的问题。

输出：《系统测试缺陷记录》《产品单元测试报告》《集成测试报告》《系统测试报告》。

5. 项目验收阶段

进行试用，然后解决问题。

输出：《项目总结报告》《项目中无法满足功能项说明书》《维护方案》。

2.3 Android 开发代码规范

2.3.1 项目和包命名规范

包名一律小写，少用缩写和长名，采用以下规则。

➢［基本包］.［项目名］.［模块名］

➢ 包名一般不要超过三级，级别多了太复杂。

➢ 不得将类直接定义在基本包下，所有项目中的类、接口等都应当定义在各自的项目和模块包中。

例如：package com. lcw. test. util。

规范命名能够提高项目组织性，从而更好地协同开发。

2.3.2 类和接口命名方法

Android 代码一般使用驼峰式规则，用名词或名词词组命名，每个单词的首字母大写。

常用类的命名，类或接口名是个一名词，采用大小写混合的方式，每个单词的首字母大写。尽量使类名简洁而富有描述性。使用完整单词，避免用缩写词（除非该缩写词被更广泛使用，比如 URL、HTML）。

例如：class Raster、class ImageSprite、interface RasterDelegate、interface Storing。

命名采用单词组合形式，单词首字母为大写，单词之间可采用“_”下划线进行区分，也可不采用。

根据定义类型首字母加以区分，比如 Interface，命名首字母加大写的 I。再如 Abstract class，命名首字母加大写 A。

根据功能类型结尾加上功能描述字符串。

➢ Activity 类：命名以 Activity 为后缀，如 LoginActivity。

➢ Fragment 类：命名以 Fragment 为后缀，如 ShareDialogFragment。

➢ Service 类：命名以 Service 为后缀，如 DownloadService。

➢ Adapter 类：命名以 Adapter 为后缀，如 CouponListAdapter。

➢ 工具类：命名以 Util 为后缀，如 EncryptUtil。

➢ 模型类：命名以 BO 为后缀，如 CouponBO。

➢ 接口实现类：命名以 Impl 为后缀，如 ApiImpl。

需要注意的是，类命名不能使用中文字符，也不能在命名字符串中出现 0-9 的数值描述和除下划线以外的其他字符描述，命名的字母组合尽量能够让人通过本身的文字意义初步了解类的大体功能。

类的命名要达到不见注释见名知意。采用大小写混合的方式，第一个单词的首字母小写，其后单词的首字母大写。

2.3.3　变量和常量命名方法

下面介绍变量和常量的命名方法。

1. 变量

变量名不应以下划线或美元符号开头。尽量避免单个字符的变量名，除非是一次性的临时变量。临时变量通常被取名为 i、j、k、m 和 n，它们一般用于整型；c、d、e 一般用于字符型。

不建议采用匈牙利命名法则，对不易清楚识别出变量类型的变量，应使用类型名或类型名缩写作其后缀组件或部件变量，使用其类型名或类型名缩写作其后缀集合类型变量，例如数组和矢量，应采用复数命名或使用表示该集合的名词做后缀。

例如：

```
private TextView headerTitleTxt; //标题栏的标题
private Button loginBtn; // 登录按钮;
```

2. 常量

常量命名全部采用大写，单词间用下划线隔开，例如：

```
static final int MIN_WIDTH  = 4;
static final int MAX_WIDTH  = 999;
static final int GET_THE_CPU  = 1;
```

2.3.4　方法的命名方法

方法名是一个动词，采用大小写混合的方式，第一个单词的首字母小写，其后单词的首字母大写。

取值类可使用 get 前缀，设值类可使用 set 前缀，判断类可使用 is（has）前缀。

方法中一定要加上适当的非空判断，与 try catch 语句等程序健壮性的判断。

- 初始化方法，命名以 init 开头，例如 initView。
- 按钮单击方法，命名以 to 开头，例如 toLogin。
- 设置方法，命名以 set 开头，例如 setData。
- 具有返回值的获取方法，命名以 get 开头，例如 getData。
- 通过异步加载数据的方法，命名以 load 开头，例如 loadData。
- 布尔型的判断方法，命名以 is 或 has 开头，或以具有逻辑意义的单词开头如 equals，例如 isEmpty。

2.3.5　注释规范

注释是程序维护的灵魂。对已经不推荐使用的类和方法需要注明@ Deprecated，并说明替代的类或者方法；对于针对集合、开关的方法，要在方法注释中表明是否多线程安全。

1. 文件注释

对于所有源文件，都应该在开头进行注释，其中列出文件的版权声明、文件名、功能描述以及创建、修改记录。

例如：

```
/*
 * Copyright(C) 2009-2014lisi Inc. All Rights Reserved.
 * FileName:HelloWorld.java
 * @Description:简要描述本文件的内容
 * History:
 *版本号  作者  日期           简要介绍相关操作
 *  1.0  liucw 2017-04-21  Create
 *  1.1  liucw 2017-04-23  Add Hello World
 */
```

2. 类或接口注释

采用 JavaDoc 文档注释，在类、接口定义之前应当对其进行注释，包括类、接口的描述，以及最新修改者、版本号、参考链接等。

例如：

```
/**
 * 描述
 * @author liuxin
 * @version 1.0
 * @see 参考的 JavaDoc
 */
class Window extends BaseWindow
{
    ...
}
```

3. JavaDoc 文档注释

描述 Java 的类、接口、构造方法、方法以及字段。

每个文档注释都会被置于注释定界符/**...*/之中，一个注释对应一个类、接口或成员。

该注释应位于声明之前。

文档注释的第一行（/**）不需要缩进，随后的文档注释每行都缩进 1 格（使星号纵向对齐）。

方法注释：采用 JavaDoc 文档注释，在方法定义之前当对其进行注释，包括方法的描述、输入、输出及返回值说明、抛出异常说明、参考链接等。

例如：

```
/**
 * @author liucw
 * @Description: ${todo}
```

```
 * @date ${date} ${time}
 * @param 参数说明:每个参数一行,注明其取值范围等
 * @return 返回值:注释出失败、错误、异常时的返回情况
 * @exception 异常:注释出什么条件下会引发什么样的异常
 * @see 参考的 JavaDoc
 */
public char charAt(int index)
{
    ...
}
```

4. 其他注释

单行代码注释一律使用注释界定符“//”。

```
// explain what this means
  if(bar > 1)
  {
    ……
  }
  int isShow = 0;//是否显示
```

多行注释使用注释界定符“/*... */”。

```
/*
 * Here is a block comment with
 * multiple lines for text comments.
 */
```

这些命名规范和注释，看似微不足道，却是我们通往专业的重要一步。

2.4 本章小结

本章介绍了 Android 软件项目开发的整体流程及 Android 开发过程中的代码规范，为 Android 项目开发奠定基础。

第三章

应用程序用户接口——界面设计

每个 Android 应用程序首先面临的就是界面的开发。Android 系统提供了丰富的界面控件和布局，以及用户图形界面的业务处理类 Activity，Android Studio 提供了更便捷的用户界面开发方法。本章主要介绍 Android 用户界面的设计。

3.1 用户界面设计基础

用户界面（User Interface）是系统和用户之间进行信息交换的媒介，设计用户界面需要解决如下问题。

（1）界面设计与程序逻辑要完全分离，这样不仅有利于软件的并行开发，而且在后期修改界面时，不用修改程序的逻辑代码。

（2）根据不同型号手机的屏幕解析度、尺寸和纵横比，自动调整界面上控件的位置和尺寸，避免因为屏幕信息的变化而出现显示错误。

（3）能够合理利用较小的屏幕显示空间，构造出符合人机交互规律的用户界面，避免出现凌乱、拥挤的用户界面。

（4）Android 已经解决了前两个问题，使用 XML 文件描述用户界面；资源文件独立保存在资源文件夹中；对用户界面描述非常灵活，允许不明确定义界面元素的位置和尺寸，仅声明界面元素的相对位置和粗略尺寸。

Android 用户界面框架（Android UI Framework）采用 MVC（Model-View-Controler）模型。提供了处理用户输入的控制器（Controler）、显示用户界面和图像的视图（View）以及保存数据和代码的模型（Model），如图 3-1 所示。

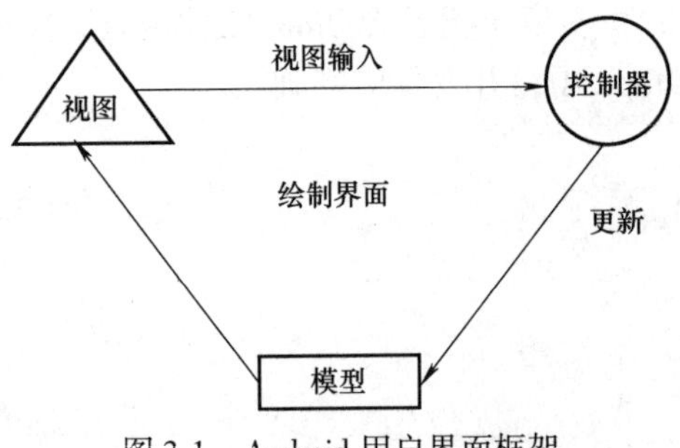

图 3-1　Android 用户界面框架

控制器（Controller）使用对立队列处理外部动作，每个外部动作作为一个对应的事件加入队列中，然后 Android 用户界面框架按照“先进先出”的规则从队列获取事件，并将这个事件分配给所对应的事件处理函数。

Android 用户界面框架中的界面元素以一种树型结构组织在一起，称为视图树，Android

系统会依据视图树的结构从上至下绘制每一个界面元素。每个元素负责对自身的绘制，如果元素包含子元素，该元素会通知其下所有子元素进行绘制，如图 3-2 所示。

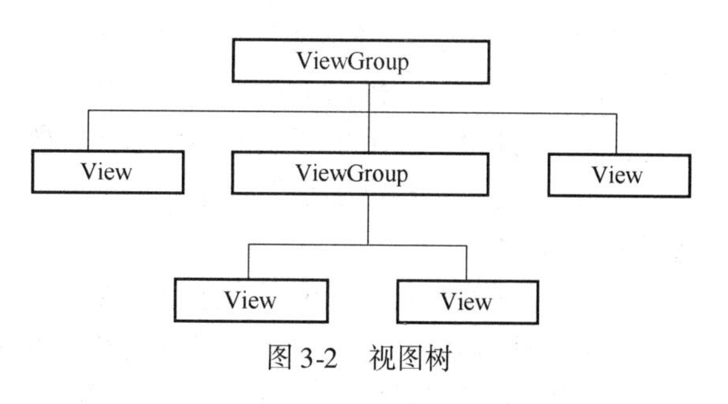

图 3-2　视图树

在一个 Android 应用程序中，用户界面通过 View 和 ViewGroup 对象构建，其中 ViewGroup 对象可以理解为一种容器，用于容纳其他的控件对象，并使这些控件对象按照特定的规则进行排列，即按照某种布局排列。Android 的布局 Layout 是 ViewGroup 的子类，能够提供各种不同的布局结构，如线性布局、表格布局和相对布局等。

视图组件 View 对象是 Android 平台上用户界面的基础单元，也可称为控件。View 控件放在 ViewGroup 容器中，Android 系统提供了许多类型的 View，例如 TextView 和 Button 等类，它们都是 View 类的子类。

View 和 ViewGroup 之间采用了组合设计模式。ViewGroup 作为布局容器类的最上层，负责对添加进 ViewGroup 的这些 View 进行布局，布局容器里面又可以有 View 和 ViewGroup。当然，一个 ViewGroup 也可以加入到另一个 ViewGroup 里。因为 ViewGroup 也是继承于 View. ViewGroup 类，在每个 ViewGroup 类中都会有一个嵌套类，这个嵌套类的属性中定义了子 View 的位置和大小。

Android Studio 2. 3 提供了界面方便操作的布局管理器，如图 3-3 所示。

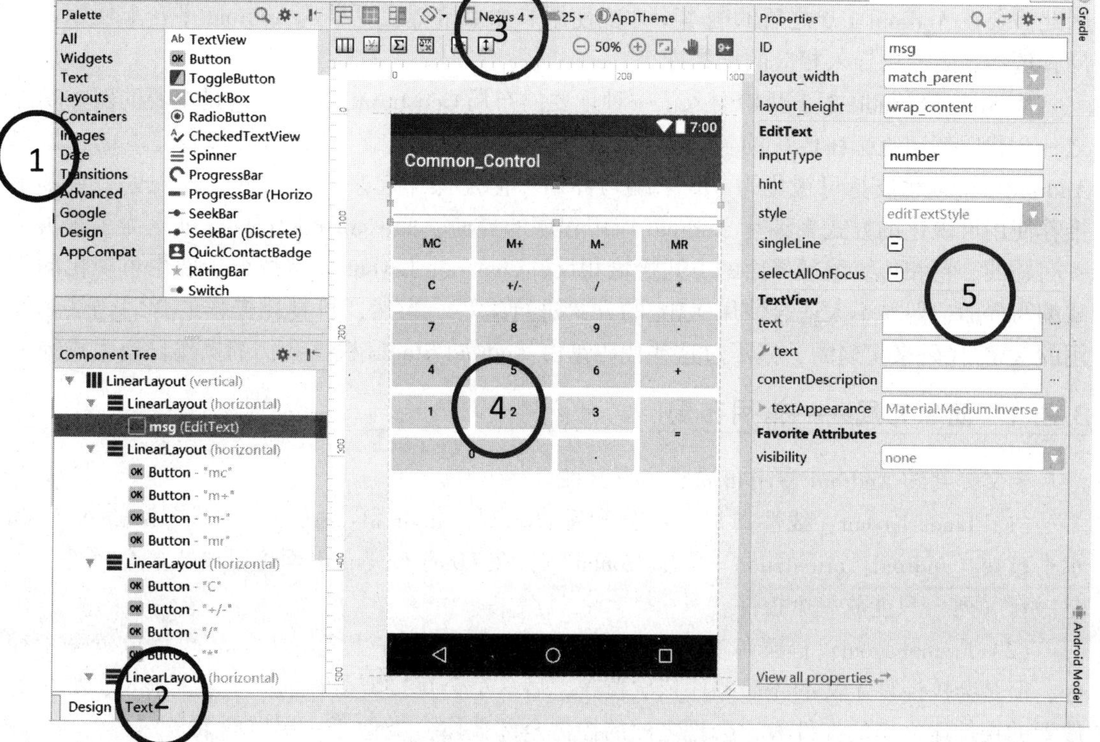

图 3-3　布局管理器

与图 3-3 中的数字相对应，管理器中的各部分功能如下。

（1）组件列表：提供了常用的 UI 组件，可以拖拽到布局编辑器中的控件列表。

（2）组件树：显示了布局的层次结构，既可以管理控件，又可以调整控件的位置。

（3）工具栏：提供了很多预览工具，可以查看布局在不同分辨率、不同 API 和不同主题上的显示效果。

（4）设计编辑器：显示所有组件的布局效果，并提供一个设计模型图。

（5）属性界面：可改变当前所选的控件的属性。

在图 3-10 的设计编辑器中添加一个按钮。在图 3-10 的最下方有两个选项卡，左边是 Design，右边是 Text。Design 是当前的可视化布局管理器，在这里不仅可以预览当前的布局，还可以通过拖放的方式编辑布局；而 Text 则通过 XML 文件的方式来编辑布局。

3.2 界面最外层设计——布局

Android 常用的布局有以下五种：线性布局（LinearLayout）、框架布局（FrameLayout）、表格布局（TableLayout）、相对布局（RelativeLayout）和绝对布局（AbsoluteLayout）。

这几种布局满足了绝大部分界面设计要求，不过，细心的读者会发现，只有 LinearLayout 支持使用 layout_weight 属性按比例指定控件大小，其他布局不支持此功能。为此，Android 引入了一种全新的布局方式来解决此问题——百分比布局，2015 年，Google 正式提供百分比布局支持库（android-support-percent-lib）。

从 Android 4.0 开始，新增了网格布局 GridLayout，这是 Android 4.0 新增的布局管理器，因此需要在 Android 4.0 之后的版本中才能使用。如果希望在更早的 Android 平台上使用该布局管理器，则需要导入相应的支撑库。

从 Android Studio 2.2 开始引入了一种重要的布局 ConstraintLayout，这是主要的新增功能之一。我们都知道，在传统的 Android 开发中，界面基本是靠编写 XML 代码完成的，虽然 Android Studio 也支持可视化的方式来编写界面，但是操作起来并不方便，我们也一直都不推荐使用可视化的方式来编写 Android 应用程序的界面，而 ConstraintLayout 就是为了解决这一问题的。它和传统编写界面的方式恰恰相反，ConstraintLayout 非常适合于使用可视化的方式编写界面，但并不太适合使用 XML 的方式进行编写。当然，可视化操作的背后仍然是使用的 XML 代码来实现的，只不过这些代码是由 Android Studio 根据我们的操作自动生成的。

3.2.1 简单布局——常用布局

上文提到的 Android 常用布局有如下五种。

（1）LinearLayout：线性布局，可分为垂直布局（android：orientation =" vertical"）和水平布局（android：orientation =" horizontal"），在 LinearLayout 里面可以放置多个控件，但是一行（列）只能放一个控件。

（2）FrameLayout：框架布局，所有控件都放置在屏幕左上角（0,0），可以放置多个控件，但是会按控件定义的先后顺序依次覆盖，后一个会直接覆盖在前一个之上显示，如果后放置的控件比之前的控件大，会把之前的控件全部盖住（类似于一层层的纸张）。

（3）AbsoluteLayout：绝对布局，可以直接指定子控件的绝对位置（例如 android：layout

_x = " 60px" android：layout_y = " 32px"），这种布局简单直接，但是由于手机的分辨率大小不统一，绝对布局的适应性比较差。

（4）RelativeLayout：相对布局，其子控件根据所设置的参照控件进行布局，设置的参照控件可以是父控件，也可以是其他的子控件。

（5）TableLayout：表格布局，以行列的形式管理子控件，在表格布局中的每一行可以是一个 View 控件或者是一个 TableRow 控件。而 TableRow 控件中还可以添加子控件。

利用这五种布局，可以在屏幕中随心所欲摆放控件，而且控件的大小和位置会随着屏幕大小的变化做出相应的调整。这五个布局在 View 的继承体系中的关系如图 3-4 所示。

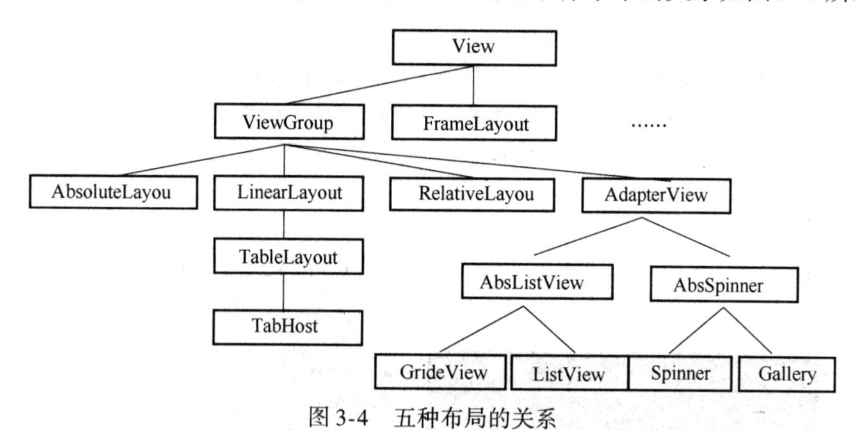

图 3-4　五种布局的关系

目前，在这 5 种布局中，线性布局、相对布局和表格布局使用较广泛，框架布局和绝对布局已经很少使用了。

1. 线性布局

线性布局（LinearLayout）是使用比较多的布局类型之一。线性布局的作用就像其名字一样，根据设置的垂直或水平的属性值，将所有的子控件按垂直或水平方式进行组织排列。当布局设置为垂直时，布局里所有子控件被组织在同一列中；当布局设置为水平时，布局里所有子控件被组织在同一行中，设置线性布局方向的属性为：android：orientation，其值可以是 horizontal 或 vertical，分别代表水平或垂直方向。

线性布局中，有如下四个非常重要的参数，直接决定元素的布局和位置。

➢ android：layout_gravity：是相对于它的父元素而言的，说明元素显示在父元素的什么位置。

➢ android：gravity：是对元素本身而言的，元素本身的文本显示在什么地方靠该属性设置，若不设置，则默认是在左侧。

➢ android：orientation：线性布局以列或行显示内部子元素。

➢ android：layout_weight：线性布局内子元素对未占用空间水平或垂直分配权重值，其值越小，权重越大。

下面是一个线性布局的例子，XML 的代码如下。

```
<LinearLayout xmlns:android = "http://schemas.android.com/apk/res/android"
xmlns:tools = "http://schemas.android.com/tools"
    android:layout_width = " match_parent"
    android: layout_height = " match_parent"
```

```
android: orientation = " vertical"
tools: context = " .LinearLayoutOneActivity" >
 <EditText android: layout_width = " fill_parent"
    android: layout_height = " wrap_content" />
 <LinearLayout android: layout_width = " fill_parent"
    android: layout_height = " wrap_content"
    android: orientation = " horizontal" >
   <Button android: layout_width = " wrap_content"
      android: layout_height = " wrap_content"
      android: text = " 登陆" />
   <Button android: layout_width = " wrap_content"
        android: layout_height = " wrap_content"
        android: text = " 退出" />
 </LinearLayout>
</LinearLayout>
```

运行效果图如图 3-5 所示。外面的 LinearLayout 的方向是垂直的，内嵌的 LinearLayout 是水平的。

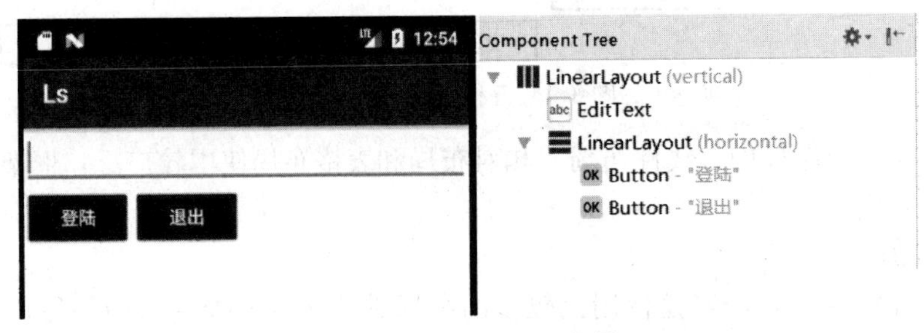

图 3-5　LinearLayout 布局

2. 相对布局

相对布局（RelativeLayout）中，控件的位置按照相对位置计算，后一个控件在什么位置依赖于前一个控件的基本位置，是布局最常用，也是最灵活的一种布局。可以使用右对齐，或上下对齐，或置于屏幕中央等形式排列元素。布局中的控件是按顺序排列的，如果第一个元素在屏幕的中央，那么相对于这个元素的其他元素将以屏幕中央的相对位置来排列。如果使用 XML 布局文件来定义这种布局，之前被关联的元素必须定义。

相对布局的相关属性如表 3-1 所示。

表 3-1　相对布局的相关属性

属　性	描　述
android：layout_above	将该控件的底部置于给定 ID 的控件之上
android：layout_below	将该控件的底部置于给定 ID 的控件之下
android：layout_toLeftOf	将该控件的右边缘与给定 ID 的控件左边缘对齐
android：layout_toRightOf	将该控件的左边缘与给定 ID 的控件右边缘对齐
android：layout_alignBaseline	将该控件的 baseline 与给定 ID 的 baseline 对齐
android：layout_alignTop	将该控件的顶部边缘与给定 ID 的顶部边缘对齐

（续）

属　　性	描　　述
android：layout_alignBottom	将该控件的底部边缘与给定 ID 的底部边缘对齐
android：layout_alignLeft	将该控件的左边缘与给定 ID 的左边缘对齐
android：layout_alignRight	将该控件的右边缘与给定 ID 的右边缘对齐
android：layout_alignParentTop	如果为 true，将该控件的顶部与其父控件的顶部对齐
android：layout_alignParentBottom	如果为 true，将该控件的底部与其父控件的底部对齐
android：layout_alignParentLeft	如果为 true，将该控件的左部与其父控件的左部对齐
android：layout_alignParentRight	如果为 true，将该控件的右部与其父控件的右部对齐
android：layout_centerHorizontal	如果为 true，将该控件的置于水平居中
android：layout_centerVertical	如果为 true，将该控件的置于垂直居中
android：layout_centerInParent	如果为 true，将该控件的置于父控件的中央
android：layout_marginTop	上偏移的值
android：layout_marginBottom	下偏移的值
android：layout_marginLeft	左偏移的值
android：layout_marginRight	右偏移的值

下面是采用 RelativeLayout 布局的例子，XML 的代码如下。

```
<RelativeLayout xmlns:android="http://schemas.android.com/apk/res/android"
        android:layout_width=" fill_parent"
        android: layout_height=" wrap_content"
        android: padding=" 10px" >
    <TextView android: id=" @ +id/label"
        android: layout_width=" fill_parent"
        android: layout_height=" wrap_content"
        android: text=" Type here:" />
    <EditText android: id=" @ +id/entry"
        android: layout_width=" fill_parent"
        android: layout_height=" wrap_content"
        android: background=" @android: drawable/editbox_background"
        android: layout_below=" @id/label" />
     <Button android: id=" @ +id/ok"
        android: layout_width=" wrap_content"
        android: layout_height=" wrap_content"
        android: layout_below=" @id/entry"
        android: layout_alignParentRight=" true"
        android: layout_marginLeft=" 10px"
        android: text=" OK" />
    <Button android: layout_width=" wrap_content"
        android: layout_height=" wrap_content"
        android: layout_toLeftOf=" @id/ok"
        android: layout_alignTop=" @id/ok"
        android: text=" Cancel" />
```

```
</RelativeLayout>
```

运行效果图如图 3-6 所示。

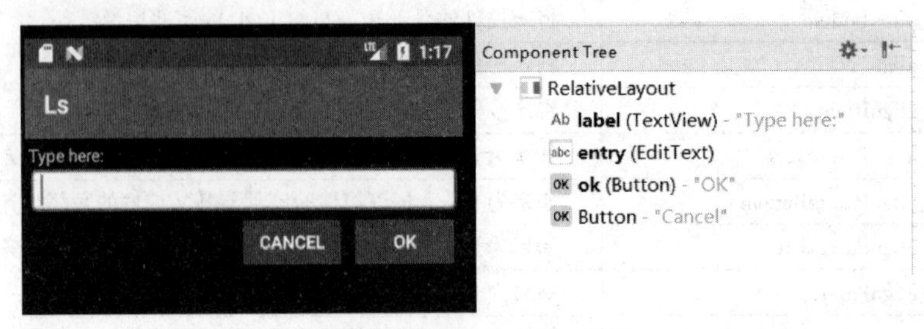

图 3-6　RelativeLayout 布局

3. 表格布局

表格布局（TableLayout）把用户界面按表格形式划为行和列，然后把控件分配到指定的行或列中，一个表格布局由许多的 TableRow 组成，每个 TableRow 定义一行 Row。表格布局容器不会显示行、列或单元格 Cell 的边框线。每行可有 0 个或多个 Cell；每个 Cell 能容纳一个 View 对象。表格允许 Cell 为空，但 Cell 不能跨列。

总体来说，这个 TableLayout 的属性和 html 中 Table 标签的属性差不多。TableLayout 可设置的属性包括全局属性及单元格属性。

（1）全局属性即列属性，有以下 3 个参数。

- android：stretchColumns：设置可伸展的列。该列可以向行方向伸展，最多可占据一整行。
- android：shrinkColumns：设置可收缩的列。当该列子控件的内容太多，已经挤满所在行时，该子控件的内容将往列方向显示。
- android：collapseColumns：设置要隐藏的列。

例如：

android：stretchColumns =" 0"　　　第 0 列可伸展

android：shrinkColumns =" 1，2"　　　第 1、2 列皆可收缩

android：collapseColumns =" *"　　　隐藏所有行

列可以同时具备 stretchColumns 及 shrinkColumns 属性，那么当该列的内容太多时，将“多行”显示其内容（这里不是真正的多行，而是系统根据需要自动调节该行的 layout_height）。

（2）单元格属性，有以下两个参数。

- android：layout_column：指定该单元格在第几列显示。
- android：layout_span：指定该单元格占据的列数（未指定时，为 1）。

例如：android：layout_column =" 1"　　　该控件显示在第 1 列

android：layout_span =" 2"　　　该控件占据 2 列

另外，一个控件也可以同时具备这两个特性。

下面是采用 TableLayout 布局的例子，XML 的代码如下。

```
<?xml version="1.0" encoding="utf-8"?>
<TableLayout xmlns:android="http://schemas.android.com/apk/res/android"
    android:layout_width=" match_parent"
```

```
android: layout_height = " match_parent"
android: stretchColumns = " 1, 2"  >
<TableRow >
    <TextView
        android: layout_width = " wrap_content"
        android: layout_height = " wrap_content"
        android: text = " 用户名:" / >
<EditText
        android: layout_width = " match_parent"
        android: layout_height = " wrap_content"
        android: hint = " 请输入用户名"
        android: layout_span = " 2" / >
</TableRow >
<TableRow >
    <TextView
        android: layout_width = " wrap_content"
        android: layout_height = " wrap_content"
        android: text = " 密码:" / >
<EditText
        android: layout_width = " match_parent"
        android: layout_height = " wrap_content"
        android: hint = " 请输入密码"
        android: password = " true"
        android: layout_span = " 2" / >
</TableRow >
<TableRow >
    <Button
        android: layout_width = " 0dp"
        android: layout_height = " wrap_content"
        android: text = " 登陆"
        android: layout_weight = " 1" / >
    <Button
        android: layout_width = " 0dp"
        android: layout_height = " wrap_content"
        android: text = " 注册"
        android: id = " @ +id/button2"
        android: layout_weight = " 1" / >
    <Button
        android: layout_width = " 0dp"
        android: layout_height = " wrap_content"
        android: text = " 取消"
        android: layout_weight = " 1" / >
</TableRow >
```

```
</TableLayout>
```

运行效果图如图 3-7 所示。

图 3-7　TableLayout 布局

3.2.2　百分比布局

Android 引入了一种全新的布局方式来解决按比例指定控件大小的功能——百分比布局。

由于 RelativeLayout 和 FrameLayout 布局没有按比例分配的属性功能，所以在兼容库 com. android. support：percent 中提供了 PercentFrameLayout 和 PercentRelativeLayout 这两个全新的布局。新的布局增加了以下几个属性。

- layout_widthPercent
- layout_heightPercent
- layout_marginPercent
- layout_marginLeftPercent
- layout_marginRightPercent
- layout_marginBottomPercent
- layout_marginStartPercent
- layout_marginEndPercent

这几个属性含义是宽高和边缘按照百分比计算。

Android 团队将百分比布局定义在了 Support 库当中，只需要在项目的 build. gradle 中添加百分比布局库的依赖，就能保证百分比布局在 Android 所有系统版本中的兼容性。

需要注意的是，Android Studio 新建的项目有两个 build. gradle 文件，这里是内层的 build. gradle 文件。

打开项目内层 build. gradle 文件，在 dependencies 闭包中添加如下内容。

```
dependencies {
compile fileTree(dir:'libs', include: ['* .jar'])
  compile 'com.android.support:appcompat-v7:25.3.1'
  compile 'com.android.support.constraint:constraint-layout:1.0.2'
  testCompile 'junit:junit:4.12'
  compile 'com.android.support:percent:25.3.1'}
```

需要注意的是，Android Studio 会弹出一个提示，如图 3-8 所示。

Gradle files have changed since last project sync. A project sync may be necessary for the IDE to work properly.　　Sync Now

图 3-8　Gradle 文件修改提示

这里单击 Sync Now 即可，之后 Gradle 即会开始同步，把新添加的百分比库引入到项目中。

下面是采用百分比布局的例子，XML 的代码如下。

```
<? xml version = "1.0" encoding = "utf-8"? >
<android.support.percent.PercentRelativeLayout
    xmlns:android = "http://schemas.android.com/apk/res/android"
    xmlns:tools = "http://schemas.android.com/tools"
    xmlns:app = "http://schemas.android.com/apk/res-auto"
    android:id = "@ +id/activity_main"
    android: layout_width = " match_parent"
    android: layout_height = " match_parent"    >
    <Button
        app: layout_widthPercent = " 60 %"
        app: layout_heightPercent = " 20 %"
        app: layout_marginLeftPercent = " 10 %"
        android: layout_alignParentLeft = " true"
        android: text = " Hello"
        />
    <Button
        app: layout_widthPercent = " 30 %"
        app: layout_heightPercent = " 40 %"
        android: layout_alignParentRight = " true"
        android: text = " Hello"
        />
</android.support.percent.PercentRelativeLayout >
```

运行效果图如图 3-9 所示。

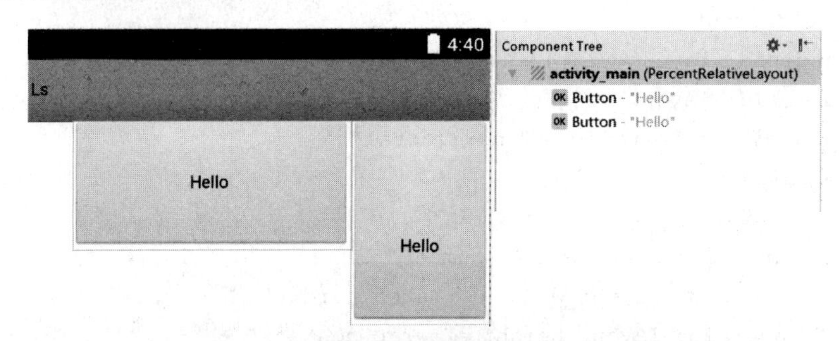

图 3-9　百分比布局

第 1 个控件的宽度：app：layout_widthPercent = " 60% "；左边的间距：app：layout_marginLeftPercent = " 10% "；第 2 个控件的宽度：app：layout_widthPercent = " 30% "。

PercentFrameLayout 的用法与此类似。

3.2.3 复杂布局——布局嵌套

在 Android 中实现复杂的界面时，仅有一种布局是很难实现的，需要综合使用多种布局。下面以计算器界面为例说明。计算器使用布局 LinearLayout 嵌套实现，在 Android Studio 中的组件树可以看到布局情况，效果如图 3-10 所示。

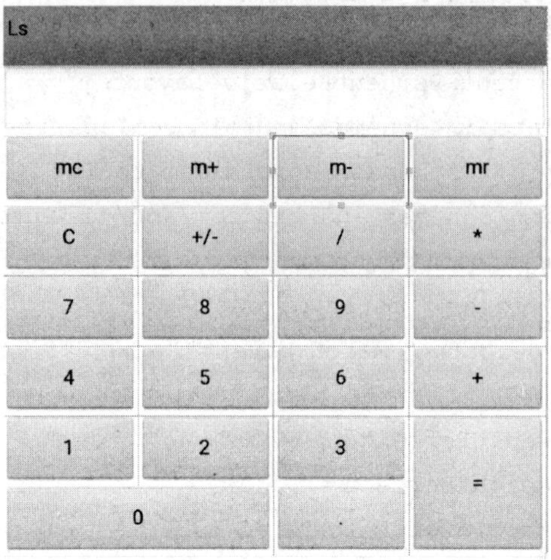

图 3-10 布局嵌套——计算器

对应 XML 的代码如下。

```
<LinearLayout xmlns:android="http://schemas.android.com/apk/res/android"
    xmlns:tools="http://schemas.android.com/tools"
    android:orientation="vertical"
    android:layout_width=" match_parent"
    android: layout_height=" match_parent"
    android: background=" #FFFFFF"
    tools: context=" .MainActivity" >
      //这里第一行显示标签为一个水平布局
    <LinearLayout
        android: layout_width=" match_parent"
        android: layout_height=" wrap_content"
        android: orientation=" horizontal" >
        <EditText
            android: id=" @ +id/msg"
            android: inputType=" number"
            android: layout_width=" match_parent"
            android: layout_height=" wrap_content"
            android: text=""  >
        </EditText>
    </LinearLayout>

    //第二行为 " mc"、" m+"、" m-"、" mr" 四个 Button 构成一个水平布局
```

```
<LinearLayout
    android: layout_width =" match_parent"
    android: layout_height =" wrap_content"
    android: orientation =" horizontal" >
    <Button
        android: layout_width =" match_parent"
        android: layout_height =" wrap_content"
        android: text =" mc" android: layout_weight =" 1" >
    </Button >
    <Button
        android: layout_width =" match_parent"
        android: layout_height =" wrap_content"
        android: text =" m+" android: layout_weight =" 1" >
    </Button >
    <Button
        android: layout_width =" match_parent"
        android: layout_height =" wrap_content"
        android: text =" m-" android: layout_weight =" 1" >
    </Button >
    <Button
        android: layout_width =" match_parent"
        android: layout_height =" wrap_content"
        android: text =" mr" android: layout_weight =" 1" >
    </Button >
</LinearLayout >

//同上、" C"、" +/-"、" / * " 四个 Button 构成一个水平布局
<LinearLayout
    android: layout_width =" match_parent"
    android: layout_height =" wrap_content"
    android: orientation =" horizontal" >
    <Button
        android: layout_width =" match_parent"
        android: layout_height =" wrap_content"
        android: layout_weight =" 1"
        android: text =" C" >
    </Button >
    <Button
        android: layout_width =" match_parent"
        android: layout_height =" wrap_content"
        android: layout_weight =" 1"
        android: text =" +/-" >
    </Button >
```

```
    <Button
        android:layout_width="match_parent"
        android:layout_height="wrap_content"
        android:layout_weight="1"
        android:text="/" >
    </Button>
    <Button
        android:layout_width="match_parent"
        android:layout_height="wrap_content"
        android:layout_weight="1"
        android:text=" * " >
    </Button>
</LinearLayout>

<LinearLayout
  android:layout_width="match_parent"
  android:layout_height="wrap_content"
  android:orientation="horizontal" >
   <Button
      android:layout_width="match_parent"
      android:layout_height="wrap_content"
      android:text="7" android:layout_weight="1" >
   </Button>
   <Button
      android:layout_width="match_parent"
      android:layout_height="wrap_content"
      android:text="8" android:layout_weight="1" >
   </Button>
   <Button
      android:layout_width="match_parent"
      android:layout_height="wrap_content"
      android:text="9" android:layout_weight="1" >
   </Button>
   <Button
      android:layout_width="match_parent"
      android:layout_height="wrap_content"
      android:text="-" android:layout_weight="1" >
   </Button>
</LinearLayout>

<LinearLayout
   android:layout_width="match_parent"
   android:layout_height="wrap_content"
```

```
android: orientation = " horizontal" >
 <Button
    android: layout_width = " match_parent"
    android: layout_height = " wrap_content"
    android: layout_weight = " 1"
    android: text = " 4" >
 </Button>
 <Button
    android: layout_width = " match_parent"
    android: layout_height = " wrap_content"
    android: layout_weight = " 1"
    android: text = " 5" >
 </Button>
 <Button
    android: layout_width = " match_parent"
    android: layout_height = " wrap_content"
    android: layout_weight = " 1"
    android: text = " 6" >
 </Button>
 <Button
    android: layout_width = " match_parent"
    android: layout_height = " wrap_content"
    android: layout_weight = " 1"
    android: text = "  +" >
 </Button>
</LinearLayout>

//最外层是一个水平布局，由左边上面一行 1、2、3 三个 Button，下面一行的 0 和" . " 两个
  Button 和 右边的"  =" 构成
<LinearLayout android: orientation = " horizontal"
   android: layout_width = " match_parent"
   android: layout_height = " wrap_content" >
   //这里 1、2、3 和 下面的 0、" . " 构成一个垂直布局
   <LinearLayout android: orientation = " vertical"
      android: layout_weight = " 3"
      android: layout_width = " wrap_content"
      android: layout_height = " wrap_content" >
      //这里的 1、2、3 构成一个水平布局
      <LinearLayout android: orientation = " horizontal"
         android: layout_width = " match_parent"
         android: layout_height = " wrap_content" >
         <Button
            android: layout_width = " wrap_content"
```

```
            android: layout_height = " wrap_content"
            android: layout_weight = " 1"
            android: text = " 1"  > </Button >
        <Button
            android: layout_width = " wrap_content"
            android: layout_height = " wrap_content"
            android: layout_weight = " 1"
            android: text = " 2"  > </Button >
        <Button
            android: layout_width = " wrap_content"
            android: layout_height = " wrap_content"
            android: layout_weight = " 1"
            android: text = " 3"  > </Button >
    </LinearLayout >
    //这里的 0 和" . " 构成一个水平布局，注意这里的 android_weight 参数设置
    <LinearLayout android: orientation = " horizontal"
        android: layout_width = " match_parent"
        android: layout_height = " wrap_content"  >
        <Button
            android: layout_width = " 0px"
            android: layout_height = " wrap_content"
            android: layout_weight = " 2"
            android: text = " 0"  > </Button >
        <Button
            android: layout_width = " 0px"
            android: layout_height = " wrap_content"
            android: layout_weight = " 1"
            android: text = " . "  > </Button >
    </LinearLayout >
  </LinearLayout >
  //这里一个单独 Button 构成垂直布局
  <LinearLayout android: orientation = " vertical"
      android: layout_weight = " 1"
      android: layout_width = " wrap_content"
      android: layout_height = " match_parent"  >
      <Button
          android: layout_width = " match_parent"
          android: layout_height = " match_parent"
          android: text = "  = "  > </Button >
  </LinearLayout >
</LinearLayout >

</LinearLayout >
```

3.2.4　Android 新布局 ConstraintLayout

Android 布局嵌套层级直接影响 UI 界面绘制的效率，如果 UI 嵌套层级太多会导致界面有性能问题，目前对于复杂的界面，使用 RelativeLayout 也无法解决。所以 Android UI 团队于 2016 年 Google I/O 开发者大会上发布了一个新的布局控件：ConstraintLayout。

ConstraintLayout 可以看作 RelativeLayout 的升级版，提供更多的手段来控制子 View 的布局，所以对于复杂的布局，用 ConstraintLayout 一个布局容器即可实现。

ConstraintLayout 可以有效地解决布局嵌套过多的问题。我们平时编写界面时，复杂的布局总会伴随着多层的嵌套，而嵌套越多，程序的性能也就越差。ConstraintLayout 则是使用约束的方式来指定各个控件的位置和关系，有点类似于 RelativeLayout，但远比 RelativeLayout 强大。

ConstraintLayout 属于 Android Studio 2.2 的新特性，ConstraintLayout 是一个新的 Support 库，支持 Android 2.3（API level 9）以及之后的版本。

新建一个 ConstraintLayoutTest 项目。另外，确保 Android Studio 是 2.2 或以上版本，为了使用 ConstraintLayout，需要在 app/build.gradle 文件中添加 ConstraintLayout 的依赖，如下所示。

```
dependencies {
  compile fileTree(dir:'libs', include: ['*.jar'])
  compile 'com.android.support:appcompat-v7:25.3.1'
  compile 'com.android.support.constraint:constraint-layout:1.0.2'
  testCompile 'junit:junit:4.12'
}
```

现在打开 res/layout/activity_main.xml 文件，由于这是一个新建的空项目，Android Studio 会自动帮我们创建一个布局，这是一个 ConstraintLayout 布局，它的代码如下。

```
<?xml version="1.0" encoding="utf-8"?>
<android.support.constraint.ConstraintLayout
xmlns:android=http://schemas.android.com/apk/res/android
   xmlns:app=http://schemas.android.com/apk/res-auto
   xmlns:tools=http://schemas.android.com/tools
   android:layout_width="match_parent"
   android:layout_height="match_parent"
   tools:context="com.example.hefugui.constraintlayouttest.MainActivity">
  <TextView
     android:layout_width="wrap_content"
     android:layout_height="wrap_content"
     android:text="Hello World!"
     app:layout_constraintBottom_toBottomOf="parent"
     app:layout_constraintLeft_toLeftOf="parent"
     app:layout_constraintRight_toRightOf="parent"
     app:layout_constraintTop_toTopOf="parent"/>
</android.support.constraint.ConstraintLayout>
```

ConstraintLayout 的基本用法很简单，现在向布局中添加一个按钮，从左侧的 Palette 区域拖一个 Button 进去即可。

虽然已经添加 Button 到界面中了，但是由于还没有为 Button 添加任何约束，因此 Button 并不知道自己应该出现在什么位置。现在在预览界面中看到的 Button 位置并不是它最终运行的实际位置，如果一个控件没有添加任何约束，则在运行之后会自动位于界面的左上角。

1. 基本约束

下面为 Button 添加约束，每个控件的约束都分为垂直和水平两类，一共可以在四个方向上为控件添加约束。

Button 的上下左右各有一个圆圈，这些圆圈就是用来添加约束的，如图 3-11 所示。

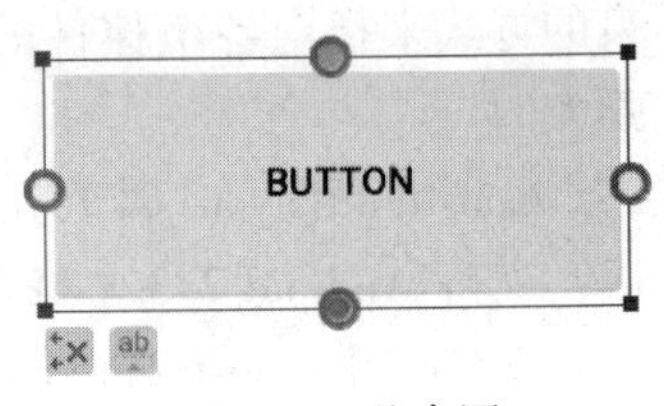

图 3-11　约束圈

可以将约束添加到 ConstraintLayout，也可以将约束添加到另一个控件。比如说，想让 Button 位于布局的左边和上边，则用鼠标按住约束圈往边界拖就可以添加约束，我们为 Button 的左边和上边添加了约束，因此 Button 会将自己定位到布局的左下角，如图 3-12 所示。

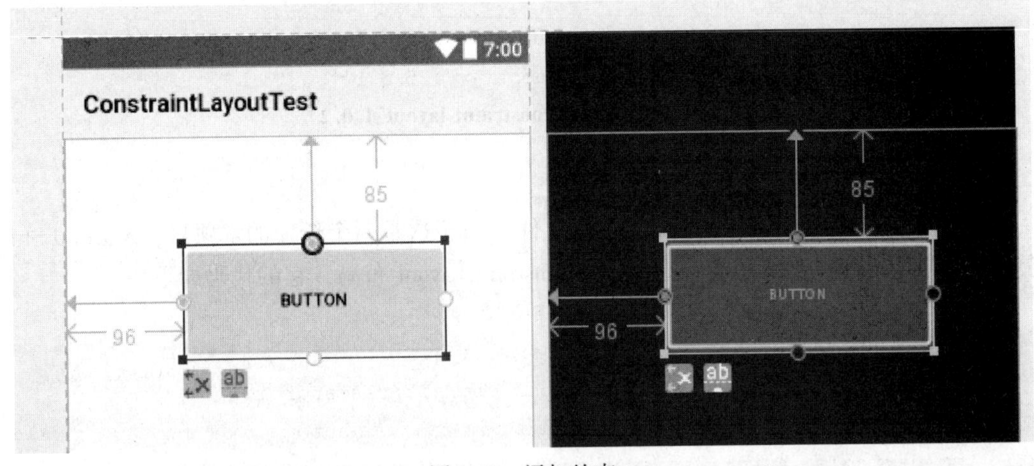

图 3-12　添加约束

这就是添加约束最基本的用法。

除此之外，我们还可以使用约束让一个控件相对于另一个控件进行定位。比如说，我们希望再添加一个 Button，让它位于第一个 Button 的正下方，间距 64dp，则用鼠标按住 Button2 的上约束圈往 Button1 边界拖动即可，如图 3-13 所示。

现在我们已经学习了添加约束的方式，那么该怎样删除约束呢？其实也很简单，删除约束的方式一共有三种。

（1）若要删除一个单独的约束，则将鼠标光标悬浮在某个约束的圆圈上，然后该圆圈会变成红色，这时单击一下即可删除。

（2）若要删除某一个控件的所有约束，则选中一个控件，它的左下角会出现一个删除约束的图标 X，单击该图标，即可删除当前控件的所有约束，如图 3-14 所示。

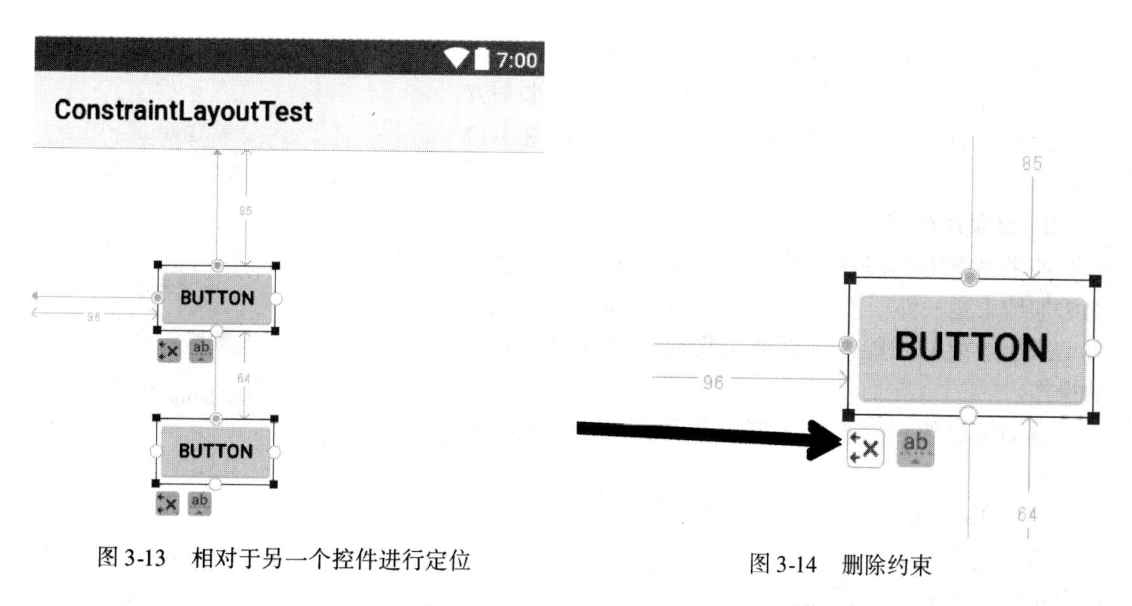

图 3-13　相对于另一个控件进行定位　　　　图 3-14　删除约束

（3）若要删除当前界面中的所有约束，则单击界面工具栏中的删除约束按钮即可，如图 3-15 所示。

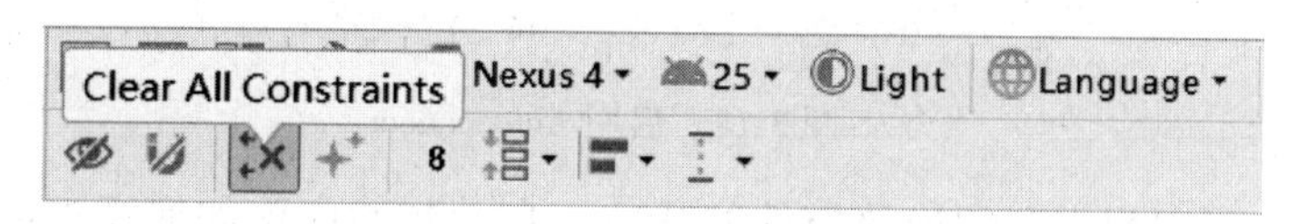

图 3-15　删除所有约束

现在已经学习了 ConstraintLayout 基本操作，并且能使用 ConstraintLayout 编写一些简单的界面了。不过目前还有一个问题可能比较麻烦，刚才我们已经实现了让一个按钮居中对齐，如果我们想让两个按钮共同居中对齐，该怎么实现呢？

其实这种情况很常见，比如在应用的登录界面中会有一个登录按钮和一个注册按钮，不管它们是水平居中还是垂直居中，肯定是两个按钮共同居中，这将用到下面要介绍的 Guideline 约束类型。

2. Guideline 约束

上文提到，想要实现两个按钮共同居中，需要用到 ConstraintLayout 中的一个新的功能 Guideline。

首先需要在界面中添加 Guideline，在界面工具栏中单击添加按钮，选择下拉列表中 Add Vertical Guideline 或者 Add Horizontal Guideline 选项，如图 3-16 所示。

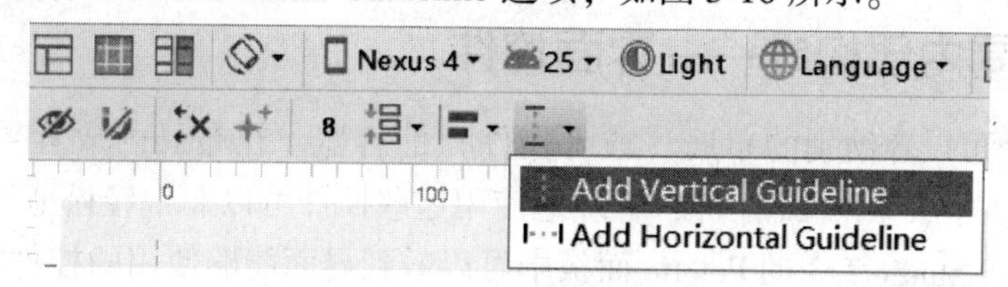

图 3-16　添加 Guideline

我们希望让这两个按钮在垂直方向上对齐，在水平方向上距离 Guideline 左部 52dp，需要先添加一个垂直方向上的 Guideline，每个按钮的左边向 Guideline 添加约束，这样就能实现两个按钮在水平方向上距离左部 Guideline 52dp 距离的要求，如图 3-17 所示。

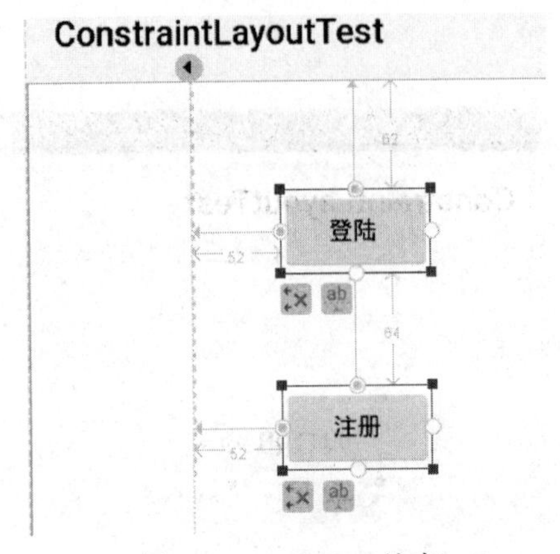

图 3-17　Guideline 约束

3. 自添加约束

如果界面中的内容比较复杂，为每个控件一个个地添加约束是一件很繁琐的事情。ConstraintLayout 支持自动添加约束的功能，可以极大程度上简化那些繁琐的操作。

自动添加约束的方式主要有两种：Autoconnect 和 Inference。

（1）Autoconnect

若要使用 Autoconnect，首先需要在工具栏中启用这个功能，默认情况下 Autoconnect 是不启用的，如图 3-18 所示。

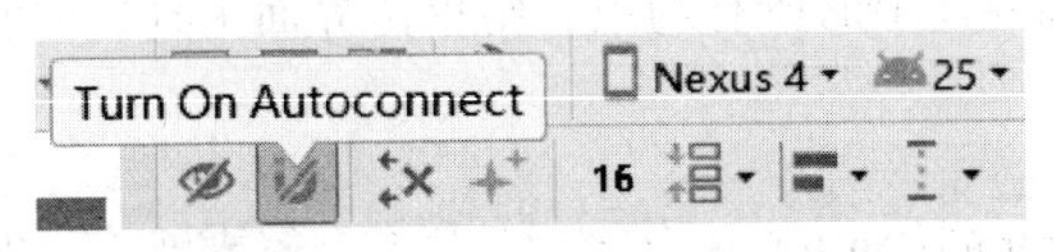

图 3-18　启用 Autoconnect

Autoconnect 可以根据我们拖放控件的状态，自动判断应该如何添加约束，比如我们将 Button 放到界面的正中央，那么它的上下左右会自动添加约束。

（2）Inference

接下来，我们看一下 Inference 的用法。Inference 也用于自动添加约束，它比 Autoconnect 功能更强大，因为 AutoConnect 只能为当前操作的控件自动添加约束，而 Inference 能为当前界面中的所有元素自动添加约束。因而 Inference 比较适合用来实现复杂度比较高的界面，它可以一键自动生成所有的约束。工具栏中 Inference 约束工具按钮如图 3-19 所示。

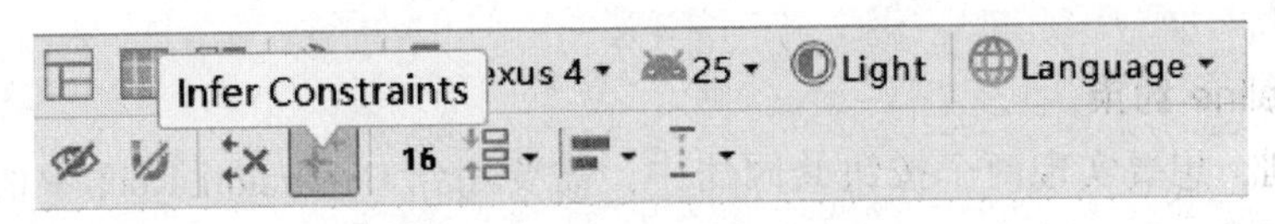

图 3-19　Inference 约束工具

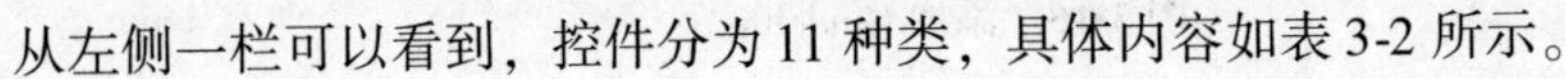

3.3 布局内部构成——界面控件

布局是一个可以容纳别的布局（或者控件）的容器。在布局内放置界面需要的各种控件。Android 提供了大量的 UI 控件，合理地使用这些控件可以非常轻松地编写出相当不错的界面。在 Android Studio 2.3 的 Palette 面板中可以看到界面能够使用的控件，如图 3-20 所示。

从左侧一栏可以看到，控件分为 11 种类，具体内容如表 3-2 所示。

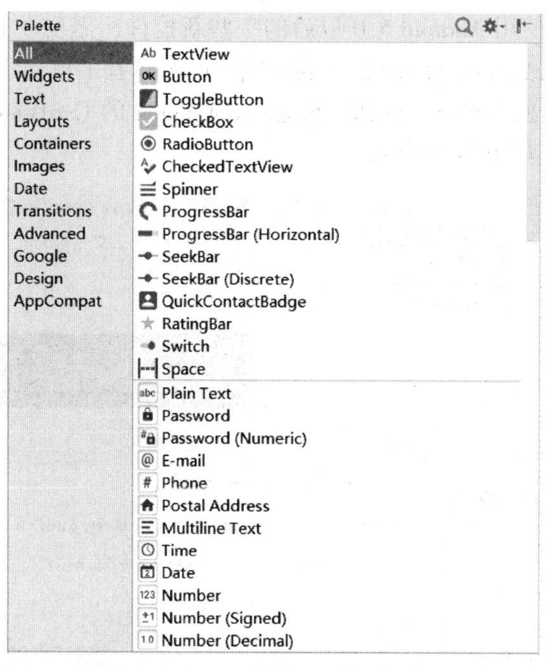

图 3-20 Android 界面可使用的控件

表 3-2 Android 控件分类

序号	控件种类	说明
1	Widgets	常用控件。包括 Button（按钮）、CheckBox（复选框）、ProgressBar（进度条）、SeekBar（拖动条）、RatingBar（评分控件）等
2	Text	文本输入类。包括 EditText（输入文本）、AutoCompleteTextView（自动完成）、MultiAutoCompleteTextView（支持选择多个值）等
3	Layouts	布局类。包括 ConstraintLayout、GridLayout、FrameLayout、LinearLauout、RelativeLayout、TableLayout 等
4	Containers	容器类。包括 RadioGroup（单选组）、ListView（列表视图）、GridView（网格视图）、ExpandableListView（扩展列表视图）、ScrollView（滚动视图）、TabHost（选项卡）、WebView（网络视图）等
5	Image	图像类。包括 ImageButton（图像按钮）、ImageView（图像视图）、VidoView（视频视图）
6	Date	日期类。包括 TimePicker（时间选择）、DatePicker（日期选择）、CalendarView（日历视图）、Chronometer（倒计时）、TextClock（时间控件）
7	Transitions	切换类。包括 ImageSwitcher（图像切换）、AdapterViewFlipper（视图切换）、StackView（堆视图）、TextSwitcher（文本切换器）、ViewAnimator（视图切换动画）、ViewFlipper（视图幻灯片）、ViewSwitcher（视图切换）
8	Advanced	高级类。包括 ViewStub（轻量级的 View）、TextureView（渲染视图）、SurfaceView（绘制视图）、NumberPicker（预定义数字控件）等
9	Google	Google 类。包括 Adview（广告视图）、MapView（地图视图）
10	Design	设计类。包括 CoordinatorLayout（协调布局）、AppBarLayout（顶部栏）、TabLayout（选项卡布局）、TabItem（表项）、NestesScrollView（嵌套滚动视图）、FloatingActionButton（悬浮按钮）、TextInputLayout（输入布局）
11	AppCompat	版本适配类。包括 CardView（卡片视图）、GridLayout（网格布局）、RecyclerView（循环器试图）、Toolbar（工具栏）

如果在应用程序中使用 Android 5.0 以后的一些新控件，需要在 app/build.gradle 文件中添加相应类的支持，在 Android Studio 2.3 的环境中已经具有了自动添加的功能。

（1）每个控件需要独立的类。例如，添加 AppCompat 的 CardView 控件时，系统会询问是否添加 CardView 类，如图 3-21 所示。

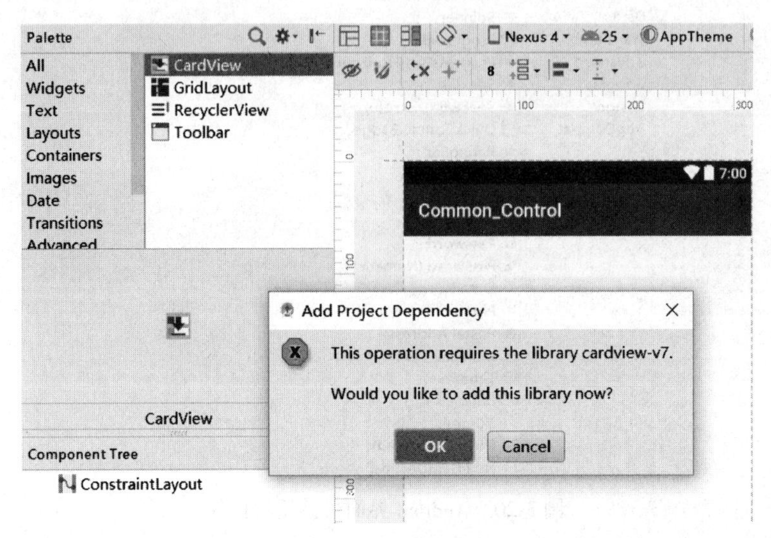

图 3-21 Android Studio 2.3 中添加 CardView 时弹出的提示

单击 OK 按钮后，自动添加相应的类 app/build.gradle 文件中，如下所示。

```
dependencies {
    compile fileTree(dir:'libs', include: ['*.jar'])
    compile'com.android.support:appcompat-v7:25.3.1'
    compile'com.android.support.constraint:constraint-layout:1.0.2'
    compile'com.android.support:cardview-v7:25.3.1'
}
```

添加 AppCompat 的 GridLayout 控件时，系统会询问是否添加 GridLayout 类。

```
dependencies {
    compile fileTree(dir:'libs', include: ['*.jar'])
    compile'com.android.support:appcompat-v7:25.3.1'
    compile'com.android.support.constraint:constraint-layout:1.0.2'
    compile'com.android.support:cardview-v7:25.3.1'
    compile'com.android.support:gridlayout-v7:25.3.1'
}
```

（2）一个类需要相同的支持类，例如 Design 类控件，在 app/build.gradle 文件中统一添加 Design 类的支持。

```
dependencies {
    compile fileTree(dir:'libs', include: ['*.jar'])
    compile'com.android.support:appcompat-v7:25.3.1'
    compile'com.android.support.constraint:constraint-layout:1.0.2'
    compile'com.android.support:design:25.3.1'
}
```

另外，还有一些控件未显示于 Android Studio 2.3 的 Palette 面板中，例如 AlertDialogLayout，这些可以在界面的 XML 文件中直接添加，直接输入开头几个字母“ <aler”会出现提示，如图 3-22 所示。

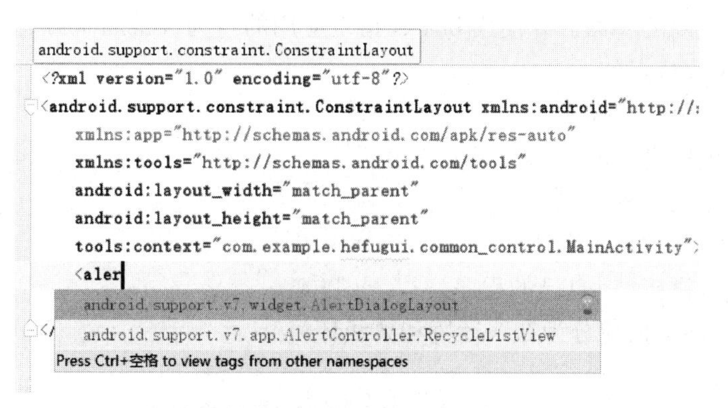

图 3-22　添加 AlertDialogLayout

界面控件的另一种添加方法是在界面的 XML 文件中直接添加，如图 3-23 所示。

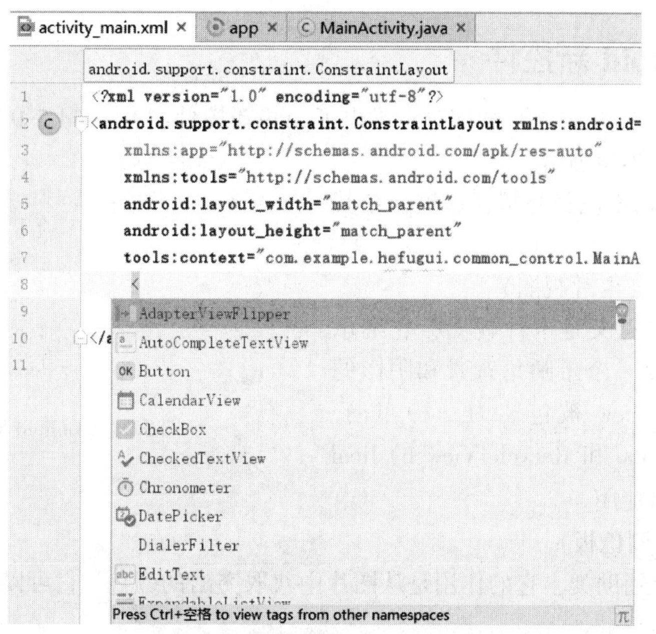

图 3-23　在界面的 XML 中直接添加控件

界面控件的设计一般是通过 Design 模式下 Palette 面板操作和 XML 文件操作结合完成的，这两种操作方法都要熟练掌握。

3.4 界面设计助手——辅助设计工具

Android 将程序中的 UI 界面布局与程序应用逻辑实现代码严格区分开，并分别放在 res 和 src 目录中。Android 的 UI 用户界面布局开发，除了在开发环境中制作之外，还可以采用辅助设计工具来实现。

1. Google App Inventor

App Inventor 是一款 Google 公司开发的手机编程软件，号称可以让任何人创建 Android 手机应用。使用 App Inventor 无须掌握编程知识，因为您根本，不需要编写代码，只需在可视化界面上设计应用的界面，并使用 blocks 指定应用的行为（behavior）。

App Inventor 的特点如下。

- 开发过程简单，易操作。
- 可以开发创造自己的应用程序。
- 不需要太多的编程知识。
- 采用代码拼接的编码方法。
- 创意 + 代码拼接 = 自己的程序。

App Inventor 环境搭建分为在线环境和离线环境。

2. DroidDraw

DroidDraw 是一个为 Android 创建图形用户界面的 UI 设计器。它是一个独立的可执行程序，可以运行在 Mac OS X、Windows 和 Linux 系统中。

3.5 Android 新控件

Google 在推出 Android 5.0 的同时，推出了一些新控件，Android 5.0 中最常用的新控件有下面 5 种，如图 3-24 所示。

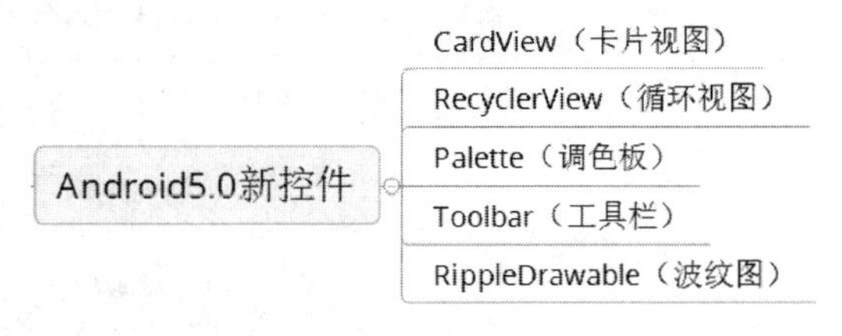

图 3-24　Android 5.0 新控件

RecyclerView 前面已经介绍，这里介绍其余 4 个控件。

（1）CardView（卡片视图）

CardView 顾名思义是卡片视图，它继承了 FrameLayout，是一个带圆角背景和阴影的 FrameLayout。CardView 被包装为一种布局，并且经常在 ListView 和 RecyclerView 的 Item 布局中，作为容器使用。

（2）Palette（调色板）

Palette 是一个辅助类，它的作用是从图片中获取突出的颜色。它可以提取下面几种特性的突出颜色。

- Vibrant：充满活力的。
- Vibrant Dark：充满活力的，黑暗的。
- Vibrant Light：充满活力的，明亮的。
- Muted：柔和的。
- Muted Dark：柔和的，黑暗的。
- Muted Light：柔和的，明亮的。

Patette 的使用方法非常简单。

```
//获取应用程序图标的 Bitmap
bitmap = BitmapFactory.decodeResource(getResources(), R.mipmap.ic_launcher);
```

```
//通过 bitmap 生成调色板 palette
Palette palette = Palette. from (bitmap). generate();
//获取 palette 充满活力色颜色
int vibrantColor = palette. getVibrantColor (Color. WHITE);
```

(3) Toolbar（工具栏）

Toolbar 顾名思义是工具栏，作为 ActionBar 的替代品，Google 推荐使用 Toolbar 替代 ActionBar。Toolbar 可以放置在任何地方，不像 ActionBar 只能放置在固定的位置。Toolbar 支持比 ActionBar 更集中的特征。

Toolbar 可能包含以下可选元素的组合。

- 导航按钮。
- 品牌的 Logo 图像。
- 标题和子标题。
- 一个或多个自定义视图。

Toolbar 的使用方法如下。

```
this. toolbar = (Toolbar) findViewById(R. id. toolbar);
this. recyclerview = (RecyclerView) findViewById(R. id. recycler_view);
this. ripplebutton = (Button) findViewById (R. id. ripple_button);
this. button = (Button) findViewById (R. id. button);
//设置 Logo
toolbar. setLogo (R. mipmap. ic_launcher);
//设置标题
toolbar. setTitle (" Android5. 0");
//设置子标题
toolbar. setSubtitle (" 新控件");
//设置 ActionBar，之后就可以获取 ActionBar 并进行操作，操作的结果会反映在 toolbar 上面
setActionBar (toolbar);
//设置了返回箭头,，相当于设置了 toolbar 的导航按钮
getActionBar(). setDisplayHomeAsUpEnabled (true);
```

(4) RippleDrawable（波纹图）

RippleDrawable 只能在 Android 5. 0 以上使用，目前还没有提供 RippleDrawable 向下兼容的支持包。RippleDrawable 可显示一个涟漪效应响应状态变化。

RippleDrawable 的使用方法如下。

定义一个 UI 的背景图片为 RippleDrawable。

```
android:background = "@ drawable/ripple"
```

在 drawable 文件夹下面定义一个 RippleDrawable 的 xml 文件

```
<? xml version = "1. 0" encoding = "utf-8"? >
<ripple xmlns: android = http://schemas. android. com/apk/res/android android:
color = "#0000FF" >
    <item >
      <shape android:shape = "rectangle" >
        <solid android:color = "#FFFFFF" / >
```

```
        <corners android:radius="4dp" />
    </shape>
  </item>
</ripple>
```

android：color：表示波纹的颜色。

<item>：表示波纹图下面的条目。

Material Design 是 Android 5.0 系统的重头戏，并在以后 App 中将成为一种设计标准，Material Design 是一种可视化、交互性、动效以及多屏幕适应的全面设计。Android 5.0 Lollipop 和已经更新的 Support Libraries 将会帮助构建 Material UI，这个在后面会详细介绍。

Android 6.0 进入 Material Design 时代，在继 Material Design 风格之后，实现很多风格上的兼容，并推出了 Android Design Support Library 库，全面支持 Material Design 设计风格的 UI 效果。该库包含了 FloatingActionButton、TextInputLayout、Snackbar、TabLayout、NavigationView、CoordinatorLayout、AppBarLayout、CollapsingToolbarLayout 八个新控件。

（1）SnackBar

SnackBar 是带有动画效果的快速提示栏，它显示在屏幕底部，是用来代替 Toast 的一个全新控件，基本上继承了 Toast 的属性和方法，用户可以单击按钮执行对应的操作，Snackbar 支持滑动消失，如果没进行任何操作，那么到一定时间自动消失。

（2）TextInputLayout

TextInputLayout 主要是用作 EditText 的容器，从而为 EditText 默认生成一个浮动的 label，当用户单击了 EditText 之后，EditText 中设置的 Hint 字符串会自动移到 EditText 的左上角。使用方法非常简单。

（3）TabLayout

TabLayout 控件用于轻松添加 Tab 分组功能，总共有两种类型可选。

固定的 Tabs：对应于 xml 配置中的 app：tabMode="fixed"。

可滑动的 Tabs：对应于 xml 配置中的 app：tabMode="scrollable"。

（4）NavigationView

以前版本中制作侧边栏可使用 SlideMenu 三方库，现在有了官方提供的 NavigationView，开发者渐渐使用此项功能。使用导航视图需要传入一组参数、一个可选的头部布局，以及一个用于构建导航选项的菜单，完成这些步骤后，只需给导航选项添加响应事件的监听器就可以了。

在使用 NavigationView 时需要提前准备好两个 XML 文件，一个是头布局，另一个是 menu 布局。

（5）FloatingActionButton

浮动操作按钮是在 Material Design 准则中新引入的组件。用于强调当前屏幕最重要、高频率的一些操作。

正常显示的情况下 FloatingActionButton 有填充颜色，并显示阴影，单击的时候会有一个 rippleColor，并且阴影的范围可以增大。

（6）CoordinatorLayout

CoordinatorLayout 是 Design 引入的一个功能强大的布局，本质上是一个增强的 FrameLayout，它可以使不同组件之间直接相互作用，并协调动画效果。我们可以定义 CoordinatorLayout 内部的视图组件如何相互作用并发生变化。

（7）CollapsingToolbarLayout

CollapsingToolbarLayout 控件可以实现当屏幕内容滚动时收缩 Toolbar 的效果，通常配合 AppBarLayout 一起使用。

Android 7.0 带来的新工具类是 DiffUtil。DiffUtil 是 support-v7：24.2.0 中的新工具类，用来比较两个数据集，寻找出旧数据集与新数据集的最小变化量。

3.6 界面背后的劳动者——Activity

本节主要介绍 Activity 的相关知识。

3.6.1 Activity 简介

说到用户界面，不得不提到 Activity，它是 Android 应用程序提供交互界面的一个重要组件，也是 Android 最重要的组件之一。

Activity 是业务类，是承载应用程序的界面以及业务行为的基础，Activity 对应 MVC 模型中的 C（Controller）。

可以这样理解：Activity 是一个工人，用来控制 Window；Window 是一面显示屏，用来显示信息；View 是要在显示屏上显示的信息，这些 View 是层层重叠在一起（通过 inflate() 和 addView()）放到 Window 显示屏上的。而 LayoutInfalter 是用来生成 View 的一个工具，XML 布局文件用来生成 View 的原料。

用户 UI 通过 Android 中布局（layout）实现，布局中包含各种控件，用户操作控件和系统的交互是通过 Activity 实现的，如图 3-25 所示。

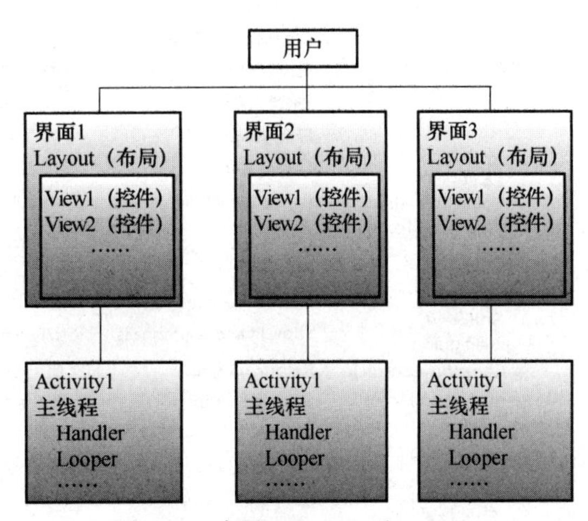

图 3-25 布局（layout）和 Activity

Activity 是 Android 的四大组件（即 Activity、Service 服务、Content Provider 内容提供者、BroadcastReceiver 广播接收器）之一。通过 Activity，用户可以与移动终端进行交互，使用 Android 应用程序进行操作，比如拨号、拍照、发送电子邮件或者浏览地图，在移动设备上实现 Activity 时，可以指定对应处理的布局 UI。Activity 的英文含义为“活动”，它是用户与应用程序进行交互的接口，同时也是一个容器，在一个 Activity 中可以放置大量的控件，这些控件决定了用户在该 Activity 中可以做什么，这也是 Activity 最关注的。

所有的 Activity 都是从 Android 提供的类 Activity 继承而来，一个 Android 应用通常由多个 Activity 构成，不同 Activity 之间采用低耦合度设计，其中某个 Activity 可以称为应用的“主 Activity”，作为在用户单击应用图标时显示的初始界面。每个 Activity 都可以触发其他

Activity 以往的某种功能。每当一个新 Activity 启动，之前的 Activity 将处于“停止”状态，但是 Android 系统会继续保留之前的 Activity 状态，这样就形成了一个 Activity 栈结构（称为 Back Stack）。新 Activity 启动后被 Android 系统推放到 Activity 栈的最前面，并且获取用户焦点（响应按键、触摸事件等），这个 Activity 栈采用“后进先出”的栈机制，因此当用户完成当前 Activity 功能后，单击“返回”键，当前 Activity 从 Activity 栈退栈并被“销毁”，之前的 Activity 变为当前 Activity，并且恢复之前的状态。

当一个 Activity 由于有新的 Activity 启动转变到“停止”状态时，Android 系统将通过 Activity 的生命周期回调函数来通知该 Activity。根据 Activity 当前状态的不同，系统将触发 Activity 多个不同的生命周期回调函数——创建、停止、恢复、销毁等。通过回调函数可以为 Activity 的不同状态添加不同的处理方法，比如当 Activity 停止时，可以释放某些系统资源，比如网络、数据库连接等，而当恢复某个 Activity 时可以重新获取这些系统资源。

3.6.2 创建 Activity 和加载布局

在 Android Studio 2.3 中，新建 Activity 是在项目中进行的，在项目中选择 app 项的 java 项目，单击鼠标右键，在弹出的菜单中选择 New > Activity 命令，选择新建的 Activity 种类，如图 3-26 所示。

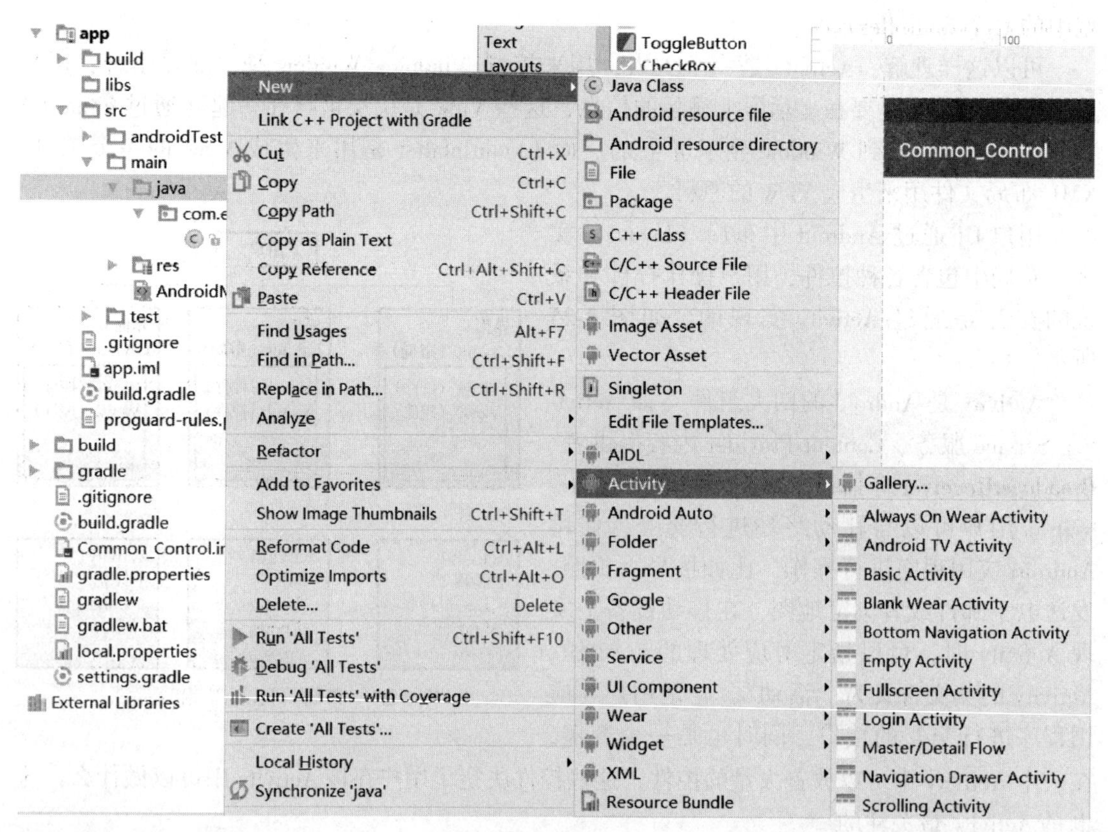

图 3-26 创建 Activity

如果要创建空白 Activity，可选择 Empty Activity 命令，弹出的窗口如图 3-27 所示，输入 Activity 的名称和对应的 Layout 名称，单击 Finish 按钮，完成 Activity 和对应的 Layout 的创建。

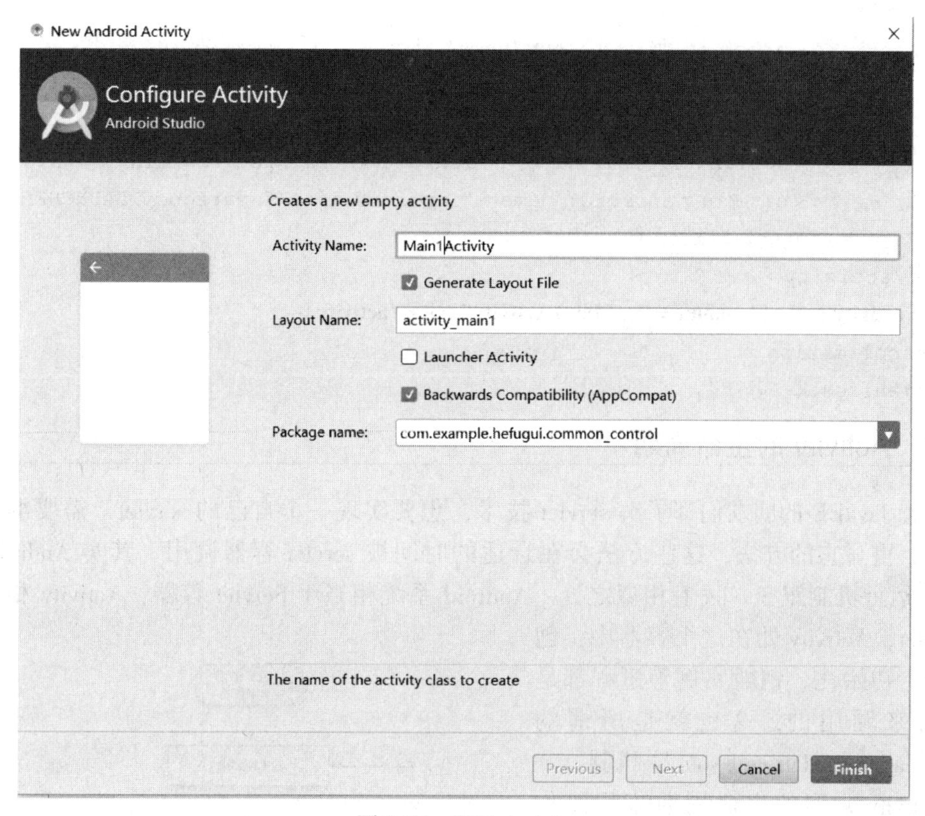

图 3-27　配置 Activity

在创建的类 Main1Activity 中，onCreate( )方法调用了 setContentView( )方法为当前的活动加载对应的布局，其中的参数为布局文件的 id。

```
public class Main1Activity extends AppCompatActivity
{
  @Override
    protected void onCreate(Bundle savedInstanceState) {
      super.onCreate(savedInstanceState);
      setContentView(R.layout.activity_main1);
  }
}
```

所有的 Activity 都要在 AndroidManifest. xml 文件中注册才能生效，在 Android Studio 2. 3 中可以自动实现注册，打开 AndroidManifest. xml 的代码如下。

```
<?xml version="1.0" encoding="utf-8"?>
<manifest xmlns:android=http://schemas.android.com/apk/res/android
  package="com.example.hefugui.common_control">
  <application
    android:allowBackup="true"
    android:icon="@mipmap/ic_launcher"
    android:label="@string/app_name"
```

```
        android: roundIcon = " @mipmap/ic_launcher_round"
        android: supportsRtl = " true"
        android: theme = " @style/AppTheme" >
        <activity android: name = " .MainActivity" >
            <intent-filter >
                <action android: name = " android. intent. action. MAIN" / >
                <category android: name = " android. intent. category. LAUNCHER" / >
            </intent-filter >
        </activity >
        <activity android: name = " . Main1Activity"  > </activity >
    </application >
    </manifest >
```

3. 6. 3 Activity 的生命周期

熟悉 JavaEE 的朋友们都了解 servlet 技术，想要实现一个自己的 servlet，需要继承相应的基类，重写它的方法，这些方法会在合适的时间被 servlet 容器调用。其实 Android 中的 Activity 运行机制跟 Servlet 有相似之处，Android 系统相当于 Servlet 容器，Activity 相当于一个 Servlet，Activity 处在这个容器中，创建实例、初始化、销毁实例等过程都是由容器来调用的，这也就是所谓的“Don't call me，I'll call you.”机制。

Activity 典型的生命周期流程图如图 3-28 所示。

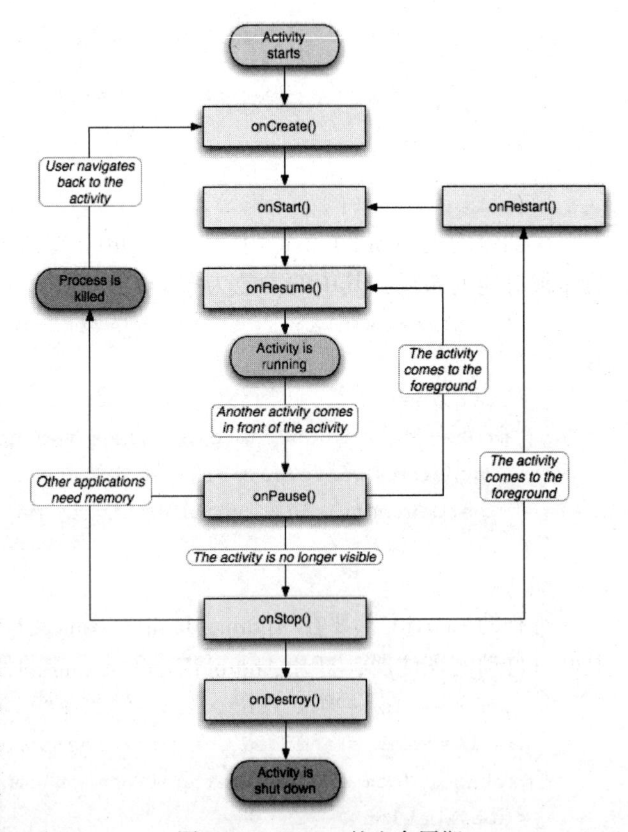

图 3-28 Activity 的生命周期

Activity 的生命周期中有 4 种状态。

1. Active/Runing

一个新 Activity 启动入栈后，显示在屏幕最前端，处于栈的最顶端（Activity 栈顶），此时它处于可见并可和用户交互的激活状态，叫作活动状态或者运行状态。

2. Paused

当 Activity 失去焦点，被一个新的非全屏的 Activity 或者一个透明的 Activity 被放置在栈顶，此时的状态叫作暂停状态（Paused）。此时它依然与窗口管理器保持连接，Activity 依然保持活力（保持所有的状态、成员信息，和窗口管理器保持连接），但是在系统内存极端低下的时候将被强行终止。所以它仍然可见，但已经失去了焦点，故不可与用户进行交互。

3. Stopped

如果一个 Activity 被另外的 Activity 完全覆盖掉，叫作停止状态（Stopped）。Activity 依然保持所有状态和成员信息，但是它不再可见，所以它的窗口被隐藏，当系统内存需要被用在其他 Activity 的时候，已经 Stopped 的 Activity 将被强行终止。

4. Killed

如果一个 Activity 处于 Paused 或者 Stopped 状态，系统可以将该 Activity 从内存中删除，Android 系统采用两种方式进行删除，要么要求该 Activity 结束，要么直接终止它的进程。当该 Activity 再次显示时，必须重新开始和重置前面的状态。

Activity 的生命周期有如下 3 个关键的循环。

（1）整个生命周期，从 onCreate（Bundle）开始到 onDestroy( )结束。Activity 在 onCreate( )设置所有的“全局”状态，在 onDestory( )释放所有的资源。例如：某个 Activity 有一个在后台运行的线程，用于从网络下载数据，则该 Activity 可以在 onCreate( )中创建线程，在 onDestory( )中停止线程。

（2）可见的生命周期，从 onStart( )开始到 onStop( )结束。在这段时间，尽管有可能不在前台，不能和用户交互，但 Activity 仍在屏幕上。在这两个接口之间，需要保持显示给用户的 UI 数据和资源等，例如：可以在 onStart( )中注册一个 IntentReceiver 来监听数据变化导致 UI 的变动，当不再需要显示时候，可以在 onStop( )中注销。onStart( )、onStop( )都可以被多次调用，因为 Activity 随时可以在可见和隐藏之间转换。

（3）前台的生命周期，从 onResume( )开始到 onPause( )结束。在这段时间里，该 Activity 处于所有 Activity 的最前面，和用户进行交互。Activity 可以经常性地在 Resumed 和 Paused 状态之间切换，例如：当设备准备休眠时，当一个 Activity 处理结果被分发时，或者是当一个新的 Intent 被分发时。所以在这些接口方法中的代码应该属于非常轻量级的。

Activity 的整个生命周期的状态转换和动作都定义在 Activity 的接口方法中，所有方法都可以被重载。

Activity 的生命是从 OnCreate( )开始的，当能够看到这个 Activity 的时候，Activity 也迈出了它人生的第一步 OnStart( )，等它成长到可以跟我们进行交互的时候，也就进入了它人生最精彩的部分 OnResume( )。当我们把注意力转移到另外的 Activity 时，Activity 进入它人生的黯淡期 OnPause( )，这个时候 Activity 有两种结果，一种是我们把注意力重新转移到它身上，它也就获得了新生 OnRestart( )，另外一种是我们不再关注这个 Activity，它从我们的视线中消失了，这个 Activity 的人生也停止了 OnStop( )，最后 Activity 寿命到了，即执行了 OnDestroy( )来结束它匆匆的一生。Activity 的人生经历了这些过程：OnCreate、OnStart、OnResume、OnPause、OnRestart、Onstop、OnDestroy。

3.6.4　使用 Intent 在 Activity 之间穿梭

一个应用通常有多个 Activity。每个 Activity 围绕一个特定的功能设计，用户可以操作它，并且可以启动其他的 Activity。例如，一个电子邮件应用可能有一个 Activity 呈现新邮件列表，当用户选择了一封邮件，会打开一个新的 Activity 来呈现邮件内容。

一个 Activity 可以启动另一个应用的 Activity。例如，如果应用想要发送 Email，您可以定义一个 Intent 来执行一个发送操作并且携带一些数据：Email 的地址、消息。一个其他应用的 Activity 需要声明可以处理这类的 Intent。在这个例子中，Intent 是要发送一封 Email，所以一个 Email 应用会启动（如果有多个 Activity 支持同一个 Intent，系统会让用户选择要使用哪一个）。当 Email 被发送出去，Activity 会恢复，看起来 Email Activity 就是应用的一部分。为维护这种无缝的用户体验，尽管 Activity 可能来自于不同的应用，Android 系统依然会将这些 Activity 保存在同一个任务中。

Android 中 Intent 实现两个不同应用 Activity 的跳转，Intent 不仅可用于应用程序之间，也可用于应用程序内部的 Activity/Service 之间的交互。因此，可以将 Intent 理解为不同组件之间通信的“媒介”，专门提供组件互相调用的相关信息。

Intent 可以启动一个 Activity，也可以启动一个 Service，还可以发起一个广播 Broadcasts，如表 3-3 所示。

表 3-3　Inten 启动组件的方法

组件名称	方法名称
Activity	startActivity( )
Service	startService( ) bindService( )
Broadcasts	sendBroadcasts( ) sendOrderedBroadcasts( ) sendStickyBroadcasts( )

Intent 是 Android 程序中各组件之间进行交互的一种重要方式，它不仅可以指明当前组件想要执行的动作，还可以在不同组件之间传递数据。Intent 一般可被用于启动活动、启动服务、以及发送广播等场景，由于服务、广播等概念暂时还未讲解，那么我们主要介绍启动活动。

Intent 切换 Activity 的方法如下。

Intent intent = new Intent（当前 Activity. this，要跳转到 Activity. class）;

startActivity（intent）;

Intent 的用法大致可以分为两种：显式 Intent 和隐式 Intent。先来看一下显式 Intent 的使用方法。

1. 显式 Intent

明确指出第一个和第二个 Activity，方法如下。

```
public void onClick(View v) {
    Intent intent = new Intent(MainActivity.this,SecondActivity.class);
    startActivity(intent);
}
```

2. 隐式 Intent

相比于显式 Intent，隐式 Intent 则含蓄得多，并不明确指出我们想要启动哪一个活动，而是指定了一系列更为抽象的 Action 和 Category 等信息，然后交由系统去分析这个 Intent，帮我们找出合适的活动并启动。

使用隐式 Intent，不仅可以启动自己程序内的 Activity，还可以启动系统内置的 Activity。例如，调用系统的拨号 Activity 的代码如下。

```
public void onClick(View v) {
        Intent intent = new Intent(Intent.ACTION_DIAL);
        intent.setData (Uri.parse (" tel: 10086"));
        startActivity (intent);
  }
```

执行结果如图 3-29 所示。

图 3-29 拨号 Activity

3.6.5 Intent 调用常见系统组件

下面介绍 Intent 调用常见系统组件的方法。

(1) 调用浏览器

```
Uri webViewUri = Uri.parse("http://blog.csdn.net/zuolongsnail");
Intent intent = new Intent(Intent.ACTION_VIEW, webViewUri);
```

(2) 调用地图

```
Uri uri = Uri.parse("geo:38.899533,-77.036476");
Intent it = new Intent(Intent.Action_VIEW, uri);
startActivity (it)
```

(3) 播放多媒体

```
Intent it = new Intent(Intent.ACTION_VIEW);
Uri uri = Uri.parse (" file: ///sdcard/song.mp3");
it.setDataAndType (uri, " audio/mp3");
startActivity (it);
Uri uri = Uri.withAppendedPath (MediaStore.Audio.Media.INTERNAL_CONTENT_URI, " 1");
Intent it = new Intent (Intent.ACTION_VIEW, uri);
```

```
startActivity (it);
```

(4) 路径规划

```
Uri uri =  Uri.parse ("http://maps.google.com/maps? f = d&saddr = startLat%20startLng&daddr = endLat%20endLng&hl = en");
Intent it = new Intent(Intent.ACTION_VIEW, URI);
startActivity (it);
```

(5) 拨打电话

```
Uri dialUri = Uri.parse("tel:10086");
Intent intent = new Intent(Intent.ACTION_DIAL, dialUri);
```

直接拨打电话，需要加上权限

```
<uses-permission id=" android.permission.CALL_PHONE" />
Uri callUri = Uri.parse (" tel: 10086");
Intent intent = new Intent (Intent.ACTION_CALL, callUri);
```

(6) 调用发送短信的程序

```
Intent it = new Intent(Intent.ACTION_VIEW);
it.putExtra (" sms_body", " The SMS text");
it.setType (" vnd.android-dir/mms-sms");
startActivity (it);
```

(7) 发送短信

```
Uri uri = Uri.parse("smsto:0800000123");
Intent it = new Intent(Intent.ACTION_SENDTO, uri);
it.putExtra (" sms_body", " The SMS text");
startActivity (it);
```

(8) 发送彩信

```
Uri uri = Uri.parse("content://media/external/images/media/23");
Intent it = new Intent(Intent.ACTION_SEND);
it.putExtra (" sms_body", " some text");
it.putExtra (Intent.EXTRA_STREAM, uri);
it.setType (" image/png");
startActivity (it);
```

(9) 发送 Email

```
Uri uri = Uri.parse("mailto:xxx@abc.com");
Intent it = new Intent(Intent.ACTION_SENDTO, uri);
startActivity (it);
Intent it = new Intent (Intent.ACTION_SEND);
it.putExtra (Intent.EXTRA_EMAIL, " me@abc.com");
it.putExtra (Intent.EXTRA_TEXT, " The email body text");
it.setType (" text/plain");
startActivity (Intent.createChooser (it, " Choose Email Client"));
Intent it=new Intent (Intent.ACTION_SEND);
String [] tos= {" me@abc.com"};
String [] ccs= {" you@abc.com"};
```

```
it.putExtra (Intent.EXTRA_EMAIL, tos);
it.putExtra (Intent.EXTRA_CC, ccs);
it.putExtra (Intent.EXTRA_TEXT, " The email body text");
it.putExtra (Intent.EXTRA_SUBJECT, " The email subject text");
it.setType (" message/rfc822");
startActivity (Intent.createChooser (it, " Choose Email Client"));
```

（10）卸载应用

```
Uri uninstallUri = Uri.fromParts("package", "com.app.test", null);
Intent intent = new Intent(Intent.ACTION_DELETE, uninstallUri);
```

（11）安装应用

```
Intent intent = new Intent(Intent.ACTION_VIEW);
intent.setDataAndType (Uri.fromFile (new File (" /sdcard/test.apk"), " appli-
cation/vnd.android.package-archive");
```

Android 的基础组件 Activity、Service 和 BroadcastReceiver，都是通过 Intent 机制激活的，不同类型的组件有不同的传递 Intent 方式。

（1）要激活一个新的 Activity，或者让一个现有的 Activity 进行新的操作，可以通过调用 Context. startActivity( )或者 Activity. startActivityForResult( )方法来实现。

（2）要启动一个新的 Service，或者向一个已有的 Service 传递新的指令，调用 Context. startService( )方法或者调用 Context. bindService( )方法将调用此方法的上下文对象与 Service 绑定。

（3）Context. sendBroadcast( )、Context. sendOrderBroadcast( )、Context. sendStickBroadcast( )这三个方法可以发送 Broadcast Intent。发送之后，所有已注册的并且拥有与之相匹配 IntentFilter 的 BroadcastReceiver 即被激活。

3.7 界面设计新体验——Material Design

长期以来，大多数人认为 Android 系统的界面并不美观，Android 标准的界面设计风格并不是特别被大众所接受，很多公司都觉得自己完全可以设计出更好看的界面，从而导致 Android 平台的界面风格长期难以得到统一。为了解决这个问题，从 Android 5.0 开始，Google 推出了一套全新的界面设计语言——Material Design。

3.7.1 什么是 Material Design

Material Design 是由 Google 的设计工程师基于传统优秀的设计原则，结合丰富的创意和科学技术所发明的一套全新的界面设计语言，包含了视觉、运动、互动效果等特性。

Google 的目标是开发一个设计系统，让所有的产品在任何平台上拥有统一的用户体验。Material Design 具有全新的设计理念，采用大胆的色彩、流畅的动画播放，以及卡片式的简洁设计。

Google 从 Android 5.0 系统开始，使用 Material Design 风格设计所有内置的应用。

为了更便捷地使用 Material Design，在 2015 年的 Google I/O 大会上推出了一个 Design Support 库，这个库将 Material Design 中最具有代表性的一些控件和效果进行了封装，带来了很多 Material Design 组件，包括 Navigation Draw View、Floating Labels、Floating Action Button、Snackbars，还有处理动作和滑屏事件的新框架。

3.7.2 Material Design 内容

Material Design 是 Google 推出的一个全新的设计语言，它的特点就是拟物扁平化。

Material Design 包含了很多内容，大致可以分为四部分。

1. 主题和布局

Material 主题只能应用在 Android L 版本。

应用 Material 主题的方法很简单，只需要修改 res/values/styles. xml 文件，使其继承 android：Theme. Material 即可，具体如下。

```
<!-- res/values/styles.xml -->
<resources>
<!-- your app's theme inherits from the Material theme -->
<style name="AppTheme" parent="android:Theme.Material">
<!-- theme customizations -->
</style>
</resources>
```

或者在 AndroidManifest. xml 中直接设置主题。

```
android:theme="@android:style/Theme.Material.Light"
```

2. 视图和阴影

View 的大小、位置都是通过（x,y）确定的，而现在有了 z 轴的概念，这个 z 值就是 View 的高度（elevation），高度决定了阴影（shadow）的大小。

3. UI 控件

新增了两个控件分别是 RecyclerView 和 CardView。

4. 动画

新增了如下几种动画。

- Touch Feedback：触摸反馈。
- Reveal Effect：揭露效果。
- Activity Transitions：Activity 转换效果。
- Curved Motion：曲线运动。
- View State Changes：视图状态改变。
- Animate Vector Drawables：可绘矢量动画。

3.8 实例：WebView 实现监控界面

WebView（网络视图）控件能加载显示网页，可以将其视为一个浏览器。它使用 WebKit 渲染引擎加载显示网页。通过它可以轻松实现显示网页功能。

WebView 控件是专门用来浏览网页的，它的使用方式与其他控件一样。我们可以通过在 XML 布局文件中添加 <WebView> 标记来完成。WebView 提供的常用方法如表 3-4 所示。

表 3-4　WebView 提供的常用方法

XML 属性	说　明
loadUrl（String url）	加载 URL 信息，Url 可以是网络地址，也可以是本地网络文件
goBack()	向后浏览历史页面
goForward()	向前浏览历史页面
loadData（String data，String mimeType，String encoding）	用于将指定的字符串数据加载到浏览器中
loadDataWithBaseURL（String baseUrl，String data，String mimeType，String encoding，String historyUrl）	用于基于 URL 加载指定的数据
stopLoading()	用于停止加载当前页面
reload()	用于刷新当前页面

其界面布局文件内容如下。

```
<RelativeLayout xmlns:android="http://schemas.android.com/apk/res/android"
    android:layout_width="fill_parent"
    android:background="@drawable/bg_environment"
    android:layout_height="fill_parent">
    <ImageView
        android:id="@+id/imageView1"
        android:layout_width="wrap_content"
        android:layout_height="wrap_content"
        android:layout_alignParentLeft="true"
        android:layout_alignParentTop="true"
        android:layout_marginLeft="20dp"
        android:layout_marginTop="30dp"
        android:src="@drawable/btn_monitoring_select"/>
    <LinearLayout
        android:layout_width="match_parent"
        android:layout_height="match_parent"
        android:background="@drawable/bg_frame_descend_setting"
        android:padding="30dip"
        android:layout_marginTop="20dp"
        android:layout_marginRight="20dp"
        android:layout_marginBottom="20dp"
        android:layout_toRightOf="@+id/imageView1"
        android:orientation="horizontal">
        <LinearLayout
            android:layout_width="0.0dp"
            android:layout_height="fill_parent"
            android:gravity="center"
            android:layout_weight="2">
        <WebView
            android:id="@+id/webView1"
            android:layout_width="match_parent"
```

```
            android: layout_height = " match_parent" / >
        < /LinearLayout >
          < LinearLayout
            android: layout_width = " 0.0dp"
            android: layout_height = " fill_parent"
            android: gravity = " center"
            android: layout_weight = " 1"
            android: orientation = " vertical" >
              < RelativeLayout
                  android: layout_width = " 150dp"
                  android: layout_height = " 120dp"
                  android: background = " @drawable/btn_direction_bg" >
                  < ImageView
                      android: id = " @ + id/imageView2"
                      android: layout_width = " 32dp"
                      android: layout_height = " 32dp"
                      android: layout_alignParentLeft = " true"
                      android: layout_centerVertical = " true"
                      android: layout_marginLeft = " 10dp"
                      android: src = " @drawable/btn_left_press" / >
                  < ImageView
                      android: id = " @ + id/imageView3"
                      android: layout_width = " 32dp"
                      android: layout_height = " 32dp"
                      android: layout_alignParentBottom = " true"
                      android: layout_centerHorizontal = " true"
                      android: layout_marginBottom = " 10dp"
                      android: src = " @drawable/btn_down_press" / >
                  < ImageView
                      android: id = " @ + id/imageView4"
                      android: layout_width = " 32dp"
                      android: layout_height = " 32dp"
                      android: layout_alignParentTop = " true"
                      android: layout_centerHorizontal = " true"
                      android: layout_marginTop = " 10dp"
                      android: src = " @drawable/btn_up_press" / >
                  < ImageView
                      android: id = " @ + id/imageView5"
                      android: layout_width = " 32dp"
                      android: layout_height = " 32dp"
                      android: layout_alignParentRight = " true"
                      android: layout_centerVertical = " true"
                      android: layout_marginRight = " 10dp"
```

```
            android: src = " @drawable/btn_right_press" />
        </RelativeLayout>
        <Button
          android: id = " @ +id/button1"
          android: layout_width = " wrap_content"
          android: layout_height = " 45dp"
          android: layout_marginTop = " 20dp"
          android: background = " @drawable/btn_page_hover"
          android: text = " 拍  照"
          android: textColor = " @color/white" />
        </LinearLayout>
    </LinearLayout>
</RelativeLayout>
```

其界面显示效果如图 3-30 所示。

图 3-30 摄像头监控界面

其中布局属性 layout_margin 为控件的外边距，是布局与布局之间的距离，Border 为控件的边距，Padding 为内边距，是布局的边与布局内部元素的距离，如图 3-31 所示。

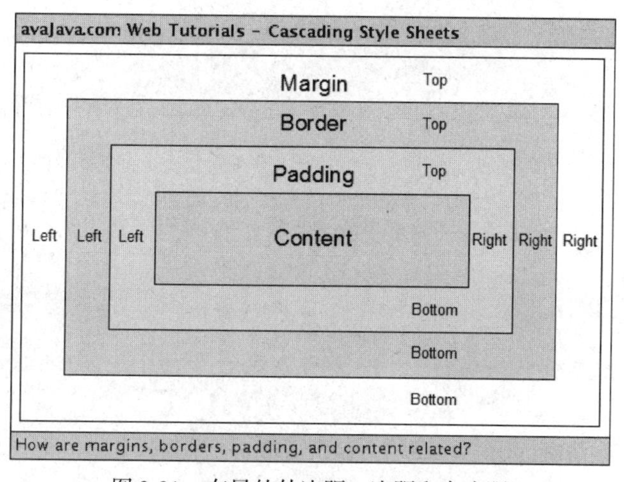

图 3-31 布局的外边距、边距和内边距

在 Android Studio 2.3 的组件树中可以看到布局情况，如图 3-32 所示。

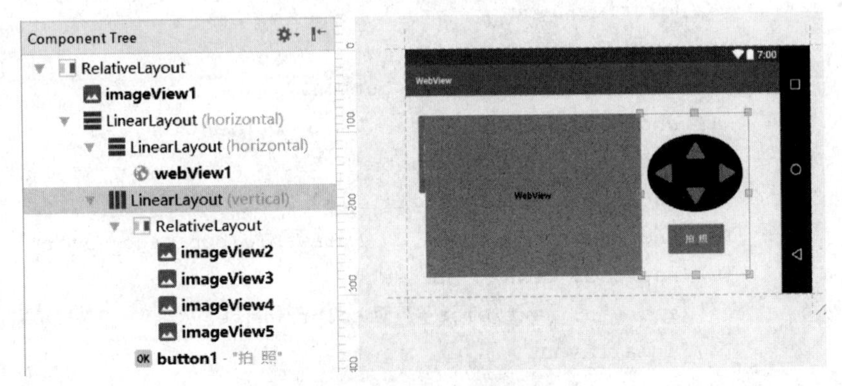

图 3-32　摄像头监控界面布局

3.9 本章小结

通过本章的学习，对 Android 有了更深的认识，实现了 Android 界面设计，了解了界面设计的三大内容：布局、控件和 Activity，掌握了 Android 界面新的设计方法：Material Design，可以制作出丰富多彩的界面。本章介绍了 UI 设计的所有重要知识点，有了这些知识的准备，下一章即可开始应用程序的开发。

第四章

应用程序的构成部件

Android 应用程序是 Android 系统智能手机的主要构成部分，实现了智能手机的多样性、多功能性，融合了办公功能、娱乐功能、生活实用功能等，广受人们的喜爱。Android 应用程序由一些松散联系的组件构成，遵守着一个应用程序清单，这个清单描述了每个组件以及它们如何交互。

4.1 应用程序架构介绍

Android 应用程序用到的各种“组件”（如 Activity、BroadcastReceiver、Service 等）会在同一个进程中执行，而且由该进程的主线程负责执行。如果有特别的指定，也可以让特定“组件”在不同的进程里执行。无论这些组件是在哪一个进程里执行，默认情况下，它们都是由该进程里面的主线程来负责执行的。

主线程除了要处理 Activity 的 UI 事件，还要处理 Service 后台服务工作，通常会忙不过来。而多线程的并行（Concurrent）可以化解主线程过于忙碌的问题，主线程可以生成多个子线程来分担它的工作，尤其是比较冗长费时的后台服务工作（如播放动画的背景音乐、从网络上下载电影等）。这样，主线程就能专心处理 UI 画面的事件了，如图 4-1 所示。

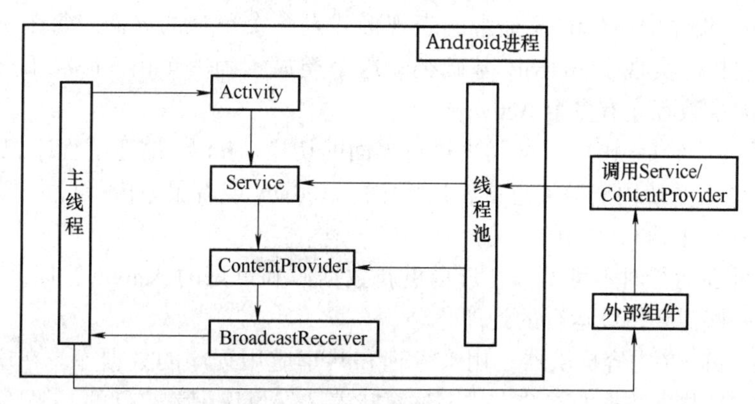

图 4-1　Android 组件和线程框架

以下六个组件构成了应用程序的基础部分。

（1）Activity：应用程序的表示层。应用程序的每个界面都是 Activity 类的扩展。Activity

用视图（View）构成 GUI，来显示信息，响应用户操作。就桌面开发而言，一个活动（Activity）相当于一个窗体。

（2）Sevvice：应用程序中的隐形工作者。Service 组件在后台运行，更新数据源和可见的 Activity，触发通知（Notification）。在应用程序的 Activity 不激活或不可见时，用于执行需要继续的长期处理。

（3）Content Provider：可共享的数据存储。Content Provider 用于管理和共享应用程序数据库，是跨应用程序边界数据共享的优先方式。这表示您可以配置自己的 Content Provider 以允许其他应用程序的访问，用他人提供的 Content Provider 来访问他人存储的数据。Android 设备包括几个本地 Content Provider，提供了如媒体库和联系人明细这样常用的数据库。

（4）Intent：一个应用程序间（inter-application）的消息传递框架。使用 Intent 可以在系统范围内广播消息或者对一个目标 Activity 或 Service 发送消息，来表示要执行一个动作。系统将辨别出相应要执行活动的目标。

（5）Broadcast Receiver：Intent 广播的消费者。如果创建并注册了一个 Broadcast Receiver，应用程序就可以监听匹配特定过滤标准的广播 Intent。Broadcast Receiver 会自动开启应用程序，以响应一个收到的 Intent，使用它们可以完美地创建事件驱动的应用程序。

Widgets 是可以添加到主屏幕界面（home screen）的可视应用程序组件。作为 Broadcast Receiver 的特殊变种，Widgets 可以为用户创建可嵌入到主屏幕界面的动态的、交互的应用程序组件。

（6）Notification：一个用户通知框架。应用 Notification，不必窃取焦点或中断当前 Activities 就能通知用户。这是在 Service 和 Broadcast Receiver 中获取用户注意的推荐技术。例如，当设备收到一条短消息或一个电话，它会通过闪光灯、发出声音、显示图标或显示消息来提醒您。您可以在应用程序中使用 Notification 触发相同的事件。

不过，不是每个程序都有这 6 个组件，可能程序只使用了其中一部分。一旦决定了程序中包含哪些组件，要在 AndroidManifest. xml 文件中将其列出，这是一个 XML 文件，包含程序所定义的组件，以及这些组件的功能和必备的条件。

Activity 中 4 个组件是最常用的：（1）Activity；（2）Intent Receiver；（3）Service；（4）Content Provider。程序中，Activity 通常的表现形式是一个单独的界面。每个 Activity 都是一个单独的类，它扩展实现了 Activity 基础类。这个类显示为一个由 Views 组成的用户界面，并响应事件。大多数程序有多个 Activity。

Android 通过一个专门的 Intent 类来进行界面的切换。Intent 描述了程序想做什么。使用 Intent，可以在整个系统内广播消息或者让特定的 Activity 或者服务执行行为意图。系统会决定哪个（些）目标来执行适当的行为。

Service 组件在运行时不可见，它负责更新数据源和可见的 Activity，以及触发通知。它们常用来执行一些需要持续运行的处理。

Content Provider（内容提供器）用来管理和共享应用程序的数据库。在应用程序之间，Content Provider 是共享数据的首选方式。这意味着，您可以配置自己的 Content Provider 以存取其他应用程序，或者通过其他应用程序暴露的 Content Provider 来存取它们的数据。

通过创建和注册 Broadcast Receiver，应用程序可以监听符合特定条件的广播的 Intent。Broadcast Receiver 会自动启动 Android 应用程序响应新来的 Intent。Broadcast Receiver 是事件驱动程序的理想手段。

4.2 应用程序并行机制——线程和线程池

在 Android 程序中，会遇到一些耗时的操作，比如网络操作、从网上抓取图片、下载文件、批量更新数据库等，这些操作对于移动终端而言，会需要很长的时间，而应用程序界面又不能等到这些操作完成后再显示，所以要让界面与这些耗时的操作并行处理，用多线程可以解决这个问题。当然还有其他解决方案，比如用 Service。

在 Android 应用程序中，至少有一个 UI 主线程、若干子线程构成，如图 4-2 所示。

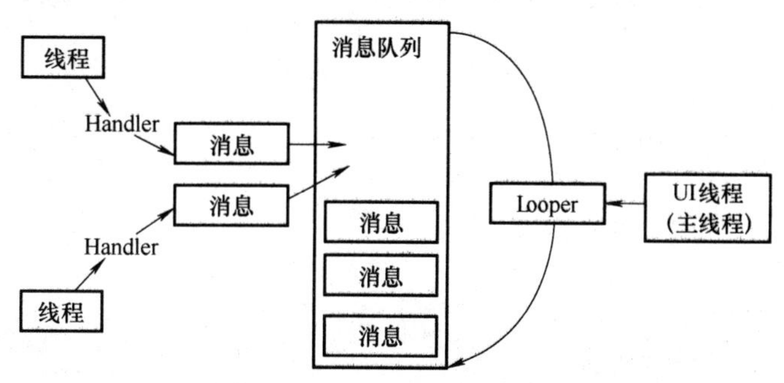

图 4-2　Android 应用程序线程

4.2.1　线程的实现方法

Android 线程的建立和 Java 线程基本相同，在 Android 中有两种实现线程的方法：（1）扩展 java. lang. Thread 类；（2）实现 Runnable 接口。

1. 扩展 java. lang. Thread 类

定义一个 Thread 类，重写父类的 run( ) 方法，如下所示。

```
class MyThread extends Thread
  {
            @Override
            public void run()
            {
            //处理具体的逻辑
       }
   }
```

使用类的方法如下。

```
MyThread  thread =new MyThread();
thread.start(); //线程启动
```

2. 实现 Runnable 接口

下面介绍第 1 种方法，定义类。

```
class MyThread implements Runnable
  {
            @Override
```

```
        public void run()
        {
            //处理具体的逻辑
        }
    }
```

使用类的方法如下。

```
MyThread  thread =new MyThread();
thread.start(); //线程启动
```

下面介绍第 2 种方法，匿名实现。

```
Thread thread=new Thread(new Runnable()
    {
        @Override
        public void run()
        {
            //处理具体的逻辑
        }
    });
thread.start(); //线程启动
```

4.2.2 Android 的线程池

多线程技术主要解决处理器单元内多个线程执行的问题，可以显著减少处理器单元的闲置时间，增加处理器单元的吞吐能力。

当同时并发多个网络线程时，要防止内存过度消耗。控制活动线程的数量，防止并发线程过多，显著减少创建线程的数目。引入线程池技术会极大地提高 App 的性能。

假设一个 App 完成一项任务的时间为 T。

- T1：创建线程的时间。
- T2：在线程中执行任务的时间，包括线程间同步所需时间。
- T3：线程销毁的时间。

显然 T = T1 + T2 + T3。这是一个简化的假设，可以看出 T1 和 T3 是多线程本身带来的开销，我们希望减少 T1 和 T3 所用的时间，从而减少 T 的时间。但一些线程的使用者并没有注意到这一点，所以在程序中频繁地创建或销毁线程，这导致 T1 和 T3 在 T 中占有相当比例。显然这是突出了线程的弱点（T1、T3），而不是优点（并发性）。

线程池的出现，恰恰能解决上面类似问题，线程池的优点如下。

（1）复用线程池中的线程，避免因为线程的创建和销毁所带来的性能开销。

（2）能够有效控制线程池的最大并发数，避免大量的线程之间因互相抢占系统资源而导致的阻塞现象。

（3）能够对线程进行简单的管理，并提供定时执行以及指定间隔循环执行等功能。

一个线程池包括以下四个基本组成部分。

（1）线程池管理器（ThreadPool）：用于创建并管理线程池，包括创建线程池、销毁线程池、添加新任务。

（2）工作线程（PoolWorker）：线程池中线程，在没有任务时处于等待状态，可以循环的执行任务。

（3）任务接口（Task）：每个任务必须实现的接口，以供工作线程调度任务的执行，它主要规定了任务的入口、任务执行完后的收尾工作，任务的执行状态等。

（4）任务队列（taskQueue）：用于存放没有处理的任务，提供一种缓冲机制。

线程池技术正是关注如何缩短或调整 T1、T3 时间的技术，从而提高服务器程序性能的。它把 T1、T3 分别安排在服务器程序的启动和结束的时间段或者一些空闲的时间段，这样在服务器程序处理客户请求时，不会有 T1、T3 的开销。线程池不仅调整 T1、T3 产生的时间段，而且还显著减少了创建线程的数目。

线程池类的构造方法如下。

ThreadPoolExecutor（int corePoolSize，int maximumPoolSize，long keepAliveTime，TimeUnit unit，BlockingQueue workQueue，RejectedExecutionHandler handler）。

其中的参数含义如下。

corePoolSize：线程池维护线程的最少数量。

maximumPoolSize：线程池维护线程的最大数量。

keepAliveTime：线程池维护线程所允许的空闲时间。

unit：线程池维护线程所允许的空闲时间的单位。

workQueue：线程池所使用的缓冲队列。

handler：线程池对拒绝任务的处理策略。

例如：private static ExecutorService exec = new ThreadPoolExecutor（8，8，0L，TimeUnit. MILLISECONDS，new LinkedBlockingQueue < Runnable >（100000），new ThreadPoolExecutor. CallerRunsPolicy( ))。

在实际使用中，建议使用较为方便的 Executors 工厂方法，通过 Executors 建立以下四种线程池。

（1）ExecutorService pool1 = Executors. newCachedThreadPool( )

缓存型池子，先查看池中有没有以前建立的线程，如果有，就重用，如果没有，就建一个新的线程加入池中。能重用的线程，必须是 timeout IDLE 内的池中线程，缺省 timeout 为 60s，超过这个 IDLE 时长，线程实例将被终止并移出池。缓存型池子通常用于执行一些生存期很短的异步型任务。

接下来通过 ExecutorService 的 execute( )方法启动线程，其参数是一个线程。例如在如下事件中启动了 9 个线程。

```
@Override
public void onClick(View view) {
  ExecutorService pool1 = Executors.newCachedThreadPool()
  for(int i = 0; i < 9; ++i) {
    final int index = i;
    pool1.execute(new Runnable() {
      @Override
       public void run() {
         try {
```

```
            //执行的代码
          Thread.sleep(2 * 1000);
          } catch(InterruptedException e) {
            e.printStackTrace();
          }
        Log.i(TAG, "Thread:" + Thread.currentThread().getId() + " activeCount:" +
Thread.activeCount() + " index:" + index);
      }
    });
  }
  }
```

这段代码执行时，建立 9 个线程。

（2） ExecutorService pool2 = Executors.newFixedThreadPool（3）

FixedThreadPool 与 cacheThreadPool 差不多，能重用就重用，但不能随时建立新的线程。其独特之处在于，任意时间点最多只能有固定数目的活动线程存在，此时如果有新的线程要建立，只能放在另外的队列中等待，直到当前的线程中某个线程终止直接被移出池子。和 cacheThreadPool 不同，fixedThreadPool 池线程数固定，所以 fixedThreadPool 多数针对一些很稳定、很固定的正规并发线程，多用于服务器。

还是以上段代码为例，其中 pool1 换成 pool2，执行以后可以看到，同时只能创建 3 个线程。

```
for(int i = 0; i < 9; ++i) {
  final int index = i;
  pool2.execute(new Runnable() {
  @Override
    public void run() {
    ......
  }
});
}
```

（3） ExecutorService pool3 = Executors.newSingleThreadExecutor()

单例线程，任意时间池中只能有一个线程，保证所有任务按照指定顺序（FIFO，LIFO，优先级）执行。

（4） ScheduledExecutorService pool4 = Executors.newScheduledThreadPool（3）

调度型线程池。这个池子里的线程可以按 schedule 依次 delay 执行，或周期执行，0 秒 IDLE（无 IDLE）。

需要注意，ScheduledExecutorService 不是使用 execute()方法，而是使用 schedule()方法，方法的定义格式为 schedule（Runnable command，long delay，TimeUnit unit）。

下面是线程建立代码。

```
for(int i = 0; i < 9; ++i) {
  final int index = i;
  pool4.schedule(new Runnable() {
```

```
            @Override
            public void run() {
              try {
                  //执行任务代码
                  Thread.sleep(2 * 1000);
              } catch(InterruptedException e) {
                  e.printStackTrace();
              }
              Log.i(TAG, "Thread:" + Thread.currentThread().getId() + " activeCount:" +
Thread.activeCount() + " index:" + index);
        }
      },2, TimeUnit.SECONDS);
    }
```

依次建立调度3个线程执行，后面的3个线程建立，前面的3个线程停止。

4.3 应用程序互动机制——事件机制

Android事件监听是整个消息的基础和关键，涉及两类对象：事件发生者和事件监听者，事件发生者是事件的起源，可以是一个按钮、编辑框等。事件监听者是事件的接受者，如果想接收某个事件，必须对该事件的发生者注册对应的事件监听器。

Android的事件处理机制有两种，基于监听接口和基于回调机制，这两种机制的原理和实现方法都有不同。

4.3.1 事件处理机制1——基于监听器的事件处理

相比于基于回调的事件处理，这是更具"面向对象"性质的事件处理方式。在监听器模型中，主要涉及三类对象。

（1）事件源Event Source：产生事件的来源，通常是各种组件，如按钮、窗口等。

（2）事件Event：事件封装了界面组件上发生的特定事件的具体信息，如果监听器需要获取界面组件上所发生事件的相关信息，一般通过事件Event对象来传递。

（3）事件监听器Event Listener：负责监听事件源发生的事件，并对不同的事件做相应的处理。

基于监听器的事件处理机制是一种委派式Delegation的事件处理方式，事件源将整个事件委托给事件监听器，由监听器对事件进行响应处理。这种处理方式将事件源和事件监听器分离，有利于增强程序的可维护性。

委托事件模型的原理如图4-3所示。外部的操作，例如按下按键、触摸屏单击或转动移动终端等动作，会触发事件源的事件。对于单击按钮的操作来说，事件源就是按钮，它会根据这个操作生成一个按钮按下的事件对象，这对于系统来说，就产生了一个事件。

事件的产生会触发事件监听器，事件本身作为参数传入到事件处理器中。事件监听器是通过代码在程序初始化时注册到事件源的，也就是说，在按钮上设置一个可以监听按钮操作的监听器，并且通过这个监听器调用事件处理器，事件处理器里有针对这个事件编写的代码。

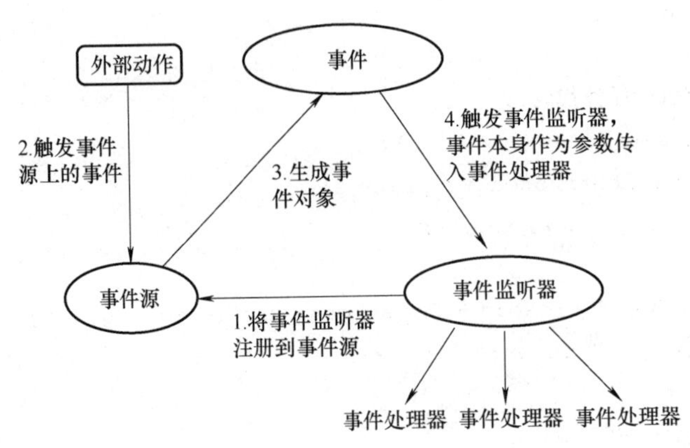

图 4-3 Android 事件监听处理机制

基于事件监听器的事件处理需要做如下 3 个工作。

(1) 定义监听器类，覆盖对应的抽象方法，在监听器中针对事件编写相应的处理代码。

(2) 创建监听器对象。

(3) 注册监听器。

4.3.2 事件处理机制 2——基于回调的事件处理

Android 的另一种事件处理机制是回调机制。

通常情况下，程序员编写程序时，需要使用系统工具提供的方法来完成某种功能，例如调用 Math. sqrt( )求取平方根。但是，某种情况下系统会反过来调用一些类的方法，例如对于用作组件或插件的类，需要编写一些共系统调用的方法，这些专门用于被系统调用的方法被称为回调方法，也就是回过来系统调用的方法。

Android 平台中，每个 View 都有自己的处理事件的回调方法，开发人员可以通过重写 View 中的这些回调方法来实现需要的响应事件。当某个事件没有被任何一个 View 处理时，便会调用 Activity 中相应的回调方法。例如：View 类实现了 KeyEvent. Callback 接口中的一系列回调函数，因此，基于回调的事件处理机制通过自定义 View 来实现，自定义 View 时重写这些事件处理方法即可。

与基于监听器的事件处理模型相比，基于回调的事件处理模型要简单一些，该模型中，事件源和事件监听器是合一的，也就是说没有独立的事件监听器存在。当用户在 GUI 组件上触发某事件时，由该组件自身特定的函数负责处理该事件。通常通过重写 Override 组件类的事件处理函数，实现事件的处理。

4.3.3 事件响应的实现

Android 系统为不同的控件和操作提供了相应的监听器，下面介绍几个常用的监听器。

- OnClickListener 接口处理的单击事件，在触控模式下，是在某个 View 上按下并抬起的组合动作，而在键盘模式下，是某个 View 获得焦点后，单击确定按钮或者按下轨迹球事件。
- OnFocusChangeListener 接口用来处理控件焦点发生改变，当某个控件失去焦点或者获得焦点时，会触发该接口中的回调方法。

➢ OnKeyListener 是对手机键盘进行监听的接口，通过对某个 View 注册该监听，当 View 获得焦点并有键盘事件时，便会触发该接口中的回调方法。

➢ OnLongClickListener 接口与之前介绍的 OnClickListener 接口相似，只是该接口为 View 长按事件的捕捉接口，即当长时间按下某个 View 时触发的事件。

➢ OnTouchListener 接口是用来处理手机屏幕事件的监听接口，当在 View 的范围内产生触摸、按下、抬起或滑动等动作时，会触发该事件。

4.3.4　实例：获取触点坐标

在智能手机上，很多应用软件需要获取用户手指操作的坐标和一些用户的操作，本应用程序通过事件获取这些信息。在 Android 2.3 中建立项目 Event_Handle。

布局文件 activity_main.xml 的代码如下。

```
<?xml version="1.0" encoding="utf-8"?>
<LinearLayout xmlns:android="http://schemas.android.com/apk/res/android"
    android:orientation="vertical"
    android:layout_width=" fill_parent"
    android: layout_height=" fill_parent" >
  <TextView
        android: id=" @+id/touch_area"
        android: layout_width=" fill_parent"
        android: layout_height=" 300dip"
        android: background=" #0FF"
        android: textColor=" #FFFFFF"
        android: text=" 触摸事件测试区" />
   <TextView
        android: id=" @+id/history_label"
        android: layout_width=" fill_parent"
        android: layout_height=" wrap_content"
        android: text=" 历史数据"   />
   <TextView
        android: id=" @+id/event_label"
        android: layout_width=" fill_parent"
        android: layout_height=" wrap_content"
        android: text=" 触摸事件:" />
</LinearLayout>
```

实现 Activity 的文件 MainActivity.java 的内容如下。

```
public class MainActivity extendsAppCompatActivity
{
    private TextView eventlable;
    private TextView histroy;
    private TextView TouchView;
    @Override
```

```
    public void onCreate(Bundle savedInstanceState) {
        super.onCreate(savedInstanceState);
        setContentView(R.layout.main);
        TouchView = (TextView) findViewById(R.id.touch_area);
        histroy = (TextView) findViewById (R.id.history_label);
        eventlable = (TextView) findViewById (R.id.event_label);
      // TouchView 控件的事件处理
        TouchView.setOnTouchListener (new View.OnTouchListener() {
            @Override
            public boolean onTouch (View v, MotionEvent event) {
                int action = event.getAction();
                switch (action) {
                //当按下的时候
                case (MotionEvent.ACTION_DOWN):
                    Display (" ACTION_DOWN", event);
                    break;
                //当按上的时候
                case (MotionEvent.ACTION_UP):
                    int historysize = ProcessHistory (event);
                    histroy.setText (" 历史数据" +historysize);
                    Display (" ACTION_UP", event);
                    break;
                //当触摸的时候
                case (MotionEvent.ACTION_MOVE):
                    Display (" ACTION_MOVE", event);
                }
                return true;
            }
        });
    }
  public void Display (String eventType, MotionEvent event) {
      //触点相对坐标的信息
      int x = (int) event.getX();
      int y = (int) event.getY();
      //表示触屏压力大小
      float pressure = event.getPressure();
      //表示触点尺寸
      float size = event.getSize();
      //获取绝对坐标信息
      int RawX = (int) event.getRawX();
      int RawY = (int) event.getRawY();
      String msg = "";
      msg += " 事件类型" +eventType + " \n";
```

```
        msg += "相对坐标" + String.valueOf(x) + "," + String.valueOf(y) + "\n";
        msg += "绝对坐标" + String.valueOf(RawX) + "," + String.valueOf(RawY)
                  + "\n";
        msg += "触点压力" + String.valueOf(pressure) + ",";
        msg += "触点尺寸" + String.valueOf(size) + "\n";
        eventlable.setText(msg);
    }
    public int ProcessHistory(MotionEvent event) {
        int history = event.getHistorySize();
        for(int i = 0; i < history; i++) {
            long time = event.getHistoricalEventTime(i);
            float pressure = event.getHistoricalPressure(i);
            float x = event.getHistoricalX(i);
            float y = event.getHistoricalY(i);
            float size = event.getHistoricalSize(i);
            }
        return history;
        }
    }
```

项目 Event_Handle 运行效果如图 4-4 所示。

图 4-4 Android 事件响应

4.4 应用程序后台劳动者——Service

Service 是 Android 系统中的四大组件（Activity、Service、BroadcastReceiver、ContentProvider）之一，跟 Activity 的级别差不多，但不能自己运行，只能后台运行，并且可以和其他组件进行交互。Service 可以在很多场合的应用中使用，比如播放多媒体的时候，用户启动了其他 Activity 时，程序要在后台继续播放，比如检测 SD 卡上文件的变化，或者在后台记录地理信息位置的改变时，总之服务是藏在后台运行的。

4.4.1 服务的创建

在 Android Studio 2.3 环境中，新建 Service 是在项目中进行的，在项目中选择 app 项的 java 项目，单击鼠标右键，在弹出的菜单中选择 New > Service 命令，选择新建的 Service 种类，共两种，如图 4-5 所示。

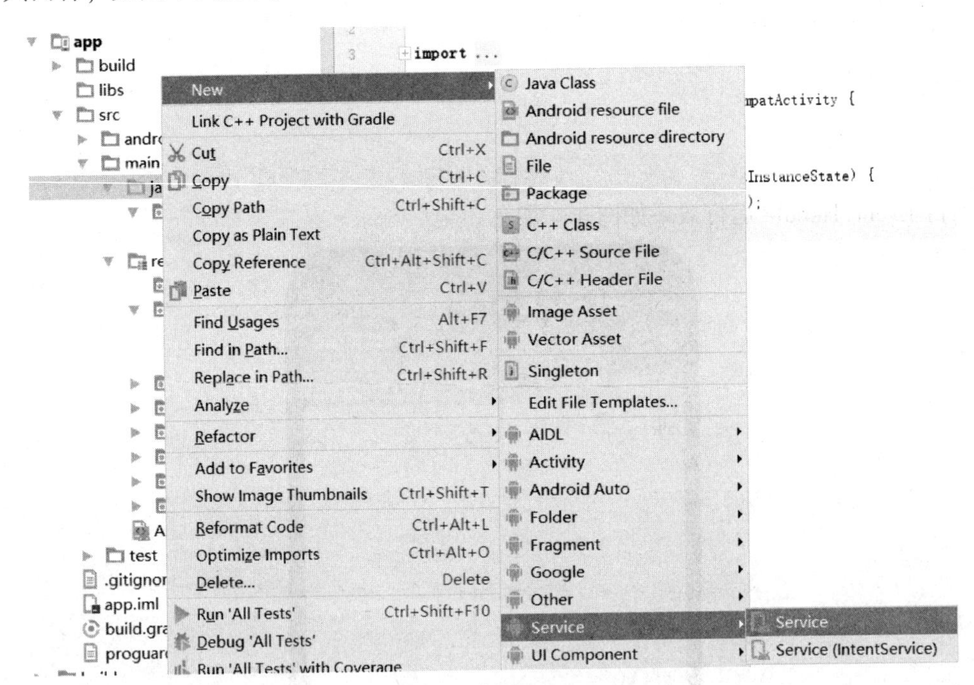

图 4-5　创建 Service

Android 的 Service 分为 IntentService 与 Service 两种。

Android 中的 Service 是用于后台服务的，当应用程序被挂到后台的时候，为了保证应用的某些组件仍然可以工作，而引入了 Service 这个概念，这里面要强调的是，Service 不是独立的进程，也不是独立的线程，它依赖于应用程序的主线程，也就是说，在更多时候不建议在 Service 中编写耗时的逻辑和操作。当我们编写的耗时逻辑不得不由 Service 管理时，就需要引入 IntentService。

IntentService 是继承并处理异步请求的一个类，在 IntentService 内有一个工作线程来处理耗时操作，启动 IntentService 的方式和启动传统的 Service 一样。同时，当任务执行完成后，IntentService 会自动停止，不需要我们手动去控制或 stopSelf()。另外，可以多次启动 Intent-

Service，每一个耗时操作会以工作队列的方式在 IntentService 的 onHandleIntent 回调方法中执行，并且，每次只会执行一个工作线程，执行完第一个再执行第二个，以此类推。

现在我们新建一个 Service，将服务命名为 MyService，如图 4-6 所示，Exported 属性用于设置是否除了当前程序之外的其他程序访问这个服务，Enabled 属性用于设置是否启动这个服务，单击 Finish 按钮完成创建。

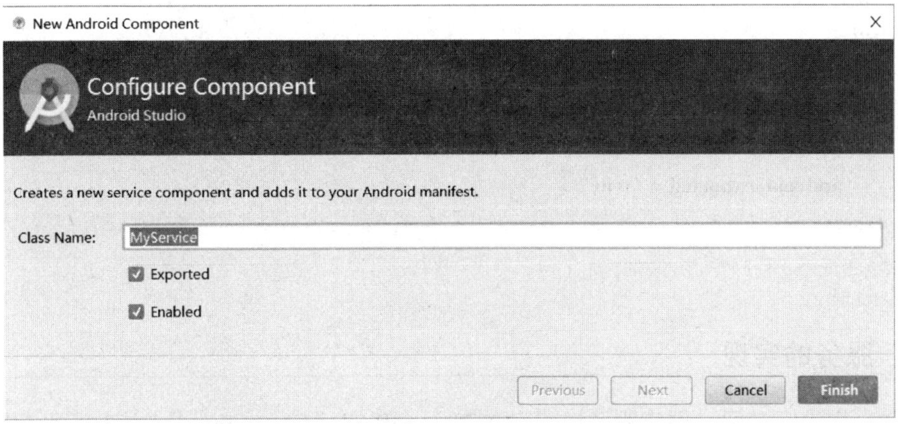

New Android Component

Configure Component
Android Studio

Creates a new service component and adds it to your Android manifest.

Class Name: MyService

Exported

Enabled

Previous　Next　Cancel　Finish

图 4-6　新建服务

下面是 MyService. java 中的代码。

```
public class MyService extends Service {
public MyService()
{
    }
    @Override
    public IBinder onBind(Intent intent) {
      // TODO: Return the communication channel to the service.
      throw new UnsupportedOperationException("Not yet implemented");
    }
}
```

可以看到 MyService 继承 Service 类，说明这是一个服务，有一个 onBind（Intent intent）方法，这个方法是 Service 中唯一的抽象方法，必须在子类里实现。

服务定义完成后，接下来我们要在 MainActivity 中启动该服务（通过 Intent 启动）。

```
public class MainActivity extends AppCompatActivity {
  @Override
  protected void onCreate(Bundle savedInstanceState) {
    super.onCreate(savedInstanceState);
    setContentView(R.layout.activity_main);
    //启动 Service
    Intent intent = new Intent (MainActivity.this, MyService.class);
    startService (intent);
  }
}
```

另外，Service 是 Android 四大组件之一，必须要在 AndroidMainfest. xml 文件中注册这个服务，打开 AndroidMainfest. xml，如下所示，发现 Android Studio 已经发生了变化，Android Studio 2. 3 已经完成了注册。

```
<manifest xmlns:android=http://schemas.android.com/apk/res/android
    package="com.example.hefugui.servicetest">
     <application
    ......
     <service
         android:name=".MyService"
         android:enabled="true"
         android:exported="true">
     </service>
     </application>
</manifest>
```

4. 4. 2　服务的实现

在上一节我们看到，创建服务时自动创建了 onBind（Intent intent）方法，那么还有其他什么方法需要实现呢？

在 Android Studio 中，在 MyService 的代码区单击鼠标右键，选择菜单中 Generate 命令，如图 4-7 所示。

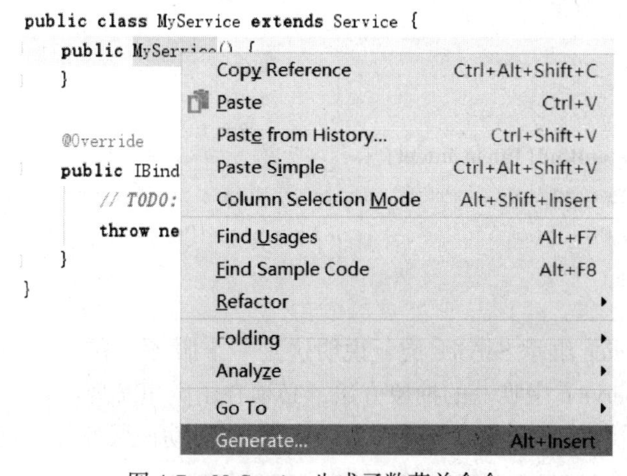

图 4-7　MyService 生成函数菜单命令

弹出的菜单窗口如图 4-8 所示，选择 Override Methods 选项。

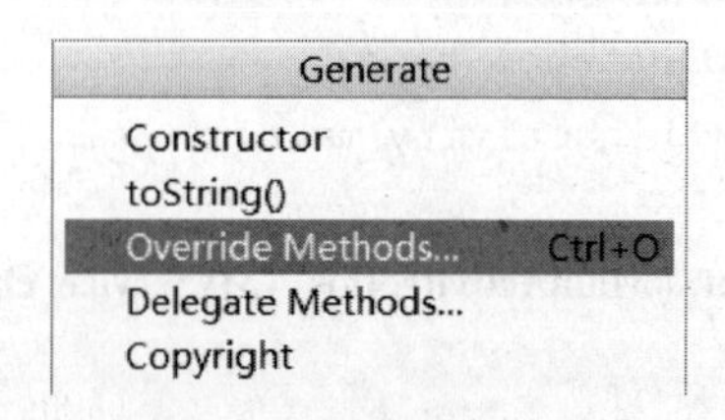

图 4-8　选择方法种类

弹出的窗口如图 4-9 所示，显示了需要覆盖的方法。

在这些方法中我们要选择哪些方法呢？

先看 Service 的生命周期，如图 4-10 所示。

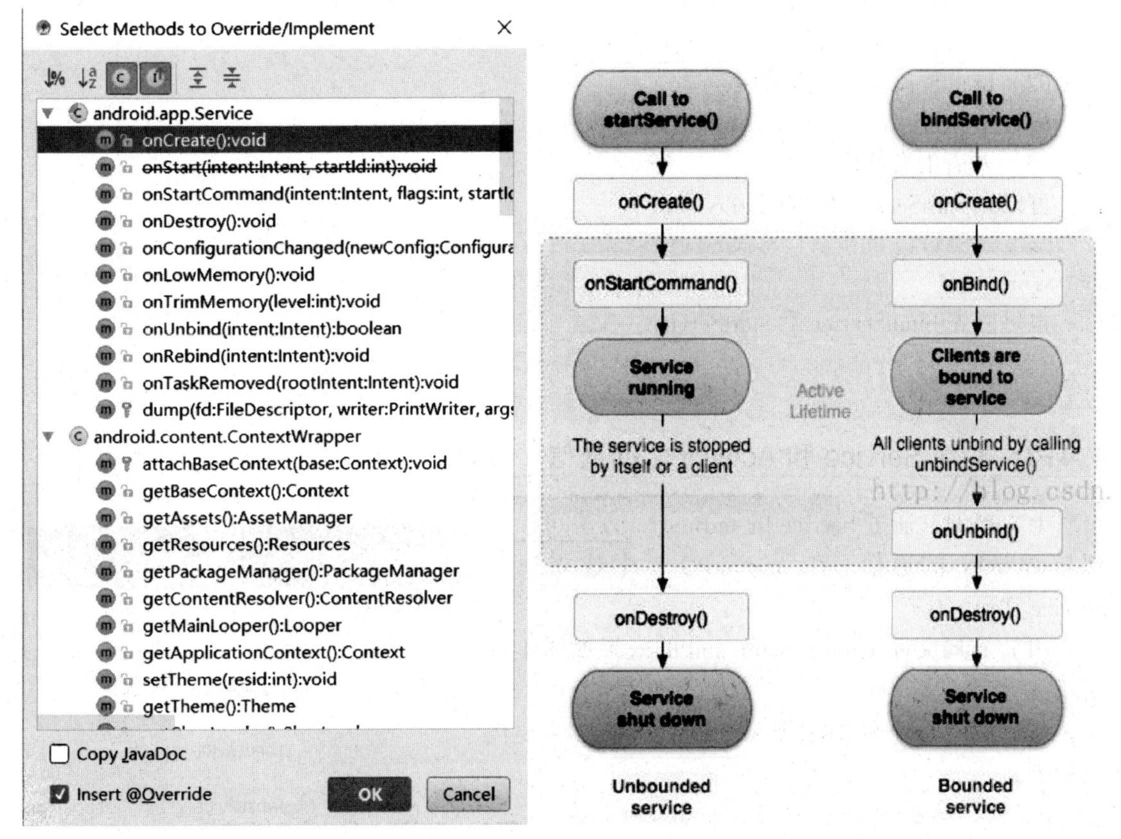

图 4-9　选择方法　　　　图 4-10　Service 的生命周期

从图中可以看出，实现 Service 的类应该实现以下方法。

（1）onCreate( )方法，当第一次启动 Service 时，先调用这个方法。

（2）onStartComman( )方法或者 onBind( )方法。

（3）如果使用 onBind( )方法，还需实现 onUnbind( )方法。要注意的是，onBind( )和 onUnbind( )方法应成对出现。

（4）onDestroy( )方法，当停止 Service 时，应执行 onDestroy( )方法。

启动 Service 有如下两种方式。

（1）startService( )：该方法启动 Service，访问者和 Service 之间没有关联，一旦启动，即使访问者退出，Service 依然运行。使用这种方法的启动顺序为 onCreate( )->onStartCommand( )->onDestory( )，如果服务已经开启，不会重复执行 onCreate( )，而是会调用 onStartCommand( )，服务停止的时候调用 onDestory( )，服务只会被停止一次。

（2）bindService( )：该方法启动 Service，访问者和 Service 绑定在一起，一旦访问者退出，Service 随即退出。使用这种方式启动的 Service 的生命周期为 onCreate( ) --->onBind( )--->onunbind( )--->onDestory( )。

Service 和 Actvity 都是从 Context 里面派生出来的，因此都可以直接调用 getResource( )、getContentResolver( )等方法。

Service 的方法调用总结如下。

（1） startService 的启动顺序

```
OnCreate()->onStartCommand()->onDestroy
```

（2） bindtService 的启动顺序

```
OnCreate()->onBind()->onServiceconnection->onUnbind()->onDestroy()
```

（3） 两者混合使用

可以先 startService 后 bindtService。

```
OnCreate()->onStartCommand()->onBind()->onServiceconnection->onUnbind()->
onDestroy()
```

也可以先 bindtService 后 startService。

```
OnCreate()->onBind()->onServiceconnection->onStartCommand()->onUnbind()->
onDestroy()
```

4.4.3 实现 Service 和 Activity 之间通信

上文提到，通过 Service 的 onBind( )方法可以实现与 Activity 的通信。在 Android 2.3 中建立项目 Service_Test。

（1） 布局文件 activity_main. xml 的效果如图 4-11 所示。

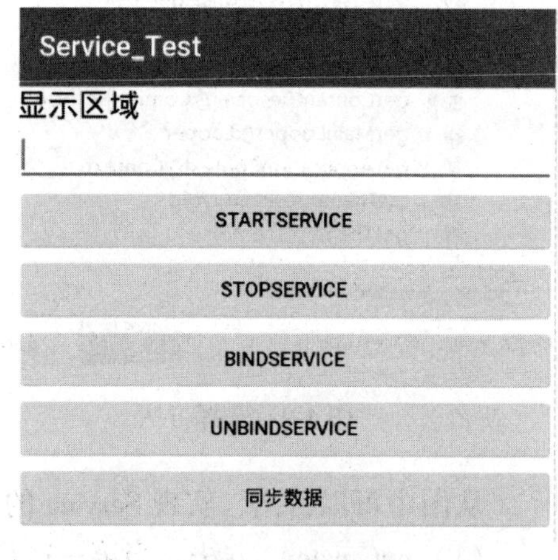

图 4-11　布局效果

（2） 创建一个 Service 类 MyService. java，主要代码如下。

```
public class MyService extends Service {
    private String data = "Service Data";
    @Override
    public IBinder onBind(Intent intent) {
        return new MyBinder();
    }
    public class MyBinder extends Binder {
        MyService getService() {
            return MyService.this;
        }
        public void setData(String data) {
            MyService.this.data = data;
        }
    }
    @Override
    public void onCreate() {
        super.onCreate();
        new Thread() {
```

```
      @Override
    public void run() {
        int n = 0;
        while(serviceRunning) {
            n++;
            String str = n+data;
            Log.d("Thread",str);
            if(dataCallback != null) {
                dataCallback.dataChanged(str);
              }
            try {
                sleep(1000);
              } catch(InterruptedException e) {
                  e.printStackTrace();
              }
            }
        };
    }.start();
}
  @Override
public int onStartCommand(Intent intent, int flags, int startId) {
          return super.onStartCommand(intent, flags, startId);
    }
    @Override
public boolean onUnbind(Intent intent) {
    return super.onUnbind(intent);
 }
@Override
  public void onDestroy() {
  super.onDestroy();
}
DataCallback dataCallback = null;
public DataCallback getDataCallback() {
      return dataCallback;
}
public void setDataCallback(DataCallback dataCallback) {
      this.dataCallback = dataCallback;
}
//通过回调机制,将 Service 内部的变化传递到外部
 public interface DataCallback {
      void dataChanged(String str);
 }
}
```

在实现的 Service 类中，在 onCreate( ) 函数启动一个线程，不断调用回调函数，把 Service 的数据传递给 Activity。

(3) 主 Activity 的类 MainActivity 的主要处理代码如下。

```
public class MainActivity extends AppCompatActivity implements View.OnClickListener {
    protected void onCreate(Bundle savedInstanceState) {
    ……
    intent = new Intent(this, MyService.class);
    myServiceConn = new MyServiceConn();
    }
  public void onClick(View v) {
      switch(v.getId()) {
        case R.id.btn_start_service:
            startService (intent);
            break;
        case R.id.btn_stop_service:
            stopService (intent);
            break;
        case R.id.btn_bind_service:
          //绑定 Service
            bindService (intent, myServiceConn, Context.BIND_AUTO_CREATE);
            break;
          case R.id.btn_unbind_service:
            unbindService (myServiceConn);
            break;
          case R.id.btn_sync_data:
          //注意：需要先绑定，才能同步数据
            if (binder != null) {
            binder.setData (et_data.getText().toString());
            }
              break;
          default:
              break;
    }
    }
  class MyServiceConn implements ServiceConnection {
        //服务被绑定成功之后执行
        @Override
        public void onServiceConnected (ComponentName name, IBinder service) {
            // IBinder service 为 onBind 方法返回的 Service 实例
            binder = (MyService.MyBinder) service;
            binder.getService().setDataCallback(new MyService.DataCallback() {
            //执行回调函数
              @Override
```

```
            public void dataChanged(String str) {
              Message msg = new Message();
              Bundle bundle = new Bundle();
              bundle.putString("str", str);
              msg.setData(bundle);
                //发送通知
                handler.sendMessage(msg);
            }
        });
    }
  Handler handler = new Handler() {
      public void handleMessage(android.os.Message msg) {
        //在 handler 中更新 UI
          tv_out.setText (msg.getData(). getString (" str"));
        };
    };
    //服务崩溃或者被杀掉执行
      @Override
      public void onServiceDisconnected (ComponentName name) {
          binder = null;
  }
  }
```

在 MainActivity 中调用 bindService( )时，第 2 个参数是一个 ServiceConnection，会运行 MyServiceConn （ServiceConnection 的实现），其中的函数 onServiceConnected( )在连接时执行，调用 Service 的回调函数，并具体实现其中的接口，在其中发送消息。

（4）运行结果如图 4-12 所示，单击 BINDSERVICE 按钮，可以看到 MyService 中的数据 Service Data 显示在 MainActivity 的界面上。

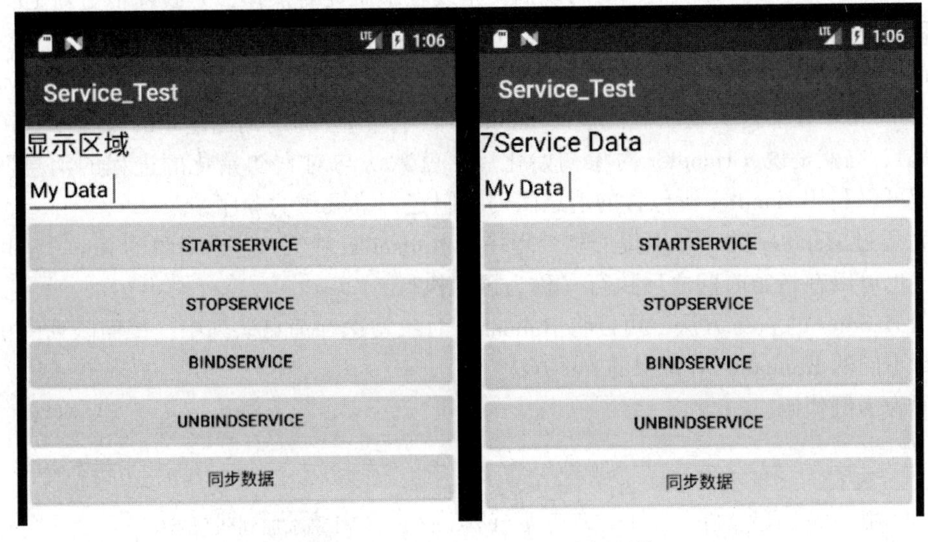

图 4-12 Service 和 Activity 之间通信

4.5 应用程序的消息处理机制——Handler

对于多线程的 Android 应用程序来说，有两类线程：一类是主线程，也就是 UI 线程；另一类是工作线程，也就是主线程或工作线程创建的线程。Android 的线程间消息处理机制主要是用来处理主线程跟工作线程间的通信。

Android 应用程序是通过消息来驱动的，即在应用程序的主线程（UI 线程）中有一个消息循环，负责处理消息队列中的消息。

线程之间和进程之间是不能直接传递消息的，必须通过对消息队列和消息循环的操作完成。Android 消息循环是针对线程的，每个线程都可以有自己的消息队列和消息循环。Android 提供了 Handler 类和 Looper 类来访问消息队列（Mesaage Queue）。

每个 Activity 是一个 UI 主线程，运行于主线程中，Android 系统在启动的时候会为 Activity 创建一个消息队列和消息循环（Looper）。

Handler 的作用是把消息加入特定的消息队列中，并分发和处理该消息队列中的消息。构造 Handler 的时候，可以指定一个 Looper 对象，如果不指定，则利用当前线程的 Looper 创建。Activity、Looper、Handler 的关系如图 4-13 所示。

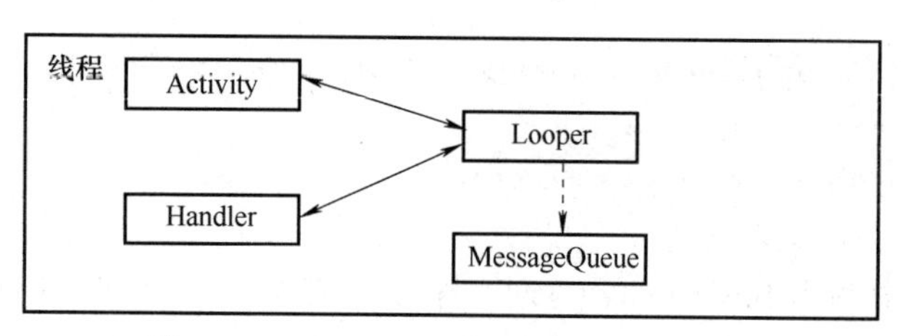

图 4-13　Activity、Looper、Handler 的关系

4.5.1　Handler 类

在 Android 开发中，有一个重要的规则：主线程不做耗时操作，子线程不更新 UI。那么在子线程需要更新 UI 界面的数据时要怎么处理呢，就要使用 Handler 来实现。

Android. os. Handler 是 Android SDK 中处理定时操作的核心类。Handler 可以分发 Message 对象和 Runnable 对象到主线程中，每个 Handler 实例，都会绑定到创建它的线程中（一般位于主线程），也就是说，Handler 对象初始化后，就默认与对它初始化的进程的消息队列绑定，因此可以利用 Handler 所包含的消息队列，制定一些操作的顺序。

（1）通过 Handler 类，可以提交和处理一个 Runnable 对象。这个对象的 run 方法可以立刻执行，也可以在指定时间之后执行（称为预约执行）。

利用 Handler 的 post 方法，可以将 Runnable 对象发送到消息队列中，按照队列的机制按顺序执行不同的 Runnable 对象中的 run 方法。

下面为示例代码。

```
    @Override
public void onClick(View v) {
    //调用 Handler 的 post 方法,将要执行的 Runnable 对象添加到队列当中
    handler.post(updateThread);
```

```
    }
      @Override
    public void onClick(View v) {
          //取消调用
          andler.removeCallbacks(updateThread);
    }
    Handler handler   = new Handler();
    //将要执行的操作写在线程对象的 run 方法当中
      Runnable updateThread =   new Runnable(){
            @Override
            public void run() {
                 System.out.println("UpdateThread");
                //在 run 方法内部,执行 postDelayed 或者 post 方法
                handler.postDelayed(updateThread, 3000);
      }
    };
```

程序的运行结果是每隔3秒钟，就会在控制台打印一行UpdateTread。这是因为实现了Runnable接口的UpdateThread对象进入了空的消息队列即被立即执行run方法，而在run方法的内部，又在3000ms之后将其再次发送进入消息队列中。

Handler延时调用Runnable对象的格式如下。

```
Handler mHandler=new Handler();
    mHandler.post(new Runnable(){
        void run(){
            //执行代码..
          }
        });
```

这个线程其实是在UI线程之内运行的，并没有新建线程。

常见的新建线程的方法如下。

```
Thread thread = new Thread();
Thread.start();
HandlerThread thread = new HandlerThread("string");
thread.start();
```

（2）传递Message。接受子线程发送的数据，并用此数据配合主线程更新UI。

在Android中，对于UI的操作通常需要放在主线程中进行。如果在子线程中有关于UI的操作，那么就需要把数据消息作为一个Message对象发送到消息队列中，然后，用Handler中的handlerMessge方法处理传过来的数据信息，并操作UI。sendMessage（Message msg）方法实现发送消息的操作。在初始化Handler对象时，重写的handleMessage（Message msg）方法接收Messgae并进行相关操作。

在线程里创建消息，并发送消息的代码示例如下。

```
Thread t1 = new Thread() {
    @Override
  public void run()
```

```
    {
        Message message = new Message(); //构造消息
        ……
        handler.sendMessage(message);   //发送消息
    }
};
```

在 Handler 内处理消息的代码示例如下。

```
Handler handler = new Handler()
{
  @Override
  public void handleMessage(Message msg)
  {
    //处理消息,更新 UI
  }
};
```

4.5.2 实例：获取当前时间

下面是 Handler 对象应用的实例，实现获取系统时间。

（1）在 Android 2.3 中创建应用项目：Handler_Test，其布局文件包含一个按钮和一个显示时间的 textView，如图 4-14 所示。

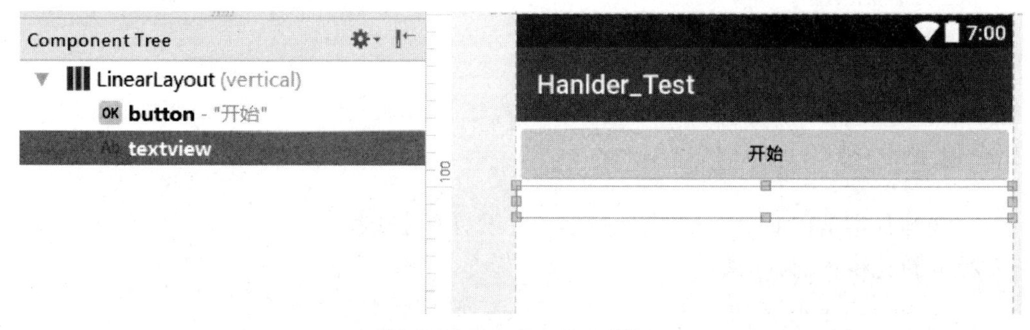

图 4-14 Handler_Test 布局

（2）其中 MainActivity 的代码如下。

```
public class MainActivity extends AppCompatActivity {
  TextView textview;
  Button button;
    @Override
  protected void onCreate(Bundle savedInstanceState) {
super.onCreate(savedInstanceState);
setContentView(R.layout.activity_main);
textview = (TextView) findViewById (R.id.textview);
button = (Button) findViewById (R.id.button);
button.setOnClickListener (new View.OnClickListener() {
    @Override
```

```
        public void onClick (View v) {
            t1.start();
        }
        });
    }
    Handler handler = new Handler()
    {
        @Override
        public void handleMessage (Message msg) {
        if (msg.what == 1) {
            textview.setText (msg.getData().getString (" time"));
          }
        }
    };
    Thread t1 = new Thread() {
        @Override
        public void run() {
        while(true) {
            try {
                Thread.sleep(1000);
                 String time = new SimpleDateFormat("yyyy/MM/dd HH:mm:ss").format
(new Date());
                System.out.println(time);
                Message message = new Message();
                Bundle bundle = new Bundle();
                bundle.putString("time", time);
                message.setData(bundle);//bundle 传值,耗时,效率低
                handler.sendMessage(message);//发送 message 信息
                message.what = 1;//标志是哪个线程传数据
                //message 有四个传值方法,
                //两个传 int 整型数据的方法 message.arg1,message.arg2
                //一个传对象数据的方法 message.obj
                //一个 bandle 传值方法
              } catch(InterruptedException e) {
                  e.printStackTrace();
              }
          }
        }
      };
    }
```

（3）运行结果如图 4-15 所示，按下按钮即开始获取时间。

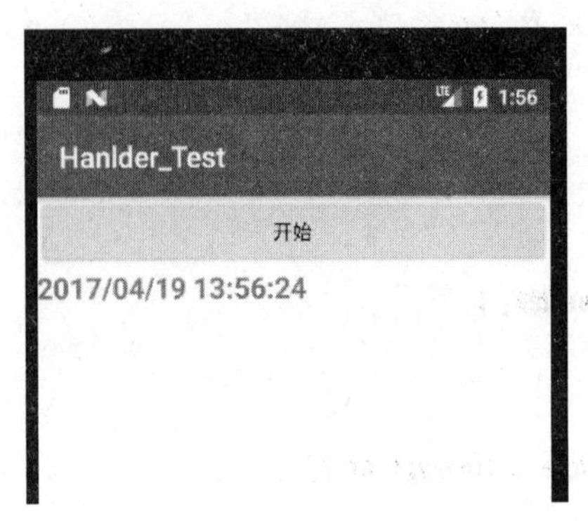

图 4-15　Handler_Test 运行结果

4.6 应用程序轻量级并行——AsyncTask 机制

用 Handler 类在子线程中更新 UI 线程，虽然避免了在主线程进行耗时计算，但费时的任务操作总会启动一些匿名的子线程，太多的子线程给系统带来巨大的负担，随之带来一些性能问题。因此，Android 提供了一个工具类 AsyncTask，顾名思义为异步执行任务。这个 AsyncTask 就是用来处理后台比较耗时的任务的，编程的语法显得优雅许多，不再需要子线程和 Handler，就可以完成异步操作并且刷新用户界面。

4.6.1 AsyncTask 抽象类

AsyncTask 是一个抽象类，使用时需要继承这个类，定义一个它的派生类并重写相关方法，然后调用 execute()方法。要注意，在继承时需要设定三个泛型 Params、Progress 和 Result 的类型，如 AsyncTask <Void，Inetger，Void>。

AsyncTask 类的声明如下。

public abstract class AsyncTask <Params，Progress，Result>

可以看到，AsyncTask 是一个泛型类，它的三个类型参数的含义如下。

- Params：是指调用 execute()方法时传入的参数类型和 doInBackground()的参数类型。
- Progress：是指更新进度时传递的参数类型，即 publishProgress()和 onProgressUpdate()的参数类型。
- Result：后台任务的返回结果类型，是指 doInBackground()的返回值类型。

AsyncTask 类主要为我们提供了如下方法。

- onPreExecute()：此方法会在后台任务执行前被调用，用于进行一些准备工作。
- doInBackground（Params... params）：此方法中定义要执行的后台任务，在这个方法中可以调用 publishProgress 来更新任务进度（publishProgress 内部会调用 onProgressUpdate 方法）。
- onProgressUpdate（Progress... values）：由 publishProgress 内部调用，表示任务进度更新。
- onPostExecute（Result result）：后台任务执行完毕后，此方法会被调用，参数即为后台任务的返回结果。
- onCancelled()：此方法会在后台任务被取消时被调用。

需要注意的是，在以上方法中，除了 doInBackground 方法由 AsyncTask 内部线程池执行外，其余方法均在主线程中执行。

为了正确使用 AsyncTask 类，必须遵守以下几条准则。

（1）AsyncTask 的实例必须在 UI 线程中创建。

（2）execute 方法必须在 UI 线程中调用。

（3）不要手动调用 onPreExecute( )、onPostExecute（Result）、doInBackground（Params...）、onProgressUpdate（Progress...）这几个方法，需要在 UI 线程中实例化这个 task 来调用。

（4）该实例只能被执行一次，否则多次调用时将会出现异常。

4.6.2　实例：实现定时器

下面是 AsyncTask 对象应用的实例，实现定时器。

（1）在 Android 2.3 中创建项目：AsyncTaskExample，其布局文件包含四个控件，输入定时时间，按下按钮开始获取定时，布局如图 4-16 所示。

图 4-16　AsyncTask 实现定时器布局

（2）其中 MainActivity 的代码如下。

```
public class AsyncTaskActivity extends AppCompatActivity {
  private EditText chronoValue;
  private TextView chronoText;
  private Button start;
  @Override
  protected void onCreate(Bundle savedInstanceState) {
    super.onCreate(savedInstanceState);
  setContentView(R.layout.main);
    //获取三个 UI 组件
  start = (Button)findViewById(R.id.start);
  chronoText = (TextView)findViewById(R.id.chronoText);
  chronoValue = (EditText)findViewById(R.id.chronoValue);
  start.setOnClickListener(new View.OnClickListener() {
```

```
            @Override
            public void onClick(View v) {
                //获取 EditText 里的数值
                int value = Integer.parseInt(String.valueOf(chronoValue.getText()));
                //验证数值是否大于零
                if(value > 0) {
                  new Chronograph().execute(value);
                }
                else {
                 Toast.makeText(AsyncTaskActivity.this, "请输入一个大于零的整数值 !",
Toast.LENGTH_LONG).show();
                }
            }
        });
    }
        private class Chronograph extends AsyncTask<Integer, Integer, Void> {
            @Override
            protected void onPreExecute() {
                super.onPreExecute();
              //在计时开始前，先使按钮和 EditText 不能用
                chronoValue.setEnabled (false);
                start.setEnabled (false);
                chronoText.setText (" 0: 0");
            }
            @Override
          protected Void doInBackground (Integer... params) {
              //计时
              for (int i = 0; i <= params [0]; i ++) {
                  for (int j = 0; j < 60; j ++) {
                      try {
                        //发布增量
                        publishProgress (i, j);
                        if (i == params [0]) {
                              return null;
                        }
                        //暂停一秒
                        Thread.sleep (1000);
                      } catch (InterruptedException e) {
                            e.printStackTrace();
                      }
                    }
                }
                if (isCancelled()) {
```

```
            return null;
        }
        return null;
    }
    @Override
    protected void onProgressUpdate (Integer... values) {
      super.onProgressUpdate (values);
        //更新 UI 界面
      chronoText.setText (values [0] +":" +values [1]);
    }
  @Override
  protected void onPostExecute (Void result) {
        super.onPostExecute (result);
      //重新使按钮和 EditText 可以使用
        chronoValue.setEnabled (true);
        start.setEnabled (true);
    }
  }
}
```

(3) 项目运行结果如图 4-17 所示。

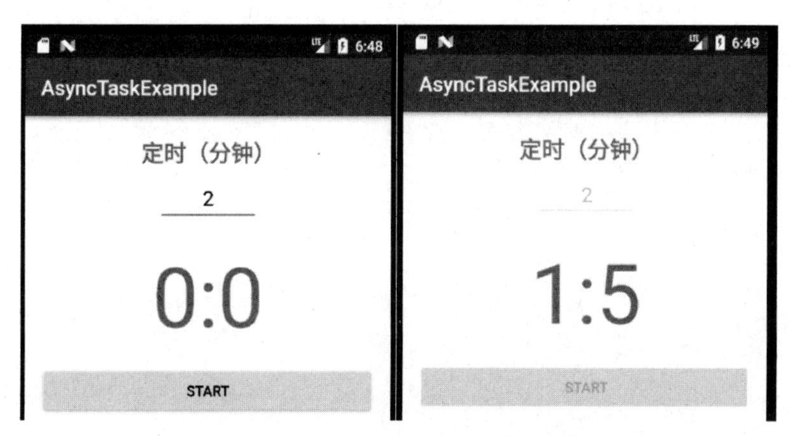

图 4-17　AsyncTaskExample 运行结果

4.7 AsyncTask 和 Handler 两种异步方式比较

本节将介绍两种异步方式的优缺点。

1. AsyncTask 优缺点

AsyncTask 是 Android 提供的轻量级的异步类，可以直接继承 AsyncTask，在类中实现异步操作，并提供接口反馈当前异步执行的程度（可以通过接口实现 UI 进度更新），最后反馈执行的结果给 UI 主线程。

(1) 优点：简单、快捷、过程可控。

（2）缺点：在使用多个异步操作并需要进行 UI 变更时，会比较复杂。

2. Handler 优缺点

在 Handler 异步实现时，涉及 Handler、Looper、Message、Thread 四个对象，实现异步的流程是主线程启动 Thread（子线程）运行，并生成 Message-Looper，获取 Message 并传递给 Handler 逐个获取 Looper 中的 Message，并进行 UI 变更。

（1）优点：结构清晰，功能定义明确；对多个后台任务时，简单清晰。

（2）使用的缺点：在单个后台异步处理时，代码过多，结构过于复杂（相对性）。

4.8 本章小结

本章对 Android 应用程序的组成部分进行了深入的研究，包括事件处理机制、Android 多线程、Android 广播组件、后台服务 Service、AsyncTask、Handler 等应用程序的主要组成部分。通过前几章和本章的学习，您已经掌握了 Android 应用程序的主要构成内容，以及它们之间的关系，完成了基本部分主要内容的学习。

第五章

界面设计更进一步——UI高级设计

一直以来，很多人认为Android系统的界面并不美观，但是随着Android版本和开发环境的不断完善，Android可以做出更好看的界面，本章介绍一些高级控件的使用方法。

5.1 自定义控件

Android虽然自带了很多控件，但有时未必能满足业务的需求，这时就需要我们自定义一些控件。

自定义控件要遵守如下要求。

（1）遵守Android标准的规范（命名、可配置、事件处理等）。

（2）在XML布局中可配置控件的属性。

（3）对交互应当有合适的反馈，比如按下、单击等。

（4）具有兼容性，Android版本很多，应该具有广泛的适用性。

自定义控件有如下两种方式。

（1）继承ViewGroup。例如：ViewGroup、LinearLayout、FrameLayout、RelativeLayout等。

（2）继承View。例如：View、TextView、ImageView、Button等。

5.1.1 自定义View类控件

自定义View的步骤如下。

- 自定义View的属性。
- 在自定义View类的构造方法中获得View属性值。
- 在自定义View类重写onMeasure（int，int）方法。
- 在自定义View类重写onDraw（Canvas canvas）方法。
- 在xml布局文件中布局自定义View类。

（1）自定义View的属性

在res/values下面新建attrs.xml属性文件。

```
<?xml version="1.0" encoding="utf-8"?>
<resources>
  <!--name是自定义属性名,format是属性的单位-->
```

```
  <attr name="titleSize" format="dimension"></attr>
  <attr name="titleText" format="string"></attr>
  <attr name="titleColor" format="color"></attr>
  <attr name="titleBackgroundColor" format="color"></attr>
  <! --name 是自定义控件的类名-->
<declare-styleable name="MyCustomView">
    <attr name="titleSize"></attr>
    <attr name="titleText"></attr>
    <attr name="titleColor"></attr>
    <attr name="titleBackgroundColor"></attr>
    </declare-styleable>
</resources>
```

自定义属性分两类：定义公共属性和定义控件的主题样式。

上面的 XML 文件第一部分是公共的属性，第二部分是自定义控件 MyCustomView 的主题样式，该主题样式里的属性必须包含在公共属性中。言外之意就是公共属性可以被多个自定义控件主题样式使用。format 字段后面的属性单位基本包括如下几个：dimension（字体大小）、string（字符串）、color（颜色）、boolean（布尔类型）、float（浮点型）、integer（整型）、enmu（枚举）、fraction（百分比）等。

（2）自定义 View 一般需要选择实现的三个构造方法

```
public classMyCustomView extends View {
  public MyCustomView(Context context, AttributeSet attrs) {
    super(context, attrs);
    ......
  }
  public MyCustomView(Context context) {
    super(context);
    ......
  }
  public MyCustomView(Context context, AttributeSet attrs, int defStyleAttr) {
    super(context, attrs, defStyleAttr);
    ......
  }
}
```

从代码中不难看出，这三个构造方法是一层调用一层的，具有递进关系，因此，我们只需要在最后一个构造方法中获得 View 的属性。

（3）自定义 View 一般需要重写 onMeasure（int，int）和 onDraw（Canvas canvas）方法

```
public classMyCustomView extends View {
  ......
  protected void onMeasure(int widthMeasureSpec, int heightMeasureSpec) {
    super.onMeasure(widthMeasureSpec, heightMeasureSpec);
    ......
  }
```

```
    protected void onDraw(Canvas canvas) {
      super.onDraw(canvas)
      ……
    }
}
```

Measure 过程用于计算视图大小，View 类 Measure 过程相关方法主要有以下三个。

- public final void measure (int widthMeasureSpec, int heightMeasureSpec)
- protected final void setMeasuredDimension (int measuredWidth, int measuredHeight)
- protected void onMeasure (int widthMeasureSpec, int heightMeasureSpec)

measure 调用 onMeasure，onMeasure 测量宽度、高度，然后调用 setMeasureDimension 保存测量结果，measure、setMeasureDimension 是 final 类型，view 的子类不需要重写，onMeasure 在 View 的子类中重写。

onDraw 过程主要利用前两步得到的参数，将视图显示在屏幕上，到这里也就完成了整个视图绘制工作。

public void draw (Canvas canvas)

protected void onDraw (Canvas canvas)

通过调用 draw 函数进行视图绘制，在 View 类中 onDraw 函数是个空函数，最终的绘制需求要在自定义的 onDraw 函数中实现。

(4) 布局中使用自定义 View

```
<LinearLayout xmlns:android=http://schemas.android.com/apk/res/android
    android:orientation="vertical" android:layout_width=" match_parent"
    android: gravity=" center"
    android: layout_height=" match_parent" >
<com.xjp.customview.MyCustomView
        android: layout_width=" match_parent"
        android: id=" @ +id/clock"
        android: layout_height=" match_parent"
custom: titleColor=" @android: color/black"
        custom: titleSize=" 25sp"
        custom: titleBackgroundColor=" #ff0000"
        custom: titleText=" 自定义的 View" />
</LinearLayout>
```

5.1.2　实例：自定义控件——走动的钟表

下面以 View 类的自定义控件为例，通过实现一个继承的 View 类实现一个走动的钟表。

下面是 View 类应用的实例，通过自定义实现 View。

(1) 在 Android 2.3 中创建应用项目：Custom_Control。其中包含实现钟表的自定义 View 类 (ClockView)、表示点的类 Point、ClockView 类的布局文件 activity_main.xml、主 Activit 类 MianActivity，其布局文件中包含一个按钮和一个显示时间的自定义控件 ClockView，如图 5-1 所示。

(2) 实现自定义控件的代码类 ClockView，代码如下。

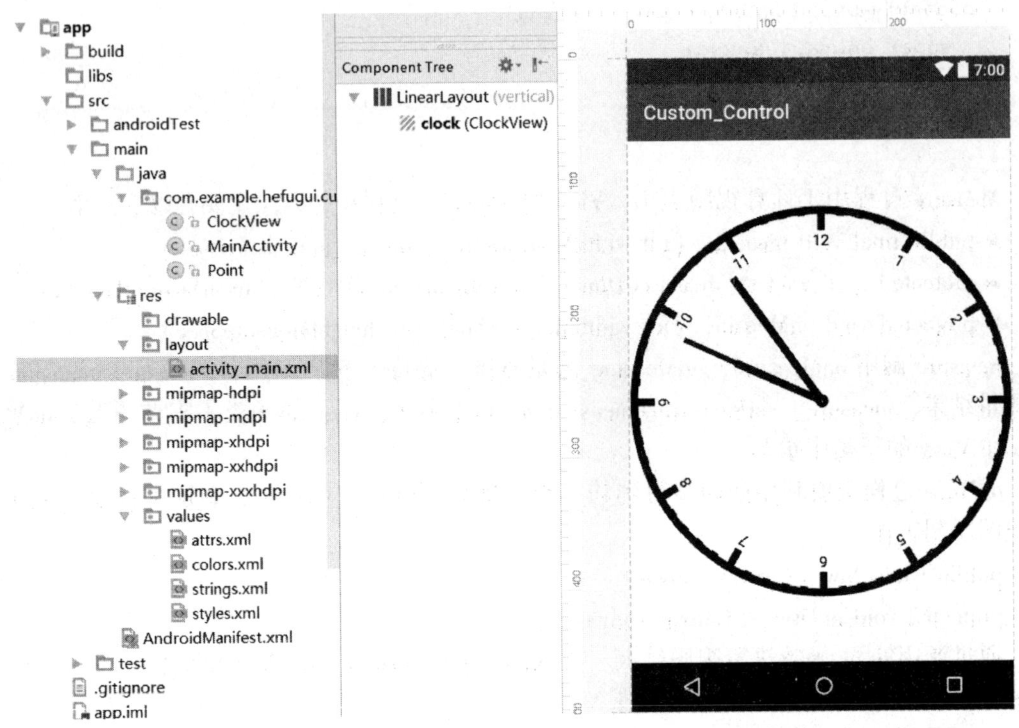

图 5-1　钟表项目 Custom_Control

```
public class ClockView extends View {
    private Paint circlePaint,dialPaint,numberPaint;
    //view 的宽高
    private float mWidth,mHeight;
    //圆的半径
    private float circleRadius;
    //圆心 X,Y 坐标
    private float circleX,circleY;
    private int second,minute;
    private double hour;
    private Handler handler = new Handler(Looper.getMainLooper()){
        @Override
    public void handleMessage(Message msg) {
            super.handleMessage(msg);
            if(msg.what = =0){
                invalidate();
            }
    }
    };
    public ClockView(Context context, AttributeSet attrs) {
        super(context, attrs);
        initPaint();
```

```
    }
    private void initPaint(){
      //刻盘圆,小时刻度,时针和分针的画笔
      circlePaint = new Paint(Paint.ANTI_ALIAS_FLAG);
      circlePaint.setColor (Color.BLACK);
      circlePaint.setStyle (Paint.Style.STROKE);
      circlePaint.setStrokeWidth (15);
      //分钟刻度的画笔
      dialPaint = new Paint (Paint.ANTI_ALIAS_FLAG);
      dialPaint.setColor (Color.BLACK);
      dialPaint.setStrokeWidth (10);
      //数字的画笔
      numberPaint = new Paint (Paint.ANTI_ALIAS_FLAG);
      numberPaint.setColor (Color.BLACK);
      numberPaint.setStrokeWidth (5);
      numberPaint.setTextSize (30);
    }
    @Override
  protected void onMeasure (int widthMeasureSpec, int heightMeasureSpec) {
    super.onMeasure (widthMeasureSpec, heightMeasureSpec);
    mWidth = getMeasuredWidth();
    mHeight = getMeasuredHeight();
    if (mWidth<mHeight) {
      //圆的半径为 view 的宽度的一半再减 9，防止贴边
      circleRadius = mWidth/2-9;
      circleX = mWidth/2;
      circleY = mHeight/2;
    } else {
      circleRadius = mHeight/2-9;
      circleX = mWidth/2;
      circleY = mHeight/2;
    }
  }
  @Override
protected void onDraw (Canvas canvas) {
    super.onDraw (canvas);
    setTimes();
    drawCirclePoint (canvas);
    drawCircle (canvas);
    drawDial (canvas);
    drawPointer (canvas);
}
/**圆心* @param canvas */
```

```
    private void drawCirclePoint (Canvas canvas) {
        canvas.drawCircle (circleX, circleY, 5, circlePaint);
    }
    private void drawCircle (Canvas canvas) {
        canvas.drawCircle (circleX, circleY, circleRadius, circlePaint);
    }
    /** 画刻度及时间 @param canvas */
    private void drawDial (Canvas canvas) {
        //时钟用长一点的刻度，画笔用画圆的画笔
        Point hourStartPoint = new Point (circleX, circleY-circleRadius);
        Point hourEndPoint = new Point (circleX, circleY-circleRadius +40);
        //分钟的刻度要稍微短一些，画笔用画圆的画笔
        Point startPoint2 = new Point (circleX, circleY-circleRadius);
        Point endPoint2 = new Point (circleX, circleY-circleRadius +10);
        //开始画刻度和数字，总共 60 个刻度，12 个时钟刻度，被 5 整除画一个时钟刻度，其余的为分针
刻度
      String clockNumber;
      for (int i =0; i <60; i ++) {
        if (i% 5 = =0) {
          if (i = =0) {
              clockNumber = " 12";
          } else {
              clockNumber = String.valueOf (i/5);
            }
            //时针刻度
            canvas.drawLine ( hourStartPoint.getX ( ), hourStartPoint.getY ( ),
hourEndPoint.getX(), hourEndPoint.getY(), circlePaint);
        //画数字,需在时针刻度末端加 30
        canvas.drawText(clockNumber,circleX-numberPaint.measureText(clockNumber)/
2,hourEndPoint.getY() +30,numberPaint);
        } else{
        //画分针刻度
         canvas.drawLine(startPoint2.getX(),startPoint2.getY(),endPoint2.getX(),
endPoint2.getY(),circlePaint);
      }
      //画布旋转 6 度
        canvas.rotate(360/60,circleX,circleY);
      }
    }
    /** 画指针 * X 点坐标 cos(弧度) * r * Y 点坐标 sin(弧度) * r
     * toRadians 将角度转成弧度 * 安卓坐标系与数学坐标系不同的地方是 X 轴是相反的,所以为了调
整方向,需要将角度 +270 度 * @param canvas */
    private void drawPointer(Canvas canvas){
```

```
    canvas.translate(circleX,circleY);
    float hourX = (float) Math.cos(Math.toRadians(hour*30+270))*circleRadius*0.5f;
    float hourY = (float) Math.sin(Math.toRadians(hour*30+270))*circleRadius*0.5f;
    float minuteX = (float) Math.cos(Math.toRadians(minute*6+270))*circleRadius*0.8f;
    float minuteY = (float) Math.sin(Math.toRadians(minute*6+270))*circleRadius*0.8f;
    float secondX = (float) Math.cos(Math.toRadians(second*6+270))*circleRadius*0.8f;
    float secondY = (float) Math.sin(Math.toRadians(second*6+270))*circleRadius*0.8f;
    canvas.drawLine(0,0,hourX,hourY,circlePaint);
    canvas.drawLine(0,0,minuteX,minuteY,circlePaint);
    canvas.drawLine(0,0,secondX,secondY,dialPaint);
      //一秒重绘一次
    handler.sendEmptyMessageDelayed(0,1000);
  }
  public void startClock(){
      setTimes();
      invalidate();
  }
  private void setTimes(){
    Date date = new Date();
    Calendar calendar = Calendar.getInstance();
    calendar.setTime(date);
    second = getTimes(date,Calendar.SECOND);
    minute = getTimes(date,Calendar.MINUTE);
    hour = getTimes(date,Calendar.HOUR) +minute/12*0.2;
  }
  private int getTimes(Date date,int calendarField){
    Calendar calendar = Calendar.getInstance();
    calendar.setTime(date);
    return calendar.get(calendarField);
  }
    public void stopClock(){
      handler.removeMessages(0);
    }
}
```

(3) 在布局文件 activity_main.xml 中包含 ClockView 类，代码如下。

```
<LinearLayout xmlns:android="http://schemas.android.com/apk/res/android"
    android:orientation="vertical" android:layout_width=" match_parent"
    android: gravity=" center"
    android: layout_height=" match_parent" >
    <com.example.customview.view.ClockView
        android: layout_width=" match_parent"
        android: id=" @+id/clock"
        android: layout_height=" match_parent" />
```

```
</LinearLayout>
```

（4）在主 Activity 中实现 ClockView 类的调用，代码如下。

```
public class MainActivity extends AppCompatActivity {
  private ClockView clockView;
  @Override
  protected void onCreate(Bundle savedInstanceState) {
    super.onCreate(savedInstanceState);
    setContentView(R.layout.clock);
    clockView = (ClockView) findViewById(R.id.clock);
  }
  @Override
  protected void onResume() {
    super.onResume();
    clockView.startClock();
  }
  @Override
  protected void onStop() {
    super.onStop();
    clockView.stopClock();
  }
}
```

Point 类实现一个平面点的操作，运行结果如图 5-2 所示。

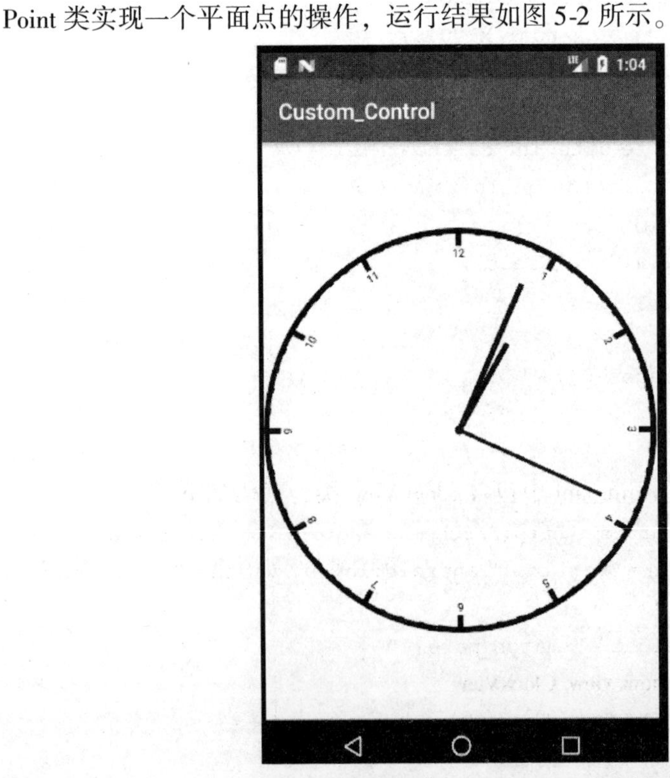

图 5-2　钟表项目 Custom_Control 运行结果

5.2 Android 适配器——BaseAdapter

在 Android 应用程序中，采用了数据和显示分开实现的数据处理方式，由于数据源形式多样，例如 ListView 所展示数据格式是有一定要求的，为了匹配这个变换，中间增加了适配器，如图 5-3 所示，图中展示了数据源、适配器、ListView 等数据显示控件之间的关系。

图 5-3　Android 的数据源、适配器、显示控件之间的关系

数据适配器正是建立了数据源与 ListView 之间的适配关系，将数据源转换为 ListView 能够显示的数据格式，从而将数据的来源与数据的显示进行解耦，降低程序的耦合性。ListView、GridView 等数据显示控件有多种数据适配器，本文讲解最通用的数据适配器——BaseAdapter。

BaseAdapter 是 Android 应用程序中经常用到的基础数据适配器，它的主要用途是将一组数据传到 ListView、Spinner、Gallery 及 GridView 等 UI 显示组件，它继承自接口类 Adapter。

BaseAdapter 使用方法比较简单，主要是通过继承此类来实现 BaseAdapter 的四个方法。

- public int getCount( )：适配器中数据集的数据个数。
- public Object getItem （int position)：获取数据集中与索引对应的数据项。
- public long getItemId （int position)：获取指定行对应的 ID。
- public View getView （int position，View convertView，ViewGroup parent)：获取每一行 Item 的显示内容。

ListView、GridView 等控件可以展示大量的数据信息。假如 ListView 可以展示 100 条信息，但是屏幕的尺寸是有限的，一屏幕只能显示 7 条。当向上滑动 ListView 时，显示项 item1 被滑出了屏幕区域，那么系统就会将 item1 回收到 Recycler 中，即 View 缓冲池中，而将要显示的 item8 从缓存池中取出布局文件，并重新设置好 item8 需要显示的数据，放入需要显示的位置。这就是 ListView 的缓冲机制，总结起来就是一句话：需要时才显示，显示完即回收到缓存。ListView、GridView 等数据显示控件采用这种缓存机制可以极大节省系统资源。

下面是一个 BaseAdapter 实现的实例。

```
public class MyAdapter extends BaseAdapter {
  private List<ItemBean> mList; //数据源 ItemBean 为数据源的内容格式
  private LayoutInflater mInflater; //布局装载器对象
  //通过构造方法将数据源与数据适配器关联起来，context:要使用当前的 Adapter 的界面对象
  public MyAdapter(Context context, List<ItemBean> list) {
      mList = list;
      mInflater = LayoutInflater.from(context);
  }
  @Override
  //ListView 需要显示的数据数量
```

```
    public int getCount() {
        return mList.size();
    }
  @Override
  //指定的索引对应的数据项
  public Object getItem(int position) {
      return mList.get(position);
  }
  @Override
  //指定的索引对应的数据项 ID
  public long getItemId(int position) {
      return position;
  }
  @Override
  //返回每一项的显示内容
  public View getView(int position, View convertView, ViewGroup parent) {
      ViewHolder viewHolder;
      //如果 view 未被实例化过,缓存池中没有对应的缓存
      if(convertView == null) {
        viewHolder = new ViewHolder();
        //由于只需要将 XML 转化为 View,并不涉及到具体的布局,所以第二个参数通常设置为 null
          convertView = mInflater.inflate(R.layout.item, null);
          //对 viewHolder 的属性进行赋值
        viewHolder.imageView = (ImageView) convertView.findViewById(R.id.iv_image);
        viewHolder.title = (TextView) convertView.findViewById(R.id.tv_title);
        viewHolder.content = (TextView) convertView.findViewById(R.id.tv_content);
        //通过 setTag 将 convertView 与 viewHolder 关联
      convertView.setTag(viewHolder);
    }else{
      //如果缓存池中有对应的 view 缓存,则直接通过 getTag 取出 viewHolder
      viewHolder = (ViewHolder) convertView.getTag();
    }
    //取出 bean 对象
    ItemBean bean = mList.get(position);
    //设置控件的数据
    viewHolder.imageView.setImageResource(bean.itemImageResId);
    viewHolder.title.setText(bean.itemTitle);
    viewHolder.content.setText(bean.itemContent);
    return convertView;
    }
    // ViewHolder 用于缓存控件,三个属性分别对应 item 布局文件的三个控件
    class ViewHolder{
      public ImageView imageView;
```

```
        public TextView title;
        public TextView content;
    }
}
```

上面代码利用了 ListView 的缓存机制，也利用 ViewHolder 类来实现显示数据视图的缓存，实现了优化程序的目的。

总体来说，BaseAdapter 类的实现步骤如下。

（1）创建 bean 对象，用于封装数据。

（2）在主 Activity 中，初始化数据对象 bean 数组。

（3）创建 ViewHolder 类，创建布局映射关系。

（4）判断 convertView，为空则创建，并设置 Tag，不为空则通过 Tag 取出 ViewHolder。

（5）为 ViewHolder 的控件设置数据。

5.3 复杂控件 ListView——实现场景对象选择

ListView 绝对可以称得上是 Android 中最常用的控件之一，几乎所有的应用程序都会用到它。由于手机屏幕空间有限，能够一次性在屏幕上显示的内容并不多，当程序中有大量的数据需要展示时，可以借助 ListView 来实现。ListView 允许用户通过手指上下滑动的方式将屏幕外的数据滚动到屏幕内，同时屏幕上原有的数据则会滚动出屏幕。

5.3.1 ListView 控件的简单应用

ListView 控件的简单使用需要以下步骤。

（1）在布局中放置 ListView 控件。

（2）准备 ListView 显示的数据，一般为数组，例如数据 data。

```
String[] data = { "Apple", "Banana", "Orange", "Watermelon","Pear", "Grape", "
Pineapple", "Strawberry", "Cherry", "Mango" };
```

（3）定义并初始化一个适配器，指定显示的上下文 MainActivity. this、显示的布局 android. R. layout. simple_list_item_1 和显示的数据 dada。

```
ArrayAdapter < String > adapter = new ArrayAdapter < String > (
MainActivity.this, android.R.layout.simple_list_item_1, data);
```

（4）获取布局的 ListView 控件。

```
ListView listView = (ListView) findViewById(R.id.list_view);
```

（5）将前面定义的适配器 adapter，利用 ListView 的函数 setAdapter( ) 设置。

```
listView.setAdapter(adapter);
```

5.3.2 ListView 控件的高级应用

只能显示一段文本的 ListView 实在太单调了，可以对 ListView 界面进行定制，让它显示更加丰富的内容。ListView 界面定制需要以下步骤。

（1）为 ListView 的显示子项指定一个自定义的布局，在 layout 目录下新建一个布局文件，例如新建一个文件 fruit_item. xml，代码如下。

```
<LinearLayout xmlns:android="http://schemas.android.com/apk/res/android"
    android:layout_width=" match_parent"
    android: layout_height=" match_parent" >
      <ImageView
       android: id=" @+id/fruit_image"
       android: layout_width=" wrap_content"
       android: layout_height=" wrap_content" />
      <TextView
       android: id=" @+id/fruit_name"
       android: layout_width=" wrap_content"
       android: layout_height=" wrap_content"
       android: layout_gravity=" center"
       android: layout_marginLeft=" 10dip" />
  </LinearLayout>
```

在这个布局中，定义了一个 ImageView 用于显示水果的图片，又定义了一个 TextView 用于显示水果的名称，ListView 的布局包含两项内容：图片和文字，显示的样式如图 5-4 所示。

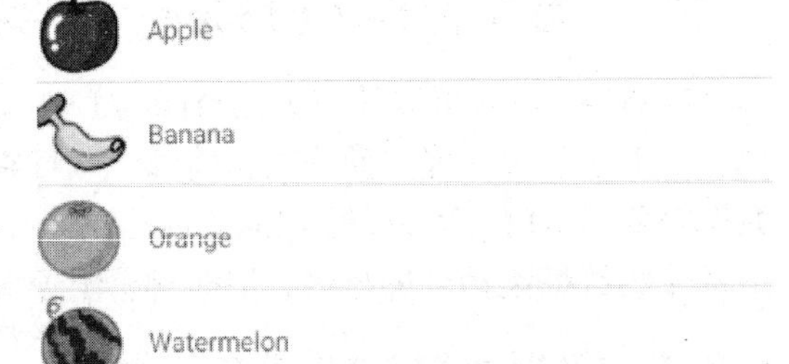

图 5-4　ListView 布局样式

(2) 接着定义一个实体类，作为 ListView 适配器的适配类型。例如新建类 Fruit，代码如下。

```
public class Fruit {
    private String name;
    private int imageId;
    public Fruit(String name, int imageId) {
        this.name = name;
        this.imageId = imageId;
    }
    public String getName() {
      return name;
    }
    public int getImageId() {
      return imageId;
    }
}
```

Fruit 类中只有两个字段，name 表示水果的名字，imageId 表示水果对应图片的资源 id。

(3) 接下来需要创建一个自定义的适配器，这个适配器继承自 ArrayAdapter，并将泛型指定为上面新建的 Fruit 类。例如新建类 FruitAdapter，代码如下。

```
public class FruitAdapter extends ArrayAdapter<Fruit> {
    private int resourceId;
    public FruitAdapter(Context context, int textViewResourceId,List<Fruit>objects){
      super(context, textViewResourceId, objects);
```

```
        resourceId = textViewResourceId;
    }
    @Override
    public View getView(int position, View convertView, ViewGroup parent) {
        Fruit fruit = getItem(position);
        View view = LayoutInflater.from(getContext()).inflate(resourceId, null);
        ImageView fruitImage =(ImageView) view.findViewById(R.id.fruit_image);
        TextView fruitName = (TextView) view.findViewById (R.id.fruit_name);
        fruitImage.setImageResource (fruit.getImageId());
        fruitName.setText (fruit.getName());
        return view;
    }
}
```

FruitAdapter 重写了父类的一组构造函数，用于将上下文、ListView 子项布局的 ID 和数据都传递进来。另外又重写了 getView( ) 方法，在每个子项被滚动到屏幕内的时候这个方法会被调用。在 getView( ) 方法中，首先通过 getItem( ) 方法得到当前项的 Fruit 实例，然后使用 LayoutInflater 来为这个子项加载我们传入的布局，接着调用 View 的 findViewById( ) 方法分别获取到 ImageView 和 TextView 的实例，并分别调用它们的 setImageResource( ) 和 setText( ) 方法来设置显示的图片和文字，最后将布局返回，这样就完成了自定义适配器。

（4）最后在 MainActivity 中实例化显示项，实例化适配器，将适配器设置到 ListView，代码如下。

```
public class MainActivity extends Activity {
        private List<Fruit> fruitList = new ArrayList<Fruit>();
        @Override
    protected void onCreate(Bundle savedInstanceState) {
        super.onCreate(savedInstanceState);
        setContentView(R.layout.activity_main);
        initFruits();    //初始化水果数据
        FruitAdapter adapter = new FruitAdapter (MainActivity.this,
                R.layout.fruit_item, fruitList); //实例化适配器
        ListView listView = (ListView) findViewById (R.id.list_view);
        listView.setAdapter (adapter);    //设置适配器
    }
    private void initFruits() {
        Fruit apple = new Fruit (" Apple", R.drawable.apple_pic);
        fruitList.add (apple);
        Fruit banana = new Fruit (" Banana", R.drawable.banana_pic);
        fruitList.add (banana);
        Fruit orange = new Fruit (" Orange", R.drawable.orange_pic);
        fruitList.add (orange);
        Fruit watermelon = new Fruit (" Watermelon", R.drawable.watermelon_pic);
```

```
        fruitList.add (watermelon);
        ……
    }
}
```

可以看到，这里添加了一个 initFruits( ) 方法，用于初始化所有的水果数据。在 Fruit 类的构造函数中将水果的名字和对应的图片 id 传入，然后把创建好的对象添加到水果列表中。接着我们在 onCreate( ) 方法中创建了 FruitAdapter 对象，并将 FruitAdapter 作为适配器传递给了 ListView。这样定制 ListView 界面的任务就完成了。

5.3.3 实例：ListView 实现场景对象选择

下面是 LsitView 控件高级应用实例，实现场景对象的选择。

(1) 在 Android 2.3 中创建应用项目：ListView_Choice。其布局文件有两个，一个是主 Activity 对应的布局文件 activity_main.xml，其中包含两个按钮和一个 ListView 控件，如图 5-5 所示。另一个是 ListView 控件的显示内容布局文件 item_layout.xml，如图 5-6 所示。

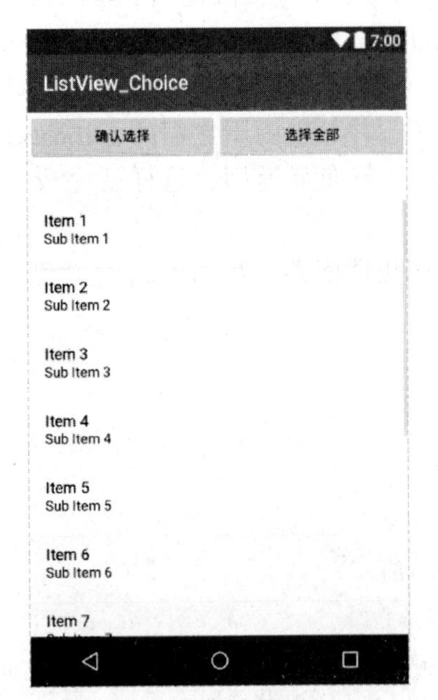

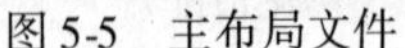

图 5-5 主布局文件

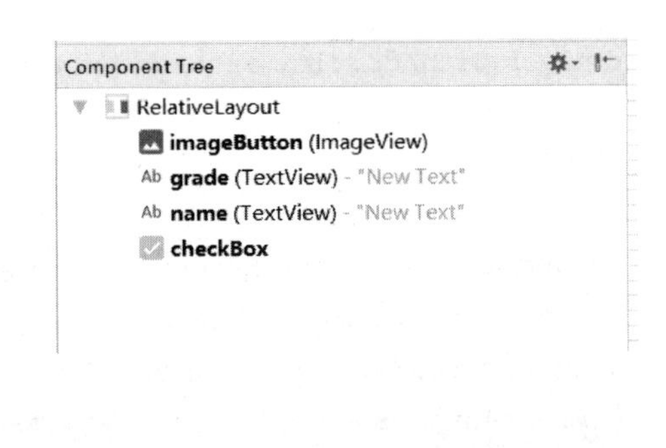

图 5-6 ListView 控件的显示内容布局

(2) ListView 控件的显示内容对应的类 ItemBean.java 的主要代码如下。

```
public class ItemBean {
    private int pictureId;  //ListView 控件的显示内容第一项图片
    private String grade; //ListView 控件的显示内容第二项等级
    private String name;  //ListView 控件的显示内容第三项名字
    private boolean isSelect; //ListView 控件的显示内容第四项是否选择
    private boolean isShowCheckBox; //ListView 控件的显示内容第四项是否显示
    ……
    Getter()和 Sertter()方法
}
```

（3）ListView 控件适配器类 ListViewWithCheckBoxAdapter. java 的代码如下。

```
public class ListViewWithCheckBoxAdapter extends BaseAdapter {
    private List<ItemBean> list;
    private LayoutInflater layoutInflater;
    public ListViewWithCheckBoxAdapter(Context context, List<ItemBean> list){
        this.list = list;
        layoutInflater = LayoutInflater.from(context);
    }
    @Override
    public int getCount() {
        return list.size();
    }
    @Override
    public Object getItem(int position) {
        return list.get(position);
    }
    @Override
    public long getItemId(int position) {
        return position;
    }
    @Override
    public View getView(int position, View convertView, ViewGroup parent) {
        if(convertView == null){
            convertView = layoutInflater.inflate(R.layout.item_layout, null);
            ViewHolder viewHolder = new ViewHolder (convertView);
            convertView.setTag (viewHolder);
        }
        ViewHolder viewHolder = (ViewHolder) convertView.getTag();
        viewHolder.imageView.setBackgroundResource (list.get (position).getPictureId());
        viewHolder.grade.setText (list.get (position).getGrade());
        viewHolder.name.setText (list.get (position).getName());
      if (list.get (position).isSelect()) {
        viewHolder.checkBox.setVisibility (View.VISIBLE);
          if (list.get (position).isSelect()) {
                viewHolder.checkBox.setChecked (true);
          } else {
              viewHolder.checkBox.setChecked (false);
          }
          return convertView;
    }
    public class ViewHolder {
        public final ImageView imageView;
        public final TextView grade;
        public final TextView name;
        public final CheckBox checkBox;
```

```
        public final View root;
        public ViewHolder (View root) {
          imageView = (ImageView) root.findViewById (R.id.imageButton);
          grade = (TextView) root.findViewById (R.id.grade);
          name = (TextView) root.findViewById (R.id.name);
          checkBox = (CheckBox) root.findViewById (R.id.checkBox);
            this.root = root;
        }
    }
}
```

(4) 主界面的 Activity 文件 MainActivity.java 的主要代码如下。

```
public class MainActivity extends Activity implements AdapterView.OnItemClickListener,
    AdapterView.OnItemLongClickListener,View.OnClickListener{
    private LinearLayout linearLayout;//按钮布局
    private Button sure,cancel;
    private ListView listView;
    private ListViewWithCheckBoxAdapter adapter;
    private List<ItemBean> list;
    private boolean isLineaLayoutVisible = false;//标记按钮布局的显示
    @Override
    protected void onCreate(Bundle savedInstanceState) {
      super.onCreate(savedInstanceState);
      setContentView(R.layout.activity_main);
      linearLayout = (LinearLayout) findViewById (R.id.linearlayout);
      sure = (Button) findViewById (R.id.sure);
      cancel = (Button) findViewById (R.id.cancel);
      sure.setOnClickListener (this);
      cancel.setOnClickListener (this);
      listView = (ListView) findViewById (R.id.listView);
      list = new ArrayList<ItemBean>();
      list.add (new ItemBean (R.drawable.apple_pic," Apple", " 1", true, false));
      //增加其他项
      adapter = new ListViewWithCheckBoxAdapter (this, list);
      listView.setAdapter (adapter);
      listView.setOnItemClickListener (this);
      listView.setOnItemLongClickListener (this);
    }
    @Override
  public void onItemClick (AdapterView<? > parent, View view, int position, long id) {
    ListViewWithCheckBoxAdapter.ViewHolder viewHolder = (ListViewWithCheckBox-
Adapter.ViewHolder) view.getTag();
   if (isLineaLayoutVisible) {//当按钮布局显示时候才有权多项选择
   list.get (position). setIsSelect (! list.get (position). isSelect()); //向表中
记录选择的 item
    adapter.notifyDataSetChanged(); //更新 ListView
```

```
    } else {
      Toast.makeText(this, list.get(position).getGrade() + "级" + list.get(posi-
tion).getName(),Toast.LENGTH_SHORT).show();
    }
    }
    @Override
    public boolean onItemLongClick(AdapterView<?> parent, View view, int posi-
tion, long id) {
      list.get(position).setIsSelect(true); //记录选择的item
        for (ItemBean itemBean : list) {
            itemBean.setIsShowCheckBox(true); //将所有的Item的CheckBox设置为选择
      }
      adapter.notifyDataSetChanged();
      linearLayout.setVisibility(View.VISIBLE); //长按item设置按钮布局为显示状态
      isLineaLayoutVisible = true;
      return true;
    }
    @Override
    public void onClick(View v) {
        switch (v.getId()) {
          case R.id.sure:
            String str = "";
            for (ItemBean itemBean : list) {
                if(itemBean.isSelect()){
                  str += itemBean.getGrade() + "级" + itemBean.getName() + "\n";
              }
            }
            if(str.equals("")){
              Toast.makeText(this,"您没有选择",Toast.LENGTH_SHORT).show();
            } else {
                Toast.makeText(this, str, Toast.LENGTH_SHORT).show();
            }
            break;
          case R.id.cancel:
            if (cancel.getText().equals("选择全部")) {
                  for (ItemBean itemBean : list) {
                  itemBean.setIsSelect(true);
              }
              cancel.setText("取消全部");
            } else {
                for (ItemBean itemBean : list) {
                  itemBean.setIsSelect(false);
                }
                cancel.setText("选择全部");
              }
```

```
                adapter.notifyDataSetChanged();
                break;
            }
        }
        @Override
        public void onBackPressed() {
            if (isLineaLayoutVisible) {
              linearLayout.setVisibility (View.INVISIBLE);
              isLineaLayoutVisible = false;
              for (ItemBean itemBean : list) {//按返回键时取消所有选择记录，同时把按钮布局设
置为不可见
                itemBean.setIsShowCheckBox (false);
                itemBean.setIsSelect (false);
              }
                adapter.notifyDataSetChanged();
            } else {
                super.onBackPressed();
          }
        }
      }
```

（5）运行结果如图 5-7 所示。

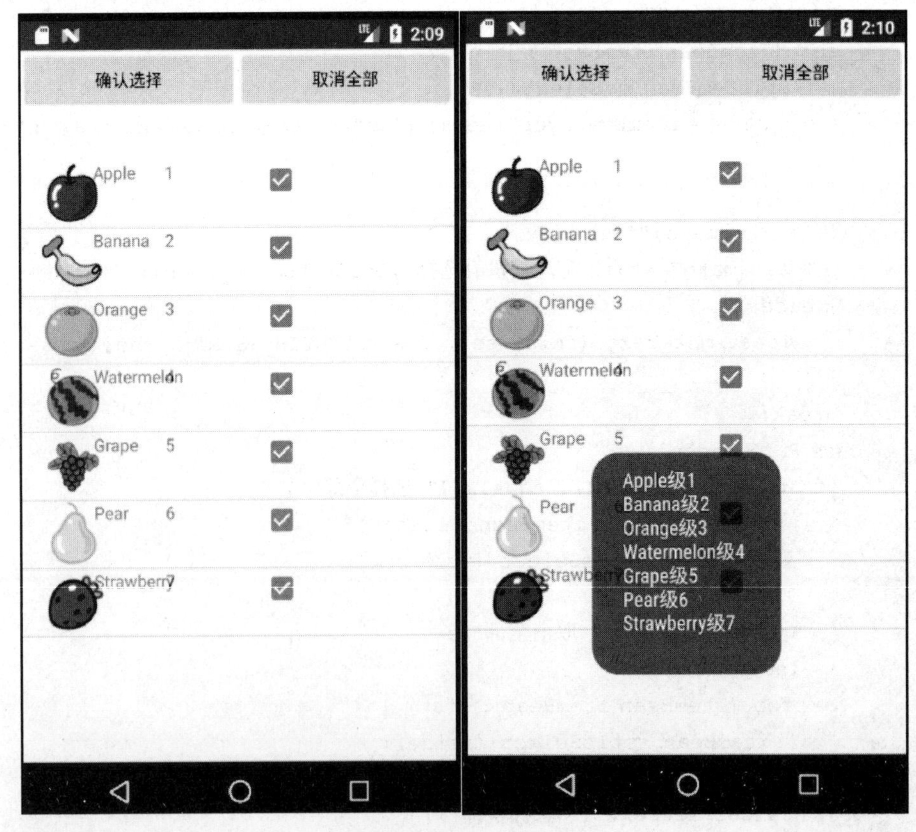

图 5-7　ListView_Choice

5.4　高级 ListView：ExpandableListView 实现商品列表折叠

ExpandableListView 是 ListView 的子类，它在普通 ListView 的基础上进行了扩展，把应用中的列表项分为几组，每组里又可包含多个列表项。ExpandableListView 的用法与普通 ListView 的用法非常相似，只是 ExpandableListView 显示的列表项应该由 ExpandableAdapter 提供。

5.4.1　ExpandableAdapter 简介

实现 ExpandableAdapter 有如下三种方式。

（1）扩展 BaseExpandableListAdpter 实现 ExpandableAdapter。

（2）使用 SimpleExpandableListAdpater 将两个 List 集合包装成 ExpandableAdapter。

（3）使用 SimpleCursorTreeAdapter 将 Cursor 中的数据包装成 SimpleCuroTreeAdapter。

一般适用于 ExpandableListView 的 Adapter 都要继承 BaseExpandableListAdapter 这个类，并且必须重载 getGroupView 和 getChildView 这两个最为重要的方法。

当扩展 BaseExpandableListAdapter 时，关键是实现如下四个方法。

（1）public abstract View getChildView（int groupPosition，int childPosition，boolean isLastChild，View　convertView，ViewGroup parent）

取得显示给定分组给定子位置的数据用的视图，参数的含义如下。

- groupPosition：包含要取得子视图的分组位置。
- childPosition：分组中子视图（要返回的视图）的位置。
- isLastChild：该视图是否为组中的最后一个视图。
- convertView：如果可能，重用旧的视图对象。使用前要保证视图对象为非空，并且是合适的类型。如果该对象不能转换为可以正确显示数据的视图，该方法将创建新视图。
- Pavent：该视图最终从属的父视图。

此方法返回指定位置相应的子视图。

（2）public abstract int getChildrenCount（int groupPosition）

取得指定分组的子元素数，参数 groupPosition 为要取得子元素个数的分组位置，返回指定分组的子元素个数。

（3）public abstract View getGroupView（int groupPosition，boolean isExpanded，View convertView，ViewGroup parent）

取得用于显示给定分组的视图。这个方法仅返回分组的视图对象，要想获取子元素的视图对象，就需要调用 getChildView（int，int，boolean，View，ViewGroup）。

参数含义如下。

- groupPosition：决定返回哪个视图的组位置。
- isExpanded：该组是展开状态还是收起状态。
- convertView：如果可能，重用旧的视图对象。使用前要保证视图对象为非空，并且是合适的类型。如果该对象不能转换为可以正确显示数据的视图，该方法将创建新视图，不保证使用先前由 getGroupView（int，boolean，View，ViewGroup）创建的视图。

➢ parent：该视图最终从属的父视图。

此方法返回指定位置相应的组视图。

（4）public abstract intgetGroupCount( )

取得分组数，返回分组数。

BaseExpandableListAdapter 的重载的其他方法如下。

（1）public abstract Object getChild（int groupPosition，int childPosition）

取得与指定分组、指定子项目关联的数据。

参数 groupPosition 包含子视图的分组的位置；childPosition 为指定分组中的子视图的位置。

此方法返回与子视图关联的数据。

（2）public abstract long getChildId（int groupPosition，int childPosition）

取得给定分组中给定子视图的 ID，该组 ID 必须在组中是唯一的，必须不同于其他所有 ID（分组及子项目的 ID）。

参数 groupPosition 包含子视图的分组的位置；childPosition 为要取得 ID 的指定分组中的子视图的位置。

此方法返回与子视图关联的 ID。

（3）public abstract longgetCombinedChildId（long groupId，long childId）

取得一览中可以唯一识别子条目的 ID（包括分组 ID 和子条目 ID）。可扩展列表要求每个条目（分组条目和子条目）具有一个可以唯一识别的 ID，该方法根据给定子条目 ID 和分组条目 ID 返回唯一识别 ID。另外，如果 hasStableIds( )为真，该函数返回的 ID 必须是固定不变的。

参数 groupId 包含子条目 ID 的分组条目 ID；childId 为子条目的 ID。

此方法返回可以在所有分组条目和子条目中唯一识别该子条目的 ID（可能是固定不变的）。

（4）public abstract Object getGroup（int groupPosition）

取得与给定分组关联的数据。

参数 groupPosition 为分组的位置。

此方法返回指定分组的数据。

（5）public abstract long getGroupId（int groupPosition）

取得指定分组的 ID。该组 ID 必须在组中是唯一的，必须不同于其他所有 ID（分组及子项目的 ID）。

参数 groupPosition 为要取得 ID 的分组位置。

此方法返回与分组关联的 ID。

（6）public abstract boolean hasStableIds( )

指定在分组视图及其子视图的 ID 对应的后台数据发生改变时，是否保持该 ID。

此方法返回是否相同的 ID 总是指向同一个对象。

（7）public abstract boolean isChildSelectable（int groupPosition，int childPosition）

指定位置的子视图是否可选择。

参数 groupPosition 包含要取得子视图的分组位置；childPosition 为分组中子视图的位置。此方法返回是否子视图可选择。

5.4.2　实例：ExpandableListView 实现商品列表折叠

下面是 ExpandableListVivew 控件应用实例，实现商品列表折叠。

（1）在 Android 2.3 中创建应用项目：ExpandableListVivew_Shopping。其布局文件有三个：一个是主 Activity 对应的布局文件 activity_main.xml，其中包含一个 ExpandableListView 控件，如图 5-8 所示；第二个布局文件是 groupitem.xml，ExpandableListView 显示项的结构，如图 5-9 所示。

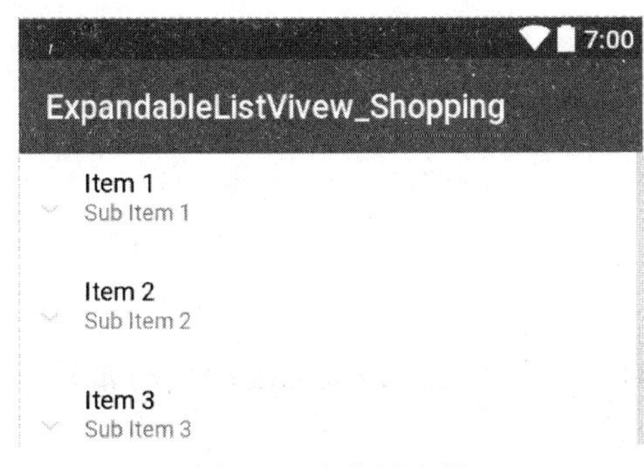

图 5-8　主布局文件

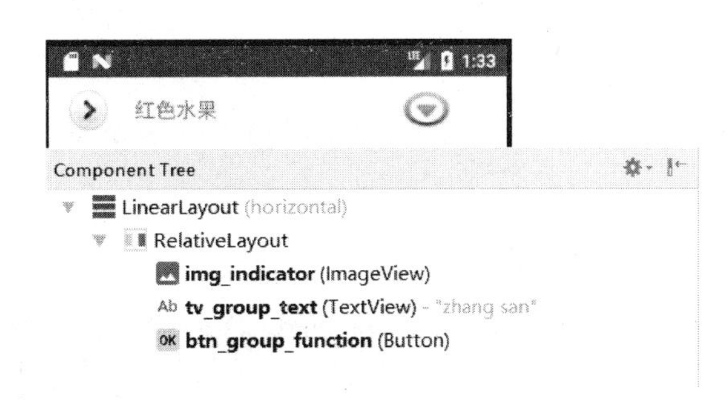

图 5-9　ExpandableListView 的显示项结构

第三个布局文件是 childitem.xml，ExpandablelistView 的显示项展开的内容结构，如图 5-10 所示。

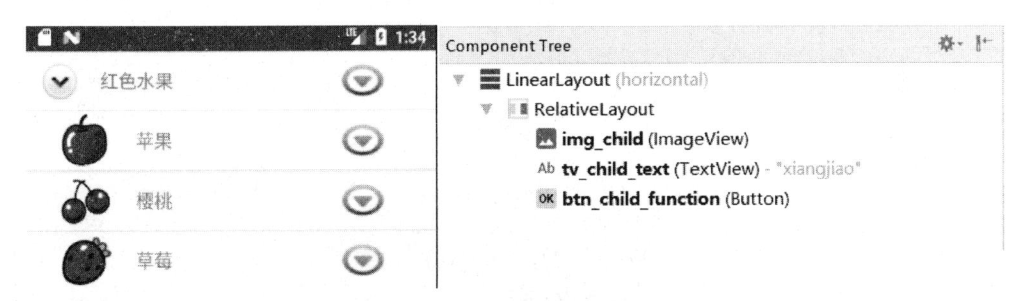

图 5-10　显示项的展开项

（2）ExpandableListVivew_Shopping 对应的处理类四个，如图 5-11 所示。

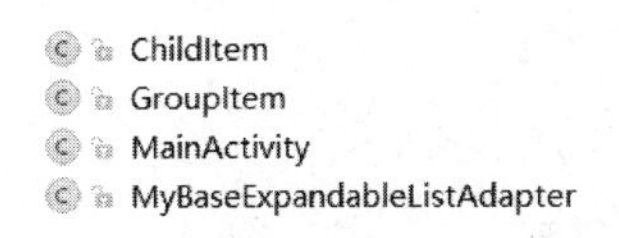

图 5-11　四个处理类

GroupItem 类，ExpandableListVivew 显示的项对应的类，GroupItem.java 的代码如下。

```
public class GroupItem {
  private String title;
  private int imageId;
  public GroupItem(String title, int imageId)
```

```
    {
        this.title = title;
        this.imageId = imageId;
    }
    public String getTitle() {
        return title;
    }
    public void setTitle(String title) {
        this.title = title;
    }
    public int getImageId() {
        return imageId;
    }
    public void setImageId(int imageId) {
        this.imageId = imageId;
    }
}
```

ChildItem 类，ExpandableListView 显示的项展开对应的类，ChildItem. java 的代码如下。

```
public class ChildItem {
    private String title; //子项显示的文字
    private int markerImgId;//每个子项的图标
    public ChildItem(String title, int markerImgId)
    {
        this.title = title;
        this.markerImgId = markerImgId;
    }
    public String getTitle() {
        return title;
    }
    public void setTitle(String title) {
        this.title = title;
    }
    public int getMarkerImgId() {
        return markerImgId;
    }
    public void setMarkerImgId(int markerImgId) {
        this.markerImgId = markerImgId;
    }
}
```

MyBaseExpandableListAdapter 类，ExpandableListView 适配器类，MyBaseExpandableListAdapter. java 的代码如下。

```
public class MyBaseExpandableListAdapter extends BaseExpandableListAdapter implements OnClickListener {
```

```
    private Context mContext;
    private List<String> groupTitle;
    //子项是一个 map,key 是 group 的 id,每一个 group 对应一个 ChildItem 的 list
    private Map<Integer, List<ChildItem>> childMap;
    private Button groupButton;//group 上的按钮
    public MyBaseExpandableListAdapter(Context context, List<String> groupTitle,
Map<Integer, List<ChildItem>> childMap) {
        this.mContext = context;this.groupTitle = groupTitle;this.childMap = childMap;
    }
    /* Gets the data associated with the given child within the given group */
    @Override
    public Object getChild(int groupPosition, int childPosition) {
      //这里返回一下每个 item 的名称,以便单击 item 时显示
        return childMap.get(groupPosition).get(childPosition).getTitle();
    }
    /*取得给定分组中给定子视图的 ID, 该组 ID 必须在组中是唯一的,必须不同于其他所有 ID(分组及
子项目的 ID)*/
    @Override
    public long getChildId(int groupPosition, int childPosition) {
        return childPosition;
    }
    /* Gets a View that displays the data for the given child within the given
group */
    @Override
    public View getChildView(int groupPosition, int childPosition, boolean isLast-
Child, View convertView,ViewGroup parent) {
    ChildHolder childHolder = null;
    if(convertView == null) {
      convertView = LayoutInflater.from(mContext).inflate(R.layout.childitem, null);
      childHolder = new ChildHolder();
      childHolder.childImg = (ImageView) convertView.findViewById(R.id.img_child);
      childHolder.childText = (TextView) convertView.findViewById (R.id.tv_child_text);
      convertView.setTag (childHolder);
    } else {
        childHolder = (ChildHolder) convertView.getTag();
    }
    childHolder.childImg.setBackgroundResource ( childMap.get ( groupPosition ).
get (childPosition). getMarkerImgId());
    childHolder.childText.setText (childMap.get (groupPosition). get (childPosi-
tion). getTitle());
    return convertView;
    }
    /*取得指定分组的子元素数*/
    @Override
```

```
    public int getChildrenCount (int groupPosition) {
      return childMap.get (groupPosition). size();
    }
    /* * 取得与给定分组关联的数据 * /
    @Override
    public Object getGroup (int groupPosition) {
        return groupTitle.get (groupPosition);
    }
    /* * 取得分组数 * /
    @Override
    public int getGroupCount() {
      return groupTitle.size();
    }
    /* * 取得指定分组的 ID,该组 ID 必须在组中是唯一的,必须不同于其他所有 ID(分组及子项目的
ID) * /
    @Override
    public long getGroupId(int groupPosition) {
      return groupPosition;
    }
    /* Gets a View that displays the given group * /
    @Override
    public View getGroupView (int groupPosition, boolean isExpanded, View convert-
View, ViewGroup parent) {
    GroupHolder groupHolder = null;
    if(convertView == null) {
        convertView = LayoutInflater.from(mContext).inflate(R.layout.groupitem, null);
        groupHolder = new GroupHolder();
      groupHolder.groupImg =(ImageView) convertView.findViewById(R.id.img_indicator);
      groupHolder.groupText = (TextView) convertView.findViewById (R.id.tv_group_
text);
      convertView.setTag (groupHolder);
    } else {
        groupHolder = (GroupHolder) convertView.getTag();
    }
    if (isExpanded) {
        groupHolder.groupImg.setBackgroundResource (R.drawable.downarrow);
    } else {
      groupHolder.groupImg.setBackgroundResource (R.drawable.rightarrow);
    }
    groupHolder.groupText.setText (groupTitle.get (groupPosition));
    groupButton = (Button) convertView.findViewById (R.id.btn_group_function);
    groupButton.setOnClickListener (this);
    return convertView;
```

```
}
@ Override
public boolean hasStableIds () {
        return true;
}
@ Override
public boolean isChildSelectable (int groupPosition, int childPosition) {
  // Whether the child at the specified position is selectable
    return true;
}
/ * show the text on the child and group item * /
private class GroupHolder
{
    ImageView groupImg;
    TextView groupText;
}
private class ChildHolder
{
    ImageView childImg;
    TextView childText;
}
@ Override
public void onClick (View v) {
    switch (v. getId ()) {
      case R. id. btn_group_function:
          Log. d (" MyBaseExpandableListAdapter", " 你单击了 group button");
      default:
        break;
    }
}
}
```

主界面的 Activity 文件 MainActivity. java 的主要代码如下。

```
public class MainActivity extends AppCompatActivity{
private ExpandableListView expandList;
private List < String > groupData;//group 的数据源
private Map < Integer, List < ChildItem > > childData;//child 的数据源
  private MyBaseExpandableListAdapter myAdapter;
  final int CONTEXT_MENU_GROUP_DELETE = 0; //添加上下文菜单时每一个菜单项的 item ID
  final int CONTEXT_MENU_GROUP_RENAME = 1;
  final int CONTEXT_MENU_CHILD_EDIT = 2;
  final int CONTEXT_MENU_CHILD_DELETE = 3;
  @ Override
  protected void onCreate (Bundle savedInstanceState) {
```

```
        super.onCreate (savedInstanceState);
        setContentView (R.layout.activity_main);
          initDatas();   initView();   initEvents();
      }
    /** group 和 child 子项的数据源 */
    private void initDatas() {
      groupData = new ArrayList<String>();
      groupData.add (" 红色水果");
      groupData.add (" 黄色水果");
      groupData.add (" 其他水果");
      List<ChildItem> childItems = new ArrayList<ChildItem>();
      ChildItem childData1 = new ChildItem (" 苹果", R.drawable.apple_pic);
      childItems.add (childData1);
      ChildItem childData2 = new ChildItem (" 樱桃", R.drawable.cherry_pic);
      childItems.add (childData2);
      ChildItem childData3 = new ChildItem (" 草莓", R.drawable.strawberry_pic);
      childItems.add (childData3);
      List<ChildItem> childItems2 = new ArrayList<ChildItem>();
      …
      List<ChildItem> childItems3 = new ArrayList<ChildItem>();
      …
      childData = new HashMap<Integer, List<ChildItem>>();
      childData.put (0, childItems);
      childData.put (1, childItems2);
      childData.put (2, childItems3);
      myAdapter = new MyBaseExpandableListAdapter (this, groupData, childData);
    }
    private void initView() {
      expandList = (ExpandableListView) findViewById (R.id.expandlist);
      //在 drawable 文件夹下新建了 indicator.xml，下面这个语句也可以实现 group 伸展收缩时
的 indicator 变化
      expandList.setGroupIndicator (null); //这里不显示系统默认的 group indicator
      expandList.setAdapter (myAdapter);
      registerForContextMenu (expandList); //给 ExpandListView 添加上下文菜单
    }
    private void initEvents() {
      //child 子项的单击事件
    expandList.setOnChildClickListener (new OnChildClickListener() {
      @Override
      public boolean onChildClick (ExpandableListView parent, View v,
int groupPosition, int childPosition, long id) {
        Toast.makeText (MainActivity.this, " 你单击了:"
+myAdapter.getChild(groupPosition, childPosition), Toast.LENGTH_SHORT).show();
```

```
            return true;
          }
        });
      }
      /*添加上下文菜单*/
      @Override
      public void onCreateContextMenu (ContextMenu menu, View v,
ContextMenuInfo menuInfo) {
      super.onCreateContextMenu (menu, v, menuInfo);
      ExpandableListView.ExpandableListContextMenuInfo info =
(ExpandableListView.ExpandableListContextMenuInfo) menuInfo;
      int type = ExpandableListView.getPackedPositionType (info.packedPosition);
      if (type == ExpandableListView.PACKED_POSITION_TYPE_GROUP) {
          menu.setHeaderTitle (" Options");
          menu.add (0, CONTEXT_MENU_GROUP_DELETE, 0, " 删除");
          menu.add (0, CONTEXT_MENU_GROUP_RENAME, 0, " 重命名");
        }
      if (type == ExpandableListView.PACKED_POSITION_TYPE_CHILD) {
          menu.setHeaderTitle (" Options");
          menu.add (1, CONTEXT_MENU_CHILD_EDIT, 0, " 编辑");
          menu.add (1, CONTEXT_MENU_CHILD_DELETE, 0, " 删除");
        }
      }
      /*每个菜单项的具体单击事件*/
      @Override
      public boolean onContextItemSelected (MenuItem item) {
          ExpandableListView.ExpandableListContextMenuInfo info =
(ExpandableListView.ExpandableListContextMenuInfo) item.getMenuInfo();
        switch (item.getItemId()) {
          case CONTEXT_MENU_GROUP_DELETE:
            Toast.makeText (this, " 这是 group 的删除", Toast.LENGTH_SHORT).show();
              break;
          case CONTEXT_MENU_GROUP_RENAME:
              Toast.makeText (this, " 这是 group 的重命名", Toast.LENGTH_SHORT).show();
              break;
          case CONTEXT_MENU_CHILD_EDIT:
              Toast.makeText (this, " 这是 child 的编辑", Toast.LENGTH_SHORT).show();
              break;
          case CONTEXT_MENU_CHILD_DELETE:
            Toast.makeText (this, " 这是 child 的删除", Toast.LENGTH_SHORT).show();
              break;
          default:
              break;
```

```
    }
  return super.onContextItemSelected (item);
  }
}
```

（3） ExpandableListVivew_Shopping 项目的运行结果如图 5-12 所示。

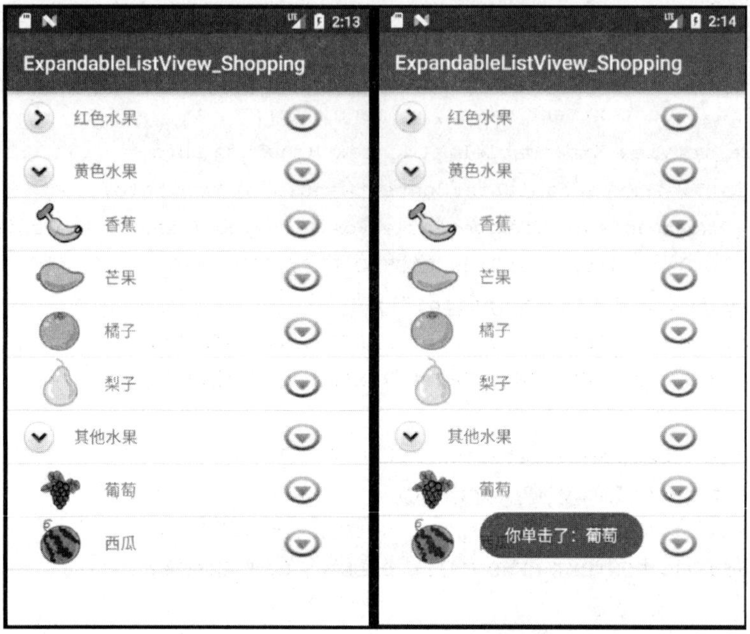

图 5-12　ExpandableListVivew_Shopping 项目的运行结果

5.5 高级控件 Camera2 + SurfaceView——实现拍照

本节我们介绍两款高级控件：Camera2 和 SurfaceView，并使用这两个控件实现拍照功能。

5.5.1　SurfaceView 简介

SurfaceView 是视图（View）的继承类，这个视图里内嵌了一个专门用于绘制的 Surface。我们可以控制这个 Surface 的格式和尺寸，SurfaceView 控制这个 Surface 的绘制位置。

Surface 是纵深排序（Z-ordered）的，这表明它总在自己所在窗口的后面。SurfaceView 提供了一个可见区域，只有在这个可见区域内的 Surface 部分内容才可见，可见区域外的部分不可见。可以通过 SurfaceHolder 接口访问 Surface，getHolder( ) 方法可以得到这个接口。

SurfaceView 变得可见时，Surface 被创建；SurfaceView 隐藏前，Surface 被销毁，这样能节省资源。如果要查看 Surface 被创建和销毁的时机，可以重载 surfaceCreated（SurfaceHolder）和 SurfaceDestroyed（SurfaceHolder）。

Surfaceview 的核心在于提供了两个线程：UI 线程和渲染线程。这里应注意以下几点。

（1）所有 SurfaceView 和 SurfaceHolder. Callback 的方法都应该在 UI 线程里调用，一般来说就是应用程序主线程。渲染线程所要访问的各种变量应该作同步处理。

（2）由于 Surface 可能被销毁，它只在 SurfaceHolder. Callback. surfaceCreated( ) 和 SurfaceHolder. Callback. surfaceDestroyed( )之间有效，所以要确保渲染线程访问的是合法有效的 Surface。

5. 5. 2　实例：Camera2 + SurfaceView——实现拍照

Android 5. 0 之后，Android. hardware. Camera 被废弃（下面称为 Camera1），由 Android. hardware. Camera2 代替。Camera2 支持 RAW 输出，可以调节曝光、对焦模式、快门等，功能比原先 Camera1 强大。

1. Camera1

Camera1 使用步骤如下。

（1）调用 Camera. open( )，打开相机，默认为后置相机，可以根据摄像头 ID 来指定打开前置还是后置相机。

（2）调用 Camera. getParameters( )得到一个 Camera. Parameters 对象。

（3）使用步骤 2 得到的 Camera. Parameters 对象，对拍照参数进行设置。

（4）调用 Camera. setPreviewDispaly （SurfaceHolder holder），指定使用哪个 SurfaceView 来显示预览图片。

（5）调用 Camera. startPreview( )方法开始预览取景。

（6）调用 Camera. takePicture( )方法进行拍照。

（7）拍照结束后，调用 Camera. stopPreview( )结束取景预览，之后再调用 release( )方法释放资源。

2. Camera2

Camear2 拍照引用了管道的概念，将安卓设备和摄像头之间联通起来，系统向摄像头发送 Capture 请求，而摄像头会返回 CameraMetadata。这一切建立在一个叫作 CameraCaptureSession 的会话中，如图 5-13 所示。

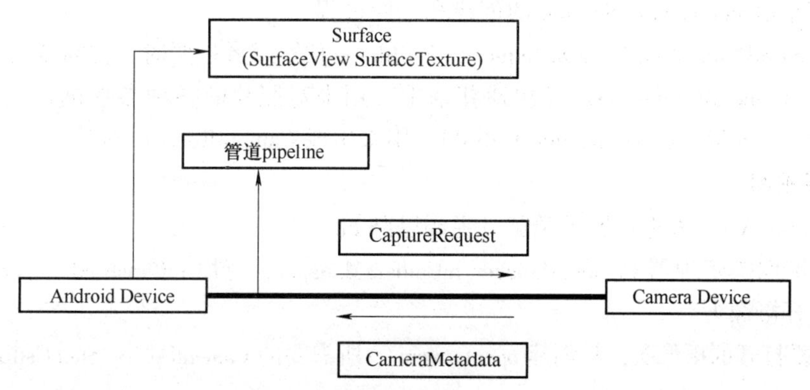

图 5-13　Camera2 拍照示意图

下面是 camera2 包中的主要类，如图 5-14 所示。

（1）CameraManager：摄像头管理器，用于检测摄像头，打开系统摄像头，调用 Camera-

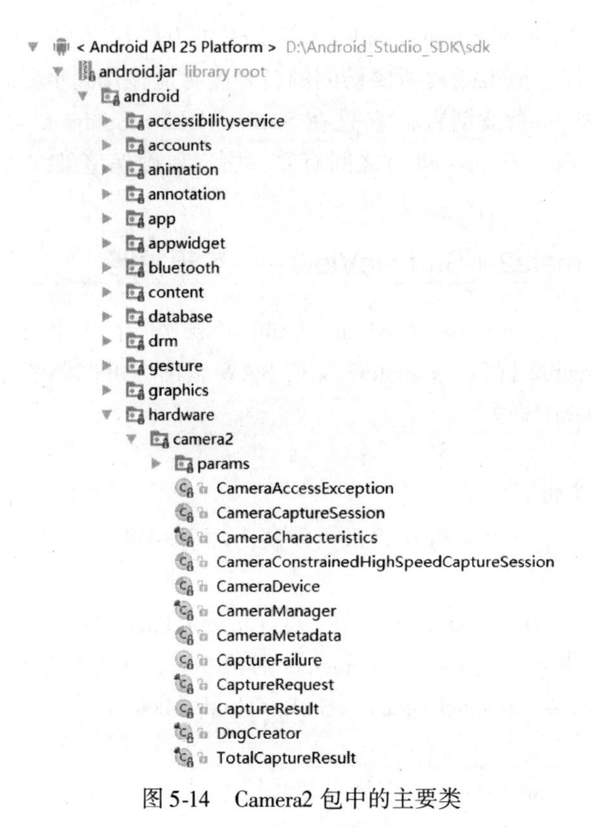

图 5-14 Camera2 包中的主要类

Manager. getCameraCharacteristics（String），可以获取指定摄像头的相关特性。

（2）CameraCharacteristics：摄像头的特性。

（3）CameraDevice：摄像头，类似于 android. hardware. Camera，也就是 Camera1 的 Camera。

（4）CameraCaptureSession：控制摄像头的预览或者拍照，setRepeatingRequest( )用于开启预览，capture( )用于拍照，CameraCaptureSession 提供了 StateCallback、CaptureCallback 两个接口来监听 CameraCaptureSession 的创建和拍照过程。

（5）CameraRequest 和 CameraRequest. Builder：预览或者拍照时，都需要一个 CameraRequest 对象。CameraRequest 表示一次捕获请求，用来对照片的各种参数进行设置，比如对焦模式、曝光模式等。CameraRequest. Builder 用来生成 CameraRequest 对象。

3. 实现拍照

使用 SurfaceView 来进行展示预览，主要思路如下。

- 获得摄像头管理器 CameraManager mCameraManager，应用 mCameraManager. openCamera( )打开摄像头。
- 指定要打开的摄像头，并创建 openCamera( )所需要的 CameraDevice. StateCallback stateCallback。
- 在 CameraDevice. StateCallback stateCallback 中调用 takePreview( )，在这个方法中，使用 CaptureRequest. Builder 创建预览需要的 CameraRequest，并初始化 CameraCaptureSession，最后调用了 setRepeatingRequest（previewRequest，null，childHandler）进行预

览，单击屏幕，调用 takePicture( )，在这个方法中，最终调用 capture (mCaptureRequest，null，childHandler)。

➢ 在 new ImageReader. OnImageAvailableListener( ) 回调方法中，展示拍照得到的图片。

具体操作步骤如下。

(1) 在 Android 2.3 中创建应用项目：SurfaceView_Camera2。其中布局文件 activity_main. xml，包含两个按钮和一个 ListView 控件。另一个布局文件是 ListView 控件的显示内容布局文件 item_layout. xml，如下所示。

```
<RelativeLayout xmlns:android="http://schemas.android.com/apk/res/android"
    android:layout_width=" match_parent"
    android: layout_height=" match_parent" >
    <SurfaceView
        android: id=" @+id/surface_view_camera2_activity"
        android: layout_width=" match_parent"
        android: layout_height=" match_parent" />
    <ImageView
        android: id=" @+id/iv_show_camera2_activity"
        android: layout_width=" 180dp"
        android: layout_height=" 320dp"
        android: visibility=" gone"
        android: layout_centerInParent=" true"
        android: scaleType=" centerCrop" />
</RelativeLayout>
```

(2) 主界面的 Activity 文件 MainActivity. java 的代码如下。

```
public class MainActivity extends AppCompatActivity implements View.OnClickListener {
  private static final SparseIntArray ORIENTATIONS = new SparseIntArray();
  ///为了使照片竖直显示
  static {
        ORIENTATIONS.append(Surface.ROTATION_0, 90);
        ORIENTATIONS.append (Surface.ROTATION_90, 0);
        ORIENTATIONS.append (Surface.ROTATION_180, 270);
        ORIENTATIONS.append (Surface.ROTATION_270, 180);
        }
  private SurfaceView mSurfaceView;
  private SurfaceHolder mSurfaceHolder;
  private ImageView iv_show;
  private CameraManager mCameraManager; //摄像头管理器
  private Handler childHandler, mainHandler;
  private String mCameraID; //摄像头 Id 0 为后，1 为前
  private ImageReader mImageReader;
  private CameraCaptureSession mCameraCaptureSession;
  private CameraDevice mCameraDevice;
  @Override
```

```
    protected void onCreate (Bundle savedInstanceState) {
        super.onCreate (savedInstanceState);
        setContentView (R.layout.activity_main);
        initVIew();
    }
    /**初始化*/
    private void initVIew() {
      iv_show = (ImageView) findViewById (R.id.iv_show_camera2_activity);
      //mSurfaceView
      mSurfaceView = (SurfaceView) findViewById(R.id.surface_view_camera2_activity);
      mSurfaceView.setOnClickListener (this);
      mSurfaceHolder = mSurfaceView.getHolder();
      mSurfaceHolder.setKeepScreenOn (true);
    // mSurfaceView 添加回调
    mSurfaceHolder.addCallback (new SurfaceHolder.Callback() {
        @Override
        public void surfaceCreated (SurfaceHolder holder) { //SurfaceView 创建
          //初始化 Camera
            initCamera2();
          }
        @Override
      public void surfaceChanged(SurfaceHolder holder, int format, int width, int height) {
        }
      @Override
      public void surfaceDestroyed(SurfaceHolder holder) { //SurfaceView 销毁
          //释放 Camera 资源
          if(null != mCameraDevice) {
                mCameraDevice.close();
                MainActivity.this.mCameraDevice = null;
              }
          }
      });
  }
  /**初始化 Camera2*/
  @RequiresApi(api = Build.VERSION_CODES.LOLLIPOP)
  private void initCamera2() {
    HandlerThread handlerThread = new HandlerThread (" Camera2");
    handlerThread.start();
    childHandler = new Handler (handlerThread.getLooper());
    mainHandler = new Handler (getMainLooper());
    mCameraID = "" +CameraCharacteristics.LENS_FACING_FRONT; //后摄像头
    mImageReader = ImageReader.newInstance (1080, 1920, ImageFormat.JPEG, 1);
    mImageReader.setOnImageAvailableListener (new
```

```
ImageReader.OnImageAvailableListener() { //可以在这里处理拍照得到的临时照片
    @Override
    public void onImageAvailable (ImageReader reader) {
      mCameraDevice.close();
      mSurfaceView.setVisibility (View.GONE);
      iv_show.setVisibility (View.VISIBLE);
      //拿到拍照照片数据
      Image image = reader.acquireNextImage();
      ByteBuffer buffer = image.getPlanes() [0].getBuffer();
      byte [] bytes = new byte [buffer.remaining()];
      buffer.get (bytes); //由缓冲区存入字节数组
      final Bitmap bitmap = BitmapFactory.decodeByteArray (bytes, 0, bytes.length);
          if (bitmap != null) {
            iv_show.setImageBitmap (bitmap);
          }
        }
      }, mainHandler);
    //获取摄像头管理
    mCameraManager = (CameraManager) getSystemService (Context.CAMERA_SERVICE);
    try {
      if(ActivityCompat.checkSelfPermission(this, Manifest.permission.CAMERA) !=
PackageManager.PERMISSION_GRANTED) {
          return;
        }
      //打开摄像头
        mCameraManager.openCamera (mCameraID, stateCallback, mainHandler);
      } catch (CameraAccessException e) {
        e.printStackTrace();
      }
    }
    /*摄像头创建监听*/
    private CameraDevice.StateCallback stateCallback = new CameraDevice.StateCallback() {
      @Override
      public void onOpened (CameraDevice camera) {//打开摄像头
        mCameraDevice = camera;
        //开启预览
        takePreview();
    }
    @Override
    public void onDisconnected (CameraDevice camera) {//关闭摄像头
        if (null != mCameraDevice) {
          mCameraDevice.close();
          MainActivity.this.mCameraDevice = null;
```

```
            }
        }
        @Override
        public void onError (CameraDevice camera, int error) {//发生错误
          Toast.makeText (MainActivity.this, " 摄像头开启失败", Toast.LENGTH_SHORT).show();
          }
        };
        /*开始预览*/
        private void takePreview() {
          try {
            //创建预览需要的 CaptureRequest.Builder
                final  CaptureRequest.Builder  previewRequestBuilder  =  mCameraDe-
vice.createCaptureRequest (CameraDevice.TEMPLATE_PREVIEW);
          //将 SurfaceView 的 surface 作为 CaptureRequest.Builder 的目标
            previewRequestBuilder.addTarget (mSurfaceHolder.getSurface());
          //创建 CameraCaptureSession，该对象负责处理预览请求和拍照请求
              mCameraDevice.createCaptureSession  ( Arrays.asList  ( mSurfaceHold-
er.getSurface  ( ),   mImageReader.getSurface  ( )),   new   CameraCaptureSes-
sion.StateCallback() {
                @Override
                public void onConfigured (CameraCaptureSession cameraCaptureSession) {
                    if (null == mCameraDevice) return;
                    //当摄像头已经准备好时，开始显示预览
                    mCameraCaptureSession = cameraCaptureSession;
                        try {
                      //自动对焦
                    previewRequestBuilder.set (CaptureRequest.CONTROL_AF_MODE, Capture-
Request.CONTROL_AF_MODE_CONTINUOUS_PICTURE);
                      //打开闪光灯
                    previewRequestBuilder.set (CaptureRequest.CONTROL_AE_MODE, Capture-
Request.CONTROL_AE_MODE_ON_AUTO_FLASH);
                    //显示预览
                  CaptureRequest previewRequest = previewRequestBuilder.build();
                      mCameraCaptureSession.setRepeatingRequest (previewRequest, null,
childHandler);
                } catch (CameraAccessException e) {
                      e.printStackTrace();
                }
            }
        @Override
        public void onConfigureFailed(CameraCaptureSession cameraCaptureSession) {
          Toast.makeText(MainActivity.this, "配置失败", Toast.LENGTH_SHORT).show();
            }
          }, childHandler);
```

```
    } catch (CameraAccessException e) {
        e.printStackTrace();
    }
    }
    /*单击事件*/
    @Override
    public void onClick (View v) {
      takePicture();
    }
    /*拍照*/
    private void takePicture() {
      if (mCameraDevice == null) return;
      //创建拍照需要的 CaptureRequest.Builder
        final CaptureRequest.Builder captureRequestBuilder;
      try {
          captureRequestBuilder = mCameraDevice.createCaptureRequest (CameraDevice.TEMPLATE_STILL_CAPTURE);
        //将 imageReader 的 surface 作为 CaptureRequest.Builder 的目标
        captureRequestBuilder.addTarget (mImageReader.getSurface());
        //自动对焦
         captureRequestBuilder.set (CaptureRequest.CONTROL_AF_MODE, CaptureRequest.CONTROL_AF_MODE_CONTINUOUS_PICTURE);
        //自动曝光
       captureRequestBuilder.set (CaptureRequest.CONTROL_AE_MODE, CaptureRequest.CONTROL_AE_MODE_ON_AUTO_FLASH);
        //获取手机方向
      int rotation = getWindowManager().getDefaultDisplay().getRotation();
        //根据设备方向计算设置照片的方向
    captureRequestBuilder.set(CaptureRequest.JPEG_ORIENTATION, ORIENTATIONS.get(rotation));
        //拍照
        CaptureRequest mCaptureRequest = captureRequestBuilder.build();
        mCameraCaptureSession.capture (mCaptureRequest, null, childHandler);
        } catch (CameraAccessException e) {
        e.printStackTrace();
        }
      }
    }
```

(3) 在 AndroidManifest.xml 文件中增加权限。

```
<uses-permission android:name="android.permission.CAMERA" />
```

(4) SurfaceView_Camera2 项目的运行结果如图 5-15 所示。

图 5-15 SurfaceView_Camera2 项目的运行结果

5.6 艺术般的控件：RecyclerView 和 CardView——实现新闻卡片

本节主要介绍 RecyclerView 和 CardView 两个控件的应用方法，并应用这两个控件实现新闻卡片效果。

5.6.1 RecyclerView 和 CardView 简介

本节介绍 RecyclerView 和 CardView 两个控件。

1. RecyclerView

RecyclerView 是 Android 5.0 之后的新控件，用来替代经典的 Listview 和 Gridview，不仅可以轻松实现 ListView 的效果，还优化了 ListView 中存在的各种不足。Android 官方推荐使用 RecyclerView，未来也会有更多的程序从 ListView 转向 RecyclerView。

相比于 ListView，RecyclerView 主要有以下特点。

- 适配器中需要提供 ViewHolder 类：ListView 中这不是必需的，但在 RecyclerView 是必需的。
- 灵活的自定义 Item 布局：ListView 只能实现垂直的线性列布局，但 RecyclerView 可以通过 RecyclerView. LayoutManager 实现 Item 任何复杂的布局。
- 处理 Item 动画：RecyclerView. ItemAnimator 可用来处理 Item 动画。
- 数据源：ListView 可以通过 ArrayAdapter、CursorAdapter 等直接提供数据源，RecyclerView 需要自定义数据源。
- Item 项：ListView 可以通过 android:dividev 等属性控制 Item 项的显示，RecyclerView 可以通过 RecyclerView. ItemDecoration 来设置。
- Item CLick Listener：ListView 提供了 AdapterView. OnItemClickListener，来实现 Item 事件的监听，RecyclerView 没有直接提供类似方法，需要自己实现。
- 提供了一种插拔式的体验，高度的解耦，异常灵活，针对一个 Item 的显示 RecylerView 专门抽取出了相应的类，来控制 Item 的显示，使其扩展性非常强。

RecyclerView 是一种更高级柔性版本的 ListView，是一个能包含很多视图的容器，能完美处理循环和滚动。

RecyclerView 使用方法很简单，因为它提供了定位 Item 的布局管理器和常见的 Item 操作默认动画。

使用 RecyclerView，必须指定一个 Adapter，并定义一个布局管理器。创建的 Adapter 必须继承自 RecyclerView. Adapter。实施的细节需要看数据类型和视图的需要，如图 5-16 所示。

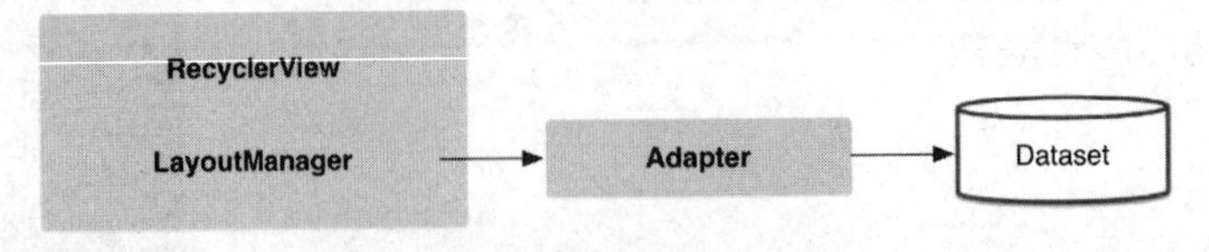

图 5-16　RecyclerView 控件

RecyclerView 提供了 LayoutManager 管理布局，RecylerView 不负责子 View 的布局，目前提供了 3 种布局管理器。

（1）LinearLayoutManager：显示垂直或水平滚动列表中的条目。

（2） GridLayoutManager：在一个网格显示项。

（3） StaggeredGridLayoutManager：在交错网格显示项。

2. CardView

CardView 是 Android 5.0 之后的新控件，CardView 继承自 FrameLayout 类，可以在一个卡片布局中一致性地显示内容，卡片可以包含圆角和阴影。可以使用 android:elevation 属性，创建带阴影的卡片。

指定 CardView 的属性的方法如下。

➢ 使用 android:cardCornerRadius 属性指定圆角半径。

➢ 使用 CardView. setRadius 设置圆角半径。

➢ 使用 android:cardBackgroundColor 属性设置卡片颜色。

5.6.2　实例：RecyclerView 和 CardView——实现新闻卡片

下面应用 RecyclerView 和 CardView 控件实现新闻卡片。

（1） 在 Android 2.3 中创建应用项目：RecyclerView_Card。需要在项目的 build. gradle（在 Project 模式下内层的 build. gradle） 添加相应的依赖库。

打开 build. gradle 文件，在 dependencies 闭包中添加如下内容。

```
dependencies {
    compile fileTree(dir:'libs', include: ['*.jar'])
    compile 'com.android.support:appcompat-v7:25.3.1'
    compile 'com.android.support.constraint:constraint-layout:1.0.2'
    compile 'com.android.support:percent:25.3.1'
    compile 'com.android.support:recyclerview-v7:25.3.1'
    compile 'com.android.support:cardview-v7:25.3.1'
}
```

（2） 主布局文件 activity_main. xml 中包含 RecyclerView 控件，代码如下。

```
<?xml version="1.0" encoding="utf-8"?>
<android.support.v7.widget.RecyclerView
    xmlns:android=http://schemas.android.com/apk/res/android
    android:id="@+id/recyclerView"
    android:layout_width="match_parent"
    android:layout_height="match_parent">
</android.support.v7.widget.RecyclerView>
```

（3） 新建 RecyclerView 显示项的布局文件 res/layout/news_item. xml，如图 5-17 所示，显示容器项是一个 CardView。

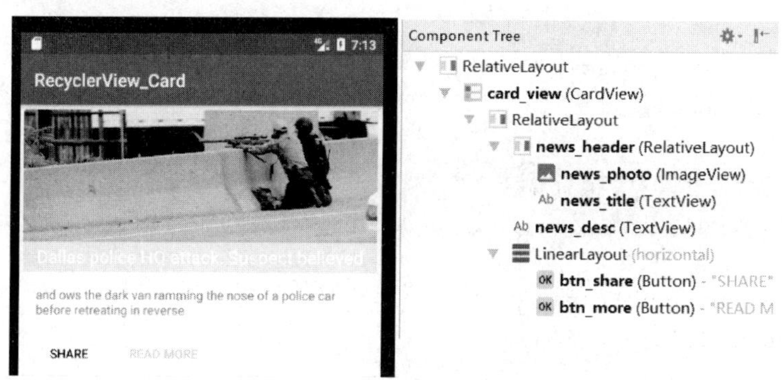

图 5-17　RecyclerView 显示项的布局文件

(4) 新建新闻卡片的布局文件 res/layout/ activity_news. xml，如图 5-18 所示，显示容器项是一个 CardView。

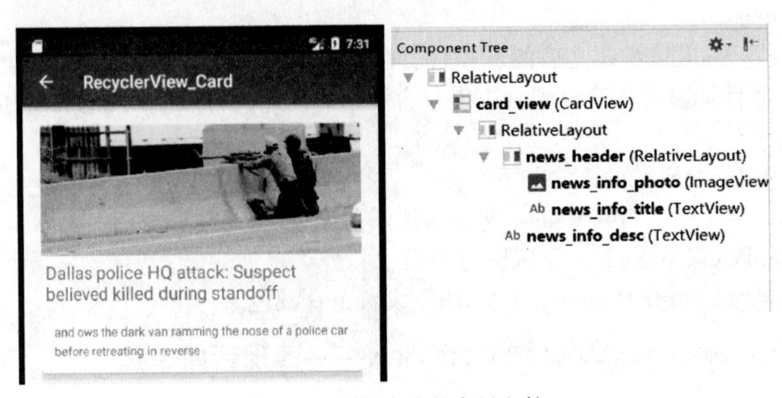

图 5-18　新闻卡片的布局文件

(5) 定义数据类。这里建立一个新闻类，implements Serialzable 是为了在 Intent 中能够直接传递 News 对象，文件名为 News. Java，代码如下。

```
public class News implements Serializable{
    //新闻标题、内容、图片
    private String title;
    private String desc;
    private int photoId;
    public News(String name, String age, int photoId) {
        this. title =name;
        this. desc =age;
        this. photoId =photoId;
    }
    public void setDesc(String desc) {
        this. desc = desc;
    }
    public void setTitle(String title) {
        this. title = title;
    }
    public void setPhotoId(int photoId) {
        this. photoId = photoId;
    }
    public String getDesc() {
        return desc;
    }
    public int getPhotoId() {
        return photoId;
    }
    public String getTitle() {
        return title;
    }
```

```
}
```

（6）自定义 Adapter。这部分内容比较重要，如果对 ListView 自定义 Adapter 比较熟悉，将很容易看懂。相比而言，RecyclerView Adapter 中必须要实现 ViewHolder 类，然后需要覆写几个方法，唯一复杂的地方是在 onBingViewHolder 方法中为两个按钮和 CardView 实现单击事件，跳转到新闻详细页面（NewsActivity）或者弹出分享。文件名为 RecyclerViewAdapter. java，代码如下。

```
public class RecyclerViewAdapter extends RecyclerView.Adapter <RecyclerViewAdapter.NewsViewHolder>{
    private List<News> newses;
    private Context context;
    public RecyclerViewAdapter(List<News> newses,Context context) {
            this.newses = newses;
            this.context=context;
    }
    //自定义 ViewHolder 类
    static class NewsViewHolder extends RecyclerView.ViewHolder{
            CardView cardView;
            ImageView news_photo;
            TextView news_title;
            TextView news_desc;
            Button share;
            Button readMore;
            public NewsViewHolder(final View itemView) {
                super(itemView);
                cardView=(CardView) itemView.findViewById(R.id.card_view);
                news_photo=(ImageView) itemView.findViewById(R.id.news_photo);
                news_title=(TextView) itemView.findViewById (R.id.news_title);
                news_desc=(TextView) itemView.findViewById (R.id.news_desc);
                share=(Button) itemView.findViewById (R.id.btn_share);
                readMore=(Button) itemView.findViewById (R.id.btn_more);
                //设置 TextView 背景为半透明
                news_title.setBackgroundColor (Color.argb (20, 0, 0, 0));
            }
        }
        @Override
        public RecyclerViewAdapter.NewsViewHolder onCreateViewHolder (ViewGroup viewGroup, int i) {
            View v=LayoutInflater.from(context).inflate(R.layout.news_item,viewGroup,false);
            NewsViewHolder nvh=new NewsViewHolder(v);
            return nvh;
        }
        @Override
        public void onBindViewHolder(RecyclerViewAdapter.NewsViewHolder personViewHolder, int i) {
            final int j=i;
```

```
        personViewHolder.news_photo.setImageResource(newses.get(i).getPhotoId());
        personViewHolder.news_title.setText(newses.get(i).getTitle());
        personViewHolder.news_desc.setText(newses.get(i).getDesc());
        //为 btn_share btn_readMore cardView 设置单击事件
        personViewHolder.cardView.setOnClickListener(new View.OnClickListener() {
            @Override
            public void onClick(View v) {
                Intent intent = new Intent(context,NewsActivity.class);
                intent.putExtra("News",newses.get(j));
                context.startActivity(intent);
            }
        });
        personViewHolder.share.setOnClickListener(new View.OnClickListener() {
            @Override
            public void onClick(View v) {
                Intent intent = new Intent(Intent.ACTION_SEND);
                intent.setType("text/plain");
                intent.putExtra(Intent.EXTRA_SUBJECT, "分享");
                intent.putExtra(Intent.EXTRA_TEXT, newses.get(j).getDesc());
                intent.setFlags(Intent.FLAG_ACTIVITY_NEW_TASK);
                context.startActivity(Intent.createChooser(intent, newses.get(j).getTitle()));
            }
        });
        personViewHolder.readMore.setOnClickListener(new View.OnClickListener() {
            @Override
            public void onClick(View v) {
                Intent intent = new Intent(context,NewsActivity.class);
                intent.putExtra("News",newses.get(j));
                context.startActivity(intent);
            }
        });
    }
    @Override
    public int getItemCount() {
    return newses.size();
    }
}
```

（7）创建 NewsActivity 类，用来显示新闻详细内容，布局使用 res/layout/activity_news.xml，文件名为 NewsActivity.java，代码如下。

```
public class NewsActivity extends AppCompatActivity {
    private ImageView newsPhoto;
    private TextView newsTitle;
    private TextView newsDesc;
    @Override
    protected void onCreate(Bundle savedInstanceState) {
```

```
        super.onCreate(savedInstanceState);
        setContentView(R.layout.activity_news);
        newsPhoto = (ImageView) findViewById (R.id.news_info_photo);
        newsTitle = (TextView) findViewById (R.id.news_info_title);
        newsDesc = (TextView) findViewById (R.id.news_info_desc);
        Intent intent =getIntent();
        News item = (News) intent.getSerializableExtra (" News");
        newsPhoto.setImageResource (item.getPhotoId());
        newsTitle.setText (item.getTitle());
        newsDesc.setText (item.getDesc());
    }
}
```

（8）主界面中的 Activity 文件 MainActivity.java 的代码如下。

```
public class MainActivity extends AppCompatActivity {
    private RecyclerView recyclerView;
    private List<News> newsList;
    private RecyclerViewAdapter adapter;
    @Override
    protected void onCreate(Bundle savedInstanceState) {
        super.onCreate(savedInstanceState);
        setContentView(R.layout.activity_main);
        LinearLayoutManager layoutManager =new LinearLayoutManager(this);
        recyclerView = (RecyclerView) findViewById(R.id.recyclerView);
        initPersonData();
        adapter =new RecyclerViewAdapter(newsList,MainActivity.this);
        recyclerView.setHasFixedSize(true);
        recyclerView.setLayoutManager(layoutManager);
        recyclerView.setAdapter(adapter);
    }
    private void initPersonData() {
      newsList =new ArrayList<>();
      newsList.add(new
News (getString (R.string.news_one_title), getString (R.string.news_one_desc),
R.mipmap.news_one));
      newsList.add(new
News (getString (R.string.news_two_title), getString (R.string.news_two_desc),
R.mipmap.news_two));
      newsList.add(new
News(getString(R.string.news_three_title),getString(R.string.news_three_desc),
R.mipmap.news_three));
      newsList.add(new
News(getString (R.string.news_four_title), getString (R.string.news_four_desc),
R.mipmap.news_four));
    }
```

```
}
```

（9）RecyclerView_Card 项目的运行结果如图 5-19 所示。

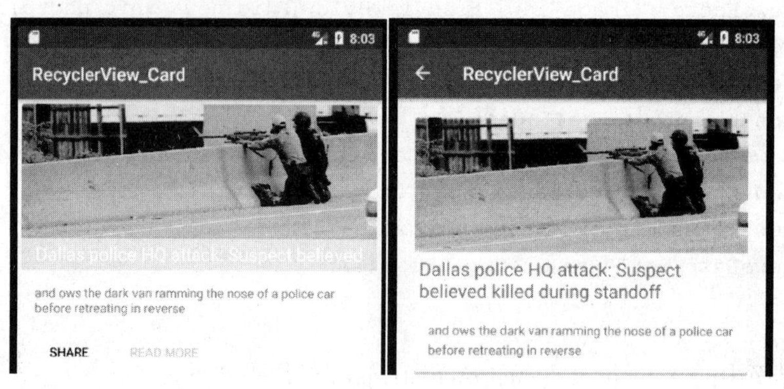

图 5-19　RecyclerView_Card 项目的运行结果

5.7 Android 7.0 新工具类：DiffUtil

DiffUtil 是 support-v7：24.2.0 中的新工具类，用来比较两个数据集，寻找出旧数据集与新数据集的最小变化量。说到数据集，相信大家已经知道它与 RecyclerView 相关。

RecyclerView 刷新时，调用 RecyclerView 的 Adapter 类的方法 notifyDataSetChanged( )，notifyDataSetChanged( )有两个缺点：（1）不会触发 RecyclerView 的动画（删除、新增、位移、change 动画）；（2）性能较低，刷新了一遍整个 RecyclerView，在极端情况下，新老数据集一模一样，效率是最低的。

使用 DiffUtil 后，改为如下代码。

```
DiffUtil.DiffResult diffResult = DiffUtil.calculateDiff(new DiffCallBack(mDatas, newDatas), true);
diffResult.dispatchUpdatesTo(mAdapter);
```

它会自动计算新老数据集的差异，并根据差异情况，自动调用以下四个方法。

```
adapter.notifyItemRangeInserted(position, count);
adapter.notifyItemRangeRemoved(position, count);
adapter.notifyItemMoved(fromPosition, toPosition);
adapter.notifyItemRangeChanged(position, count, payload);
```

显然，这个四个方法在执行时都伴有 RecyclerView 动画，且都是定向刷新方法，刷新效率大幅上升。

我们需要实现一个继承自 DiffUtil.Callback 的类，实现它的四个 Abstract 方法。虽然这个类叫 Callback，但是可以把它理解成定义了一些用来比较新老 Item 是否相等的契约（Contract）、规则（Rule）的类。

DiffUtil.Callback 抽象类如下。

```
public abstract static class Callback {
    public abstract int getOldListSize();//老数据集 size
    public abstract int getNewListSize();//新数据集 size
    public abstract boolean areItemsTheSame(int oldItemPosition, int newI-
```

```
temPosition);//新老数据集在同一个 position 的 Item 是否是一个对象?(可能内容不同,如果这里返回 true,会调用下面的方法)
        public abstract boolean areContentsTheSame(int oldItemPosition, int newItemPosition);//这个方法仅仅是上面方法返回 true 时才会调用,只有 notifyItemRangeChanged()才会调用,判断 item 的内容是否有变化
      //该方法在 DiffUtil 高级用法中用到
        @Nullable
        public Object getChangePayload(int oldItemPosition, int newItemPosition) {
            return null;
        }
}
```

下面以一个实例进行说明。

(1) 在 Android 2.3 中创建应用项目：DiffUtils，项目的构成如图 5-20 所示。

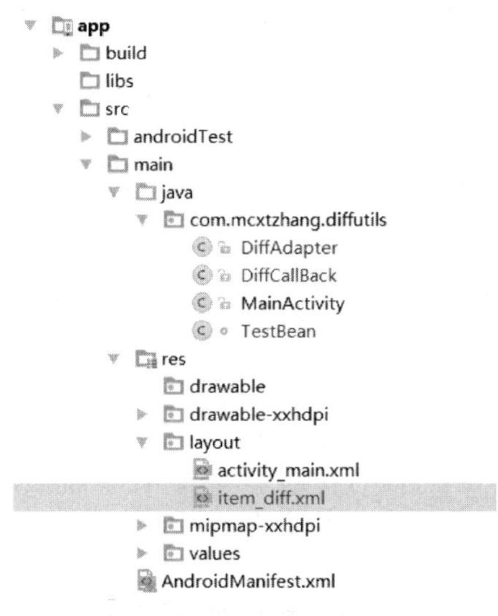

图 5-20　项目构成

(2) 其中有两个布局文件：一个是主布局文件 activity_main. xml，其中包含 RecyclerView 和按钮控件；另一个是 RecyclerView 显示项的布局文件 item_diff. xml，如图 5-21 所示。

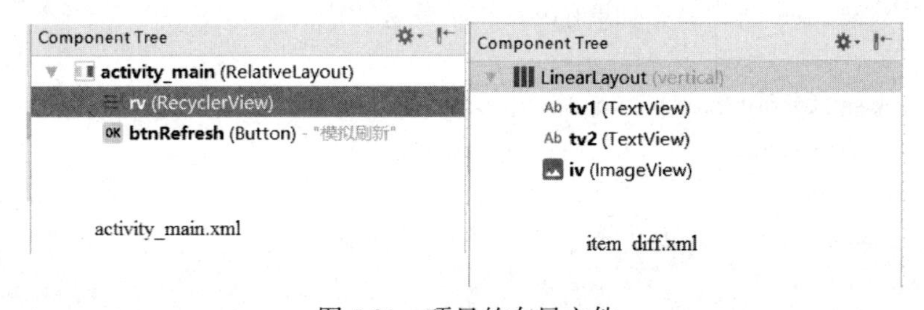

图 5-21　项目的布局文件

(3) java 文件有以下四个：MainAcvity. java——主 Activity 处理类；TestBean. java——RecyclerView 显示的数据；DiffAdapter. java——RecyclerView 的适配器类；DiffCallBack. java——DiffUtil. Callback 的实现。TestBean. java、MainAcvity. java、DiffAdapter 和前面介绍的 Recy-

clerView 相类似，这里主要介绍 DiffCallBack. java，其代码如下。

```
public class DiffCallBack extends DiffUtil. Callback {
  private List < TestBean > mOldDatas, mNewDatas;
  public DiffCallBack(List < TestBean > mOldDatas, List < TestBean > mNewDatas) {
    this. mOldDatas = mOldDatas;
    this. mNewDatas = mNewDatas;
}
//旧数据集大小
@ Override
public int getOldListSize() {
    return mOldDatas ! = null ? mOldDatas. size() : 0;
}
//新数据集大小
@ Override
public int getNewListSize() {
    return mNewDatas ! = null ? mNewDatas. size() : 0;
}
/ * * 被 DiffUtil 调用,用来判断两个对象是否是相同的 Item,本例判断 name 字段是否一致 * /
@ Override
public boolean areItemsTheSame(int oldItemPosition, int newItemPosition) {
    return mOldDatas. get(oldItemPosition).getName().
      equals(mNewDatas. get(newItemPosition).getName());
}
/ * * 被 DiffUtil 调用,用来检查两个 item 是否含有相同的数据
 * DiffUtil 用返回的信息(true/false)来检测当前 item 的内容是否发生了变化
 * DiffUtil 用这个方法替代 equals 方法检查是否相等
  @ Override
  public boolean areContentsTheSame(int oldItemPosition, int newItemPosition){
      TestBean beanOld = mOldDatas. get(oldItemPosition);
      TestBean beanNew = mNewDatas. get(newItemPosition);
      if(! beanOld. getDesc().equals(beanNew. getDesc())) {
          return false;//如果有内容不同,就返回 false
      }
      if(beanOld. getPic() ! = beanNew. getPic()) {
          return false;//如果有内容不同,就返回 false
      }
          return true; //默认两个 data 内容是相同的
      }
      / * * 返回一个代表着新老 item 的改变内容的 payload 对象
      public Object getChangePayload(int oldItemPosition, int newItemPosition) {
          TestBean oldBean = mOldDatas. get(oldItemPosition);
          TestBean newBean = mNewDatas. get(newItemPosition);
          Bundle payload = new Bundle();
```

```
            if(! oldBean.getDesc().equals(newBean.getDesc())) {
                payload.putString("KEY_DESC", newBean.getDesc());
            }
            if (oldBean.getPic() ! = newBean.getPic()) {
                payload.putInt (" KEY_PIC", newBean.getPic());
              }
            if (payload.size() = = 0) //如果没有变化，返回空值
                  return null;
              return payload;
        }
}
```

(4) DiffUtils 项目的运行结果如图 5-22 所示。

图 5-22　DiffUtils 项目的运行结果

5.8 更炫的控件：DrawerLayout——实现侧滑菜单效果

DrawerLayout 是 Support Library 包中实现侧滑菜单效果的控件，可以说 DrawerLayout 是 Google 借鉴第三方控件如 MenuDrawer 等的产物。DrawerLayout 分为侧边菜单和主内容区两部分，侧边菜单可以根据手势展开与隐藏（DrawerLayout 自身特性），主内容区的内容可以随着菜单的单击而变化。

做抽屉菜单的时候，左边滑出来的那一部分的布局是由自己来定义的，多花点时间能做出比较美观的侧拉菜单，但是要耗费很多时间，于是 Google 在 5.0 之后推出了 NavigationView，Google 最新推出规范式设计中的 NavigationView 和 DrawerLayout 结合可以实现侧滑菜

单栏效果，就是左边滑出来的那个菜单。这个菜单整体上分为两部分，上面一部分叫作 HeaderLayout，下面的那些单击项都是 menu，如图 5-23 所示。

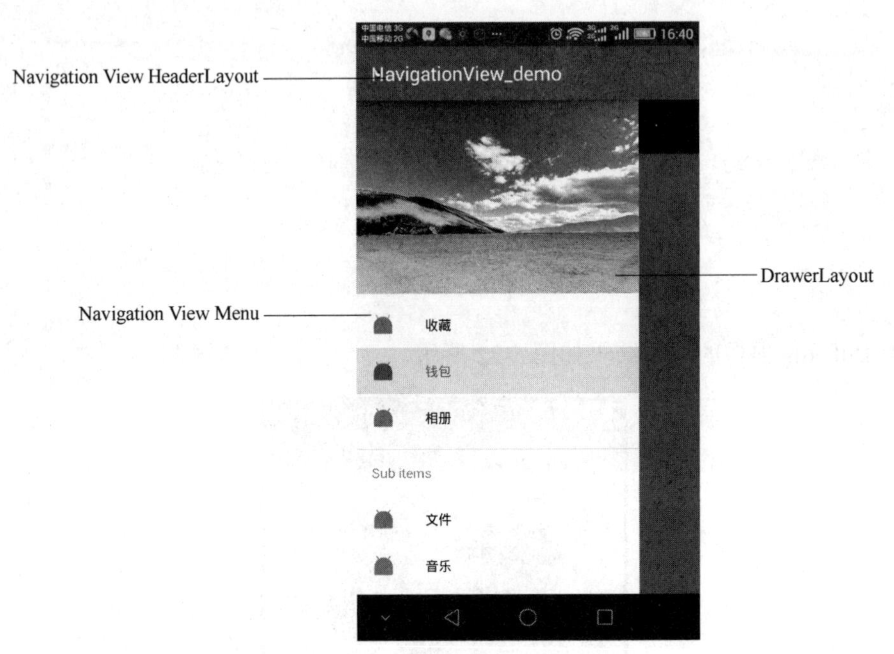

图 5-23　抽屉菜单

下面是一个采用导航菜单的 Android 应用程序的主 Activity 界面实例。

```
<? xml version = "1.0" encoding = "utf-8"? >
<android. support. v4. widget. DrawerLayout
    xmlns:android = "http://schemas. android. com/apk/res/android"
    xmlns:app = "http://schemas. android. com/apk/res-auto"
    xmlns:tools = "http://schemas. android. com/tools"
    android:layout_width = " match_parent"
    android: layout_height = " match_parent"
    tools: context = " org. mobiletrain. drawerlayout. MainActivity" >
/ * ------------------------drawerLayout 内容部分--------------- * /
<LinearLayout
        android: layout_width = " match_parent"
        android: layout_height = " match_parent"
        android: orientation = " vertical" >
        <TextView
            android: layout_width = " wrap_content"
            android: layout_height = " wrap_content"
            android: text = " 主页面" / >
    </LinearLayout >
/ * ------------------------drawerLayout 内容部分结束--------------- * /
/ * ------------------------drawerLayout 左侧菜单部分--------------- * /
<android. support. design. widget. NavigationView
```

```
    android: id = " @ + id/navigation_view"
    android: layout_width = " wrap_content"
    android: layout_height = " match_parent"
    android: layout_gravity = " left"    //左侧菜单
    android: fitsSystemWindows = " true"
    app: headerLayout = " @ layout/header_layout"    //头部（上部）
  app: menu = " @ menu/main" >  //菜单（下部）
  </android. support. design. widget. NavigationView >
  / * ------------------------drawerLayout 左侧菜单部分结束--------------- * /
  </android. support. v4. widget. DrawerLayout >
```

DrawerLayout 最好为界面的根布局，主内容区的布局代码要放在侧滑菜单布局的前面，因为 XML 顺序意味着按层叠顺序排序，并且抽屉式导航栏必须位于内容顶部；侧滑菜单部分的布局（这里是 NavigationView）必须设置 layout_gravity 属性，确定侧滑菜单是在左边还是右边，如果不设置，在打开和关闭抽屉时会报错，设置了 layout_gravity = " start/left" 的视图才会被认为是侧滑菜单。

下面通过一个实例进行讲解。在 Android Studio 中新建项目 NavigationView_demo。项目的文件结构，如图 5-24 所示。

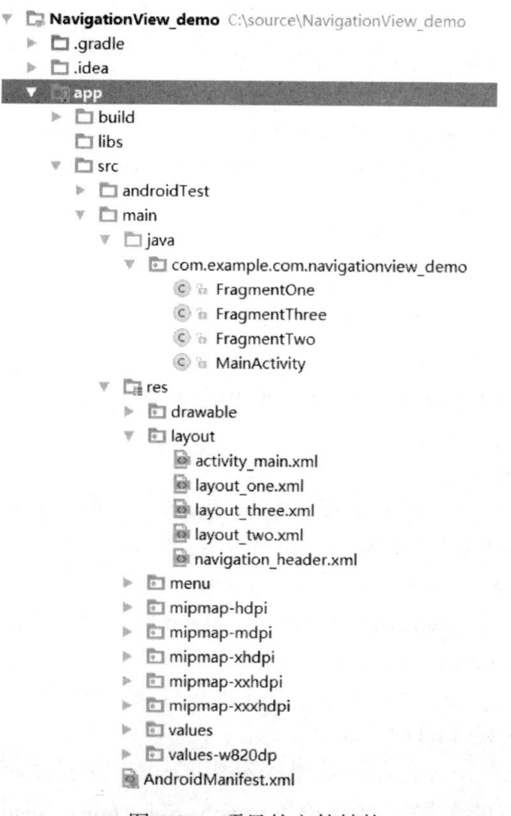

图 5-24　项目的文件结构

（1）编辑 app/build. gradle 文件，在 dependencies 闭包中增加以下内容。

```
dependencies {
```

```
    compile fileTree(dir:'libs', include: ['*.jar'])
    compile'com.android.support:appcompat-v7:25.3.1'
    compile'com.android.support:design:25.3.1'
    testCompile'junit:junit:4.12'
}
```

（2）主 Activity 对应的布局文件 activity_main.xml 的代码如下。

```
<?xml version="1.0" encoding="utf-8"?>
  <android.support.v4.widget.DrawerLayout
    xmlns:android=http://schemas.android.com/apk/res/android
    xmlns:app=http://schemas.android.com/apk/res-auto
    xmlns:tools=http://schemas.android.com/tools
    android:id="@+id/drawer_layout"
    android: layout_width=" match_parent"
    android: layout_height=" match_parent"
    tools: context=" com.example.com.navigationview_demo.MainActivity" >
     <LinearLayout
      android: layout_width=" match_parent"
      android: layout_height=" match_parent"
      android: orientation=" vertical" >
     <android.support.v7.widget.Toolbar
      android: id=" @+id/toolbar"
      android: layout_width=" match_parent"
      android: layout_height="?attr/actionBarSize"
      android: background=" #30469b" />
     <!--内容显示布局-->
     <FrameLayout
      android: id=" @+id/frame_content"
      android: layout_width=" match_parent"
      android: layout_height=" match_parent" >
     </FrameLayout>
     </LinearLayout>
     <android.support.design.widget.NavigationView
      android: id=" @+id/navigation_view"
      android: layout_width=" match_parent"
      android: layout_height=" match_parent"
      android: layout_gravity=" left"
      app: headerLayout=" @layout/navigation_header"
      app: menu=" @menu/drawer" />
     </android.support.v4.widget.DrawerLayout>
```

（3）主 Activity 对应的布局文件 activity_main.xml，app：headerLayout = "@ layout/navigation_header" 所对应的 navigation_header.xml 代码如下。

```
<?xml version="1.0" encoding="utf-8"?>
<LinearLayout xmlns:android=http://schemas.android.com/apk/res/android
```

```
        android:layout_width = " match_parent"
        android: layout_height = " 200dp"
        android: orientation = " vertical" >
        <ImageView
          android: id = " @ +id/iv"
          android: layout_width = " match_parent"
          android: layout_height = " match_parent"
          android: scaleType = " centerCrop"
          android: src = " @drawable/timg" />
  </LinearLayout>
```

(4) 主 Activity 对应的布局文件为 activity_main. xml，app：menu = " @ menu/drawer"，在 res 目录下建立 menu 目录，在 menu 目录下建立对应的 drawer. xml，代码如下。

```
<? xml version = "1.0" encoding = "utf-8"? >
<menu xmlns:android = "http://schemas.android.com/apk/res/android" >
<group android:checkableBehavior = "single" >
<item
    android:id = "@ +id/favorite"
    android:icon = "@mipmap/ic_launcher"
    android: title = " 收藏" />
<item
    android: id = " @ +id/wallet"
    android: icon = " @mipmap/ic_launcher"
    android: title = " 钱包" />
<item
    android: id = " @ +id/photo"
    android: icon = " @mipmap/ic_launcher"
    android: title = " 相册" />
</group>
<item android: title = " Sub items" >
<menu>
  <item
      android: id = " @ +id/file"
      android: icon = " @mipmap/ic_launcher"
      android: title = " 文件" />
  <item
      android: id = " @ +id/music"
      android: icon = " @mipmap/ic_launcher"
    android: title = " 音乐" />
  </menu>
</item>
</menu>
```

menu 可以分组，将 group 的 android:checkableBehavior 属性设置为 single，表示设置该组为单选。

(5) 主 Activity 对应处理文件 MainActivity. java 的代码如下。

```
public class MainActivity extends AppCompatActivity {
    @Override
    protected void onCreate(Bundle savedInstanceState) {
      super.onCreate(savedInstanceState);
      setContentView(R.layout.activity_main);
      //设置 ToolBar
      final Toolbar mToolbar = (Toolbar) findViewById (R.id.toolbar);
    mToolbar.setTitleTextColor (Color.WHITE);
      //设置抽屉 DrawerLayout
  final DrawerLayout mDrawerLayout = (DrawerLayout) findViewById (R.id.drawer_
layout);
  //ActionBarDrawerToggle 是 DrawerLayout.DrawerListener 实现
  ActionBarDrawerToggle mDrawerToggle = new ActionBarDrawerToggle (this, mDraw-
erLayout, mToolbar,          R.string.drawer_open, R.string.drawer_close);
  mDrawerToggle.syncState(); //初始化状态
  mDrawerLayout.setDrawerListener (mDrawerToggle);
  //设置导航栏 NavigationView 的单击事件
  NavigationView mNavigationView = (NavigationView) findViewById (R.id.navigation_view);
  mNavigationView.setNavigationItemSelectedListener (new NavigationView. OnNavigation-
ItemSelectedListener() {
      @Override
     public boolean onNavigationItemSelected (MenuItem menuItem) {
       switch (menuItem.getItemId())
        {
         case R.id.favorite:
  getSupportFragmentManager ( ). beginTransaction ( ). replace (R.id.frame_con-
tent, new FragmentOne()).commit();
  mToolbar.setTitle (" 我的爱好");
              break;
  case R.id.wallet:
      getSupportFragmentManager().beginTransaction().replace (R.id.frame_con-
tent, new FragmentTwo()).commit();
      mToolbar.setTitle (" 我的钱包");
  break;
  case R.id.photo:
      getSupportFragmentManager().beginTransaction().replace (R.id.frame_con-
tent, new FragmentThree()).commit();
      mToolbar.setTitle (" 照片");
  break;
  }
  menuItem.setChecked (true); //单击将其设为选中状态
  mDrawerLayout.closeDrawers(); //关闭抽屉
```

```
    return true;
      }
    });
    }
}
```

其中，getSupportFragmentManager(). beginTransaction(). replace (R. id. frame_ content, new FragmentOne()). commit()打开一个视图类 FragmentOne()，显示在 activity_main. xml 文件中控件 ID 为 frame_content，FragmentOne. java 在源代码目录下建立，其代码如下。

```
public class FragmentOne extends Fragment {
public FragmentOne() {
}
@Override
public View onCreateView(LayoutInflater inflater, ViewGroup container,
                         Bundle savedInstanceState) {
        return inflater.inflate(R.layout.layout_one, container, false);
  }
}
```

其中，layout_one 是布局文件，在 layout 目录下，layout_one. xml 的内容如下。

```
<?xml version="1.0" encoding="utf-8"?>
<RelativeLayout xmlns:android=http://schemas.android.com/apk/res/android
  android:layout_width=" match_parent"
android: layout_height=" match_parent" >
<TextView
  android: layout_width=" wrap_content"
  android: layout_height=" wrap_content"
  android: layout_centerInParent=" true"
  android: text=" 我的爱好"
  android: textSize=" 25sp" />
</RelativeLayout>
```

其他两个文件与此类似：FragmentTwo. java、FragmentThree. java、layout_two. xml、layout_three. xml。

(6) 运行结果如图 5-25 所示。

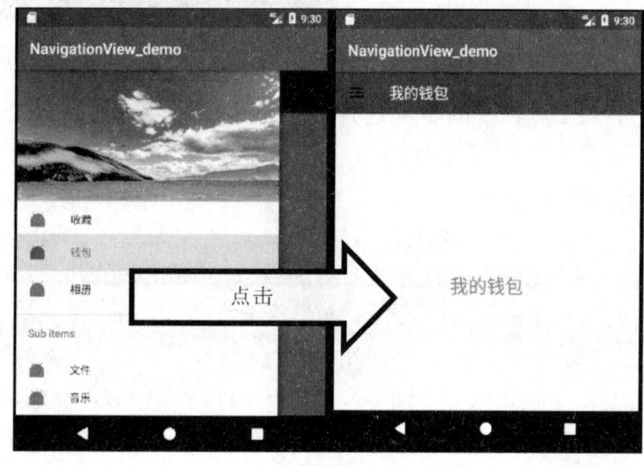

图 5-25　项目运行结果

5.9 对话框

在 Android 开发中，经常需要在 Android 界面中弹出一些对话框，比如让用户确认或者让用户选择，这些对话框称为 Android Dialog。

5.9.1 常用对话框

下面介绍一些常用的对话框。

(1) 在 Android 2.3 中创建应用项目：Dialog_Test，主布局如图 5-26 所示。

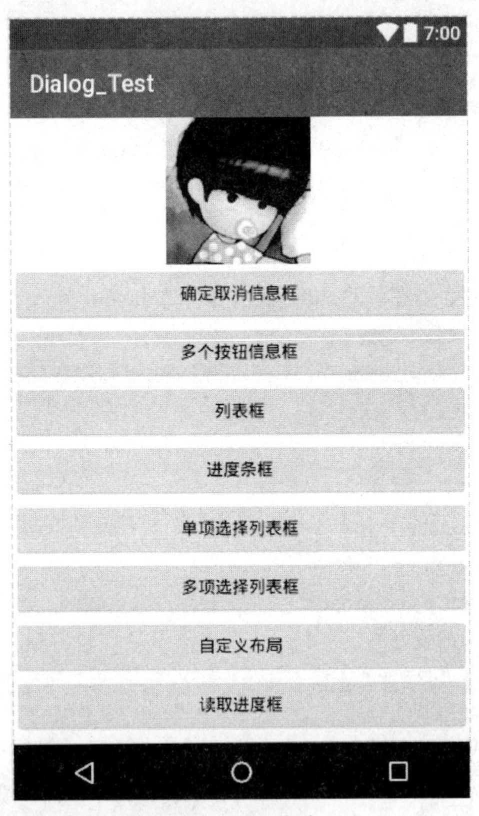

图 5-26 项目主布局

(2) 其中的单项选择列表框的代码如下。

```
final String[] mItems = {"item0","item1","itme2","item3","itme4","item5","i-
tem6"};
AlertDialog.Builder builder
builder.setIcon(R.mipmap.ic_launcher);
builder.setTitle("单项选择");
builder.setSingleChoiceItems(mItems, 0, new DialogInterface.OnClickListener() {
    public void onClick(DialogInterface dialog, int whichButton) {
        mSingleChoiceID = whichButton;
        showDialog("你选择的 id 为" +whichButton +", " +mItems[whichButton]);
```

```
}
});
builder.setPositiveButton("确定", new DialogInterface.OnClickListener() {
    public void onClick(DialogInterface dialog, int whichButton) {
        if(mSingleChoiceID > 0) {
            showDialog("你选择的是" +mSingleChoiceID);
        }
    }
});
builder.setNegativeButton("取消", new DialogInterface.OnClickListener() {
    public void onClick(DialogInterface dialog, int whichButton) {
    }
});
```

显示结果如图 5-27 所示。

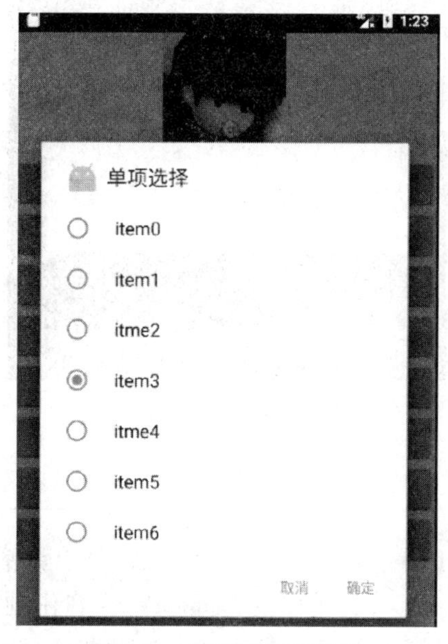

图 5-27　单项选择列表框

(3) 其中的自定义布局的代码如下。

```
LayoutInflater factory = LayoutInflater.from(this);
final View textEntryView = factory.inflate(R.layout.test, null);
builder.setIcon(R.mipmap.ic_launcher);
builder.setTitle (" 自定义输入框");
builder.setView (textEntryView);
builder.setPositiveButton (" 确定", new DialogInterface.OnClickListener() {
    public void onClick (DialogInterface dialog, int whichButton) {
        EditText userName = (EditText) textEntryView.findViewById (R.id.etUserName);
        EditText password = (EditText) textEntryView.findViewById (R.id.etPassWord);
        showDialog (" 姓名 :"  + userName.getText().toString()  + " 密码:" +
```

```
password.getText().toString());
        }
    });
    builder.setNegativeButton ("取消", new DialogInterface.OnClickListener() {
        public void onClick (DialogInterface dialog, int whichButton) {
        }
    });
```

显示结果如图 5-28 所示。

图 5-28　自定义布局

5.9.2　MDDialog

MDDialog 是一款 Material Designed 风格的 Dialog，可以灵活定制其内容以及显示方式，例如，可以添加中间的 ContentView，可以为 ContentView 自由地添加代码操作。对于多个选项风格的 Dialog，MDDialog 提供了直接设置多个选项的功能，并提供了精细的 UI 按下效果、单击回调函数等。

这款 Material Designed 风格的 Dialog 的设计灵感来自于 MD 设计理念，可以通过使用和 AlertDialog 相似的代码来构建 MDDialog。

MDDialog 具有多种有趣的属性。

(1) 可以设置显示/隐藏 title、显示/隐藏“确定/取消”按钮（或者同时隐藏两个 Button，具体 UI 效果可见微信的选择对话框）。

(2) 可以向 MDDialog 添加一个自定义的 View，同时可以在构建 MDDialog 时，使用 setContentViewOperator (...) 函数，添加操作自定义 View 的代码。

(3) 可以为 MDDialog 设置 String [ ] messages，构建 MDDialog 的 Builder 中提供了响应单击每一个 String 的回调函数，即，通过 setOnItemClickListener (...) 设置单击每一条目的回调。

（4）可以自由定制 MDialog 的四个角的半径大小。

（5）MDDialog 自动为每个 message 提供了按下效果，且按下效果会随着此 item 是否具有圆角而改变。

（6）可以通过两种方式设置 MDDialog 的宽度，可以设置宽度占整个屏幕宽度的比值，也可以设置精确的尺寸。

要使用 MDDialog，需要在项目的 build. gradle（在 Project 模式下内层的 build. gradle）添加相应的依赖库，打开 build. gradle 文件，在 dependencies 闭包中添加如下内容。

```
dependencies {
    compile fileTree(dir:'libs', include: ['*.jar'])
    compile 'com.android.support:appcompat-v7:25.3.1'
    compile 'com.android.support.constraint:constraint-layout:1.0.2'
    compile 'cn.carbs.android:MDDialog:1.0.0'
}
```

MDDialog 对话框的内容可以通过自定义视图实现，如下所示。

```
<?xml version="1.0" encoding="utf-8"?>
  <LinearLayout xmlns:android="http://schemas.android.com/apk/res/android"
  android:layout_width=" match_parent"
  android: layout_height=" match_parent"
  android: orientation=" vertical"    >
      <LinearLayout
        android: layout_gravity=" center"
        android: orientation=" horizontal"
        android: layout_width=" match_parent"
        android: layout_height=" wrap_content" >
      <TextView
        android: layout_width=" wrap_content"
        android: layout_height=" wrap_content"
        android: textSize=" 20dp"
        android: id=" @+id/messg1"
        android: text=" 输入内容 1" />
      <EditText
        android: id=" @+id/et1"
        android: layout_width=" match_parent"
        android: layout_height=" wrap_content" />
      </LinearLayout>
      <LinearLayout
        android: layout_gravity=" center"
        android: orientation=" horizontal"
        android: layout_width=" match_parent"
        android: layout_height=" wrap_content" >
       <TextView
          android: layout_width=" wrap_content"
```

```
            android: layout_height = " wrap_content"
            android: textSize = " 20dp"
            android: id = " @ + id/messg2"
            android: text = " 输入内容 2" / >
        < EditText
            android: id = " @ + id/et2"
            android: layout_width = " match_parent"
            android: layout_height = " wrap_content" / >
    < /LinearLayout >
  < /LinearLayout >
```

上面的布局文件名为 customizedview. xml，显示对话框的代码如下。

```
new MDDialog. Builder (MainActivity. this)
    . setContentView (R. layout. customizedview) //test 为自定义 view 文件名
    . setContentViewOperator (new MDDialog. ContentViewOperator () {
            @ Override
     public void operate (View contentView) {//这里的 contentView 就是上面代码中传入的
自定义的 View 或者 layout 资源 inflate 出来的 View
                EditText et  = (EditText) contentView. findViewById (R. id. et1);
                et. setHint ("hint set in operator");
            }
            })
        . setTitle ("添加")
        . setNegativeButton (new View. OnClickListener () {
            @ Override
          public void onClick (View v) {
              }
          })
          . setPositiveButton (new View. OnClickListener () {
            @ Override
            public void onClick (View v) {
                        }
          })
          . setPositiveButtonMultiListener (new MDDialog. OnMultiClickListener () {
              @ Override
              public void onClick (View clickedView, View contentView) {
            //这里的 contentView 就是上面代码中传入的自定义的 View 或者 layout 资源 in-
flate 出来的 View,目的是方便在确定/取消按键中对 contentView 进行操作,如获取数据等
                EditText et  = (EditText) contentView. findViewById (R. id. et1);
                 Toast. makeText (getApplicationContext (), "edittext 0 : " + et. getText (),
Toast. LENGTH_SHORT). show ();
                    }
              })
            . setNegativeButtonMultiListener  (new MDDialog. OnMultiClickListener () {
```

```
                @Override
                public void onClick (View clickedView, View contentView) {
                     EditText et = (EditText) contentView.findViewById (R.id.et2);
                     Toast.makeText (getApplicationContext (), " edittext 1 : " +
et.getText (), Toast.LENGTH_SHORT). show ();
                     }
                })
                .setWidthMaxDp (600)
                .setShowTitle (false) //default is true
                .setShowButtons (true) //default is true
                .create ()
                .show ();
```

对话框显示结果如图 5-29 所示。

图 5-29　MDDialog 示例

5.10 本章小结

本章针对 Android 界面的设计，在前面章节介绍内容的基础上，讲解了一些更复杂、更高级的界面设计，包括 Android 的一些新控件的使用方法，通过本章的学习，可以设计出更美观的界面。

第六章

数据持久化方案

数据存储在开发中是非常频繁的，任何应用程序都需要通过文件系统存储文件，Android提供了6种持久化应用程序数据的方式，这6种方式分别用于不同的情况下，具体选择方式取决于具体的需求。

6.1 轻量级存储：SharedPreferences——实现“记住密码”功能

SharedPreferences是Android平台上一个轻量级的存储类，主要用于保存一些常用的配置参数，采用XML文件存放数据，文件存放在/data/data <package name >/shared_prefs目录下。

SharedPreferences是一个接口，在这个接口里没有提供写入数据和读取数据的能力，它通过其Editor接口中的一些方法来操作SharedPreference，用法如下。

（1）应用Context. getSharedPreferences（String name，int mode）得到一个SharedPreferences实例。其中参数含义如下。

name：指文件名称，不需要加后缀 . xml，系统会自动添加。

mode：用于指定读写方式，其值有如下四种。

- Context. MODE_PRIVATE：为默认操作模式，代表该文件是私有数据，只能被应用本身访问，在该模式下，写入的内容会覆盖原文件的内容，如果想把新写入的内容追加到原文件中，可以使用Context. MODE_AppEND。
- Context. MODE_APPEND：模式会检查文件是否存在，若存在，则往文件追加内容，否则创建新文件。
- Context. MODE_WORLD_READABLE和Context. MODE_WORLD_WRITEABLE：用来控制其他应用是否有权限读写该文件，从API 17开始已经过期。

（2）调用SharedPreferences对象的edit()方法获取一个SharedPreferences. Editor对象。

（3）向SharedPreferences. Editor对象中添加数据，比如使用putBoolean方法添加一个布尔型数据，使用putString()方法添加一个字符串，以此类推。

（4）调用commit()方法将添加的数据提交，从而完成数据存储操作。

SharedPreferences接口非常适合用来存储零散的数据，这里用来实现记录用户名和密码的功能。也可以使用IO流来实现记住密码的功能，使用SharedPreferences接口会比用IO流更加方便，代码更加简洁，也更高效。

下面是 LsitView 控件高级应用实例，实现 SharedPreferences 记住用户名和密码。

（1）在 Android 2.3 中创建应用项目：SharedPreferences_Demo。其布局文件有 3 个，即主 Activity 对应的布局文件 login.xml、登录界面 logo.xml 和登录成功界面 welcome.xml，如图 6-1 所示。

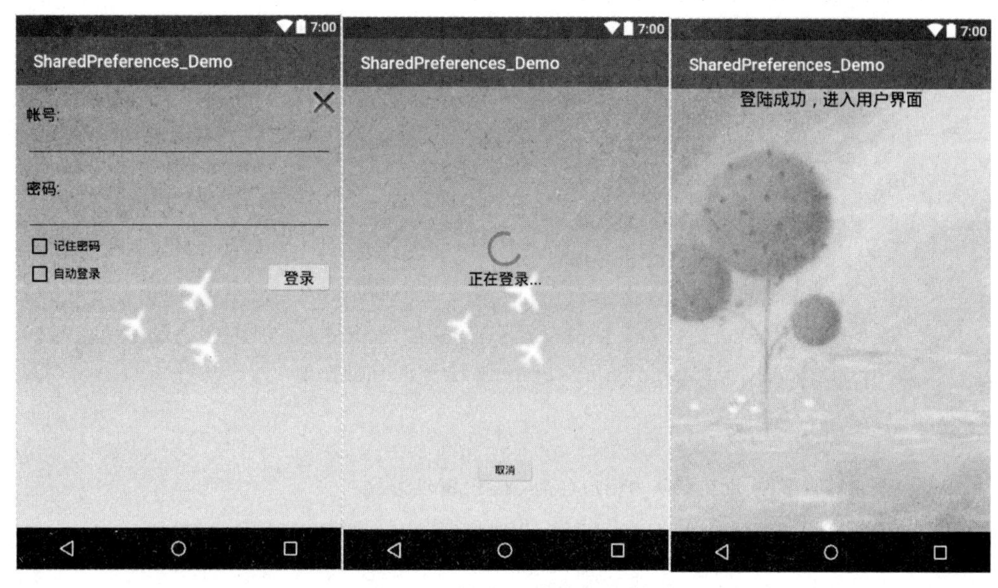

图 6-1 项目的 3 个布局文件

（2）主界面的 Activity 文件 LoginActivity.java 的代码如下。

```
public class LoginActivity extends AppCompatActivity {
  private EditText userName, password;
  private CheckBox rem_pw, auto_login;
  private Button btn_login;
  private ImageButton btnQuit;
  private String userNameValue,passwordValue;
  private SharedPreferences sp;
  public void onCreate(Bundle savedInstanceState) {
      super.onCreate(savedInstanceState);
      //去除标题
      this.requestWindowFeature(Window.FEATURE_NO_TITLE);
      setContentView(R.layout.login);
      //获得实例对象
      sp = this.getSharedPreferences("userInfo", Context.MODE_PRIVATE);
      userName = (EditText) findViewById(R.id.et_zh);
      password = (EditText) findViewById(R.id.et_mima);
      rem_pw = (CheckBox) findViewById(R.id.cb_mima);
      auto_login = (CheckBox) findViewById(R.id.cb_auto);
      btn_login = (Button) findViewById(R.id.btn_login);
```

```
        btnQuit =(ImageButton)findViewById(R.id.img_btn);
        //判断记住密码多选框的状态
        if(sp.getBoolean("ISCHECK", false))
        {
          //设置默认是记录密码状态
          rem_pw.setChecked(true);
          userName.setText(sp.getString("USER_NAME", ""));
          password.setText(sp.getString("PASSWORD", ""));
          //判断自动登录多选框状态
          if(sp.getBoolean("AUTO_ISCHECK", false))
          {
              //设置默认是自动登录状态
              auto_login.setChecked(true);
              //跳转界面
              Intent intent = new Intent(LoginActivity.this,LogoActivity.class);
              LoginActivity.this.startActivity(intent);
          }
        }
        //登录监听事件,现在默认为用户名为 user,密码为 123
        btn_login.setOnClickListener(new OnClickListener() {
          public void onClick(View v) {
              userNameValue = userName.getText().toString();
              passwordValue = password.getText().toString();
              if(userNameValue.equals("user")&&passwordValue.equals("123"))
              {
                Toast.makeText(LoginActivity.this,"登录成功", Toast.LENGTH_SHORT).show();
                //登录成功和记住密码框为选中状态才保存用户信息
                if(rem_pw.isChecked())
                {
                    //记住用户名、密码
                    Editor editor = sp.edit();
                    editor.putString("USER_NAME", userNameValue);
                    editor.putString("PASSWORD",passwordValue);
                    editor.commit();
                }
                //跳转界面
                Intent intent = new Intent(LoginActivity.this,LogoActivity.class);
                LoginActivity.this.startActivity(intent);
                //finish();
              }else{
                Toast.makeText(LoginActivity.this,"用户名或密码错误,请重新登录",
Toast.LENGTH_LONG).show();
              }
```

```
            }
        });
        //监听记住密码多选框按钮事件
        rem_pw.setOnCheckedChangeListener(new OnCheckedChangeListener() {
          public void onCheckedChanged(CompoundButton buttonView,boolean isChecked) {
              if(rem_pw.isChecked()) {
                System.out.println("记住密码已选中");
                sp.edit().putBoolean("ISCHECK", true).commit();
              }else {
                System.out.println("记住密码没有选中");
                sp.edit().putBoolean("ISCHECK", false).commit();
            }
          }
        });
        //监听自动登录多选框事件
        auto_login.setOnCheckedChangeListener(new OnCheckedChangeListener() {
          public void onCheckedChanged(CompoundButton buttonView,boolean isChecked) {
              if(auto_login.isChecked()) {
                System.out.println("自动登录已选中");
                sp.edit().putBoolean("AUTO_ISCHECK", true).commit();
              } else {
                System.out.println("自动登录没有选中");
                sp.edit().putBoolean("AUTO_ISCHECK", false).commit();
              }
          }
        });
        btnQuit.setOnClickListener(new OnClickListener() {
          @Override
          public void onClick(View v) {
              finish();
          }
        });
    }
}
```

(3) 登录界面对应的 Activity 的文件 LogoActivity. java 的代码如下。

```
public class LogoActivity extends Activity {
private ProgressBar progressBar;
private Button backButton;
    protected void onCreate(Bundle savedInstanceState) {
        super.onCreate(savedInstanceState);
        //去除标题
        this.requestWindowFeature(Window.FEATURE_NO_TITLE);
        setContentView(R.layout.logo);
```

```
        progressBar = (ProgressBar) findViewById(R.id.pgBar);
        backButton = (Button) findViewById(R.id.btn_back);
          Intent intent = new Intent(this, WelcomeAvtivity.class);
        LogoActivity.this.startActivity(intent);
          backButton.setOnClickListener(new OnClickListener() {
              @Override
            public void onClick(View v) {
                finish();
            }
        });
      }
  }
```

（4）登录成功界面对应的 Activity 的文件 WelcomeAvtivity.java 的代码如下。

```
public class WelcomeAvtivity extends Activity {
    @Override
protected void onCreate(Bundle savedInstanceState) {
    // TODO Auto-generated method stub
    super.onCreate(savedInstanceState);
    setContentView(R.layout.welcome);
    }
}
```

（5）运行结果如图 6-2 所示。

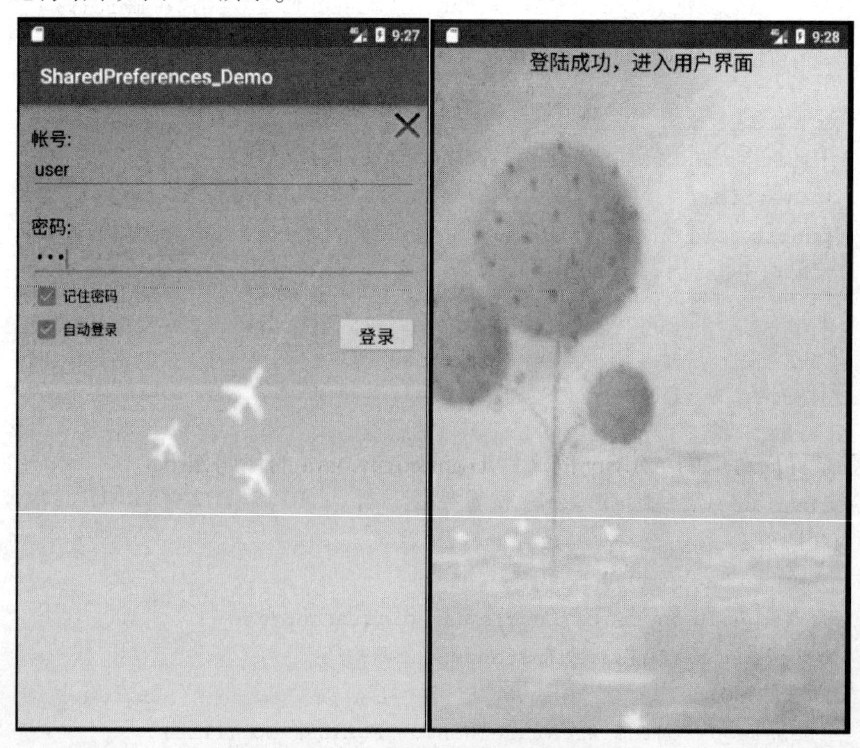

图 6-2　SharedPreferences 项目的运行结果

6.2 结构化数据存储——SQLite

Android 中通过 SQLite 数据库引擎来实现结构化数据存储，Android 提供了对 SQLite 数据的完全支持。应用中创建的任何数据库都能够通过类名来访问，下面看看 Android 的 SQLite 数据库是如何使用的。

6.2.1 SQLite 简介

SQLite 是一款开源的、轻量级的、嵌入式的、关系型的数据库，于 2000 年由 D. Richard Hipp 发布，可以支持 Java、Net、PHP、Ruby、Python、Perl、C 等几乎所有现代编程语言，支持 Windows、Linux、Unix、Mac OS、Android、IOS 等几乎所有主流操作系统平台。目前发布的版本是 SQLite3. 18. 0，简称 SQLite3，网址：http://www. sqlite. org/download. html。

SQLite 具有如下特性。

（1）事物处理原子性、一致性、独立性和持久性（ACID）。

（2）零配置，无需安装和管理配置。

（3）是储存在单一磁盘文件中的一个完整的数据库。

（4）数据库文件可以在不同字节顺序的机器间自由共享。

（5）支持数据库大小至 2TB。

（6）足够小，3 万行 C 代码约 250K。

（7）比一些流行的数据库操作速度更快。

（8）简单、轻松的 API。

（9）包含 TCL 绑定，同时通过 Wrapper 支持其他语言的绑定。

（10）良好注释的源代码，并且有 90% 以上的测试覆盖率。

（11）独立，没有额外依赖。

（12）Source 完全 Open，可以用于任何用途，包括出售。

（13）支持多种开发语言：C、PHP、Perl、Java、ASP. NET、Python。

SQLite 是一款内置到移动设备上的轻量型数据库，是遵守 ACID（原子性、一致性、隔离性、持久性）的关联式数据库管理系统，多用于嵌入式系统中。

SQLite 数据库是无类型的，可以向一个 integer 的列中添加一个字符串，但它又支持常见的类型，比如 NULL、VARCHAR、TEXT、INTEGER、BLOB、CLOB 等。

Android 系统内置了 SQLite，并提供了一系列 API 方便对其进行操作，操作步骤如下。

（1）Android 提供了一个 SQLLiteOpenHelper 的类，借助这个类可以对数据库进行创建和升级。

（2）使用 SQLLiteOpenHelper 的对象的 getWritableDatabase( ) 或 getReadableDatabase( ) 返回一个 SQLiteDataBase 的对象。

（3）Android 系统通过 SQLiteDataBase 类对 SQLite 数据库进行访问，该类封装了一些操作数据库的 API，使用该类可以完成对 SQLite 中数据库的添加（Inert）、查询（Query）、更新（Update）和删除（Delete）操作。

6.2.2 创建 SQLite 数据库

SQLiteOpenHelper 是 Android 提供的一个抽象工具类，负责管理数据库的创建、升级工作。如果想创建数据库，就需要自定义一个类继承 SQLiteOpenHelper，然后重写其中的抽象方法：onCreate( )和 onUpdate( )。应用这两个方法创建和升级数据库。

（1）在 Android Studio 中新建项目 SQLite_Test，在项目 SQLite_Test 的源代码目录中新建类文件 MySQLiteOpenHelper. java，代码如下。

```
public class MySQLiteOpenHelper extends SQLiteOpenHelper {
    private static final String DB_NAME = " user. db";
    private static final int VERSION = 1;
    public MySQLiteOpenHelper (Context context, String name, SQLiteDatabase. Cur-
sorFactory factory, int version) {
        super (context, name, factory, version);
    }
    public MySQLiteOpenHelper (Context context) {
        super (context, DB_NAME, null, VERSION);
    }
    @ Override
    public void onCreate (SQLiteDatabase db) {
        //数据库创建
  db. execSQL (" create table person (_id integer primary key autoincrement, "
+" name char (10), phone char (20), money integer (20))");
    }
    @ Override
    public void onUpgrade (SQLiteDatabase db, int oldVersion, int newVersion) {
        //数据库升级
    }
}
```

（2）其布局文件中放置一个按钮，如图 6-3 所示。

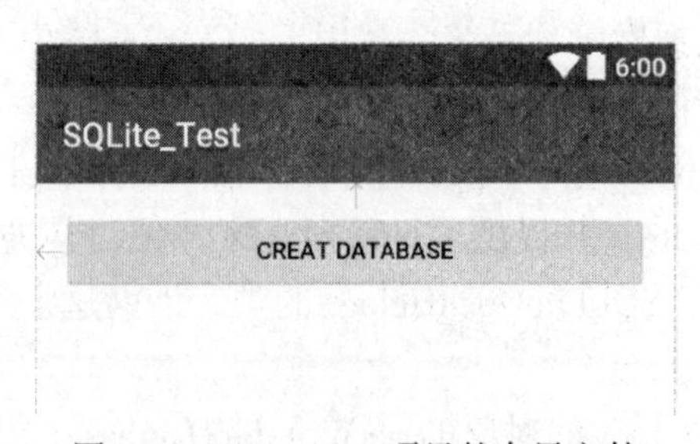

图 6-3　SQLite_Test 项目的布局文件

（3）在实现 Activity 的文件 MainActivity. java 中实现创建数据库，代码如下。

```
public class MainActivity extends AppCompatActivity {
  Button create;
  SQLiteDatabase db;  //SQLiteDatabase 对象
  public String db_name =  " user. db"; //数据库名
```

```
final MySQLiteOpenHelper helper = new MySQLiteOpenHelper (this, db_name, null, 1);
@Override
protected void onCreate (Bundle savedInstanceState) {
  super.onCreate (savedInstanceState);
  setContentView (R.layout.activity_main);
  create = (Button) findViewById (R.id.creat);
  create.setOnClickListener (new View.OnClickListener()
   {
    @Override
    public void onClick (View v) {
      db = helper.getWritableDatabase(); //从辅助类获得数据库对象
    }
  });
 }
}
```

SQLiteOpenHelper 创建数据库的方法如表 6-1 所示。

表 6-1 SQLiteOpenHelper 创建数据库的方法

方　法	说　明
getWritableDatabase()	打开可读写的数据库，没有权限或磁盘已满时会抛出异常
getReadableDatabase()	在磁盘空间不足时打开只读数据库，否则打开可读写数据库；有异常时返回一个只读数据库

（4）运行项目，单击 CREATE DATABASE 按钮，在 Android Device Monitor 窗口中，可以看到新建的数据库文件“user.db”，创建的数据库位于/data/data/包名/databases/目录中，如图 6-4 所示。

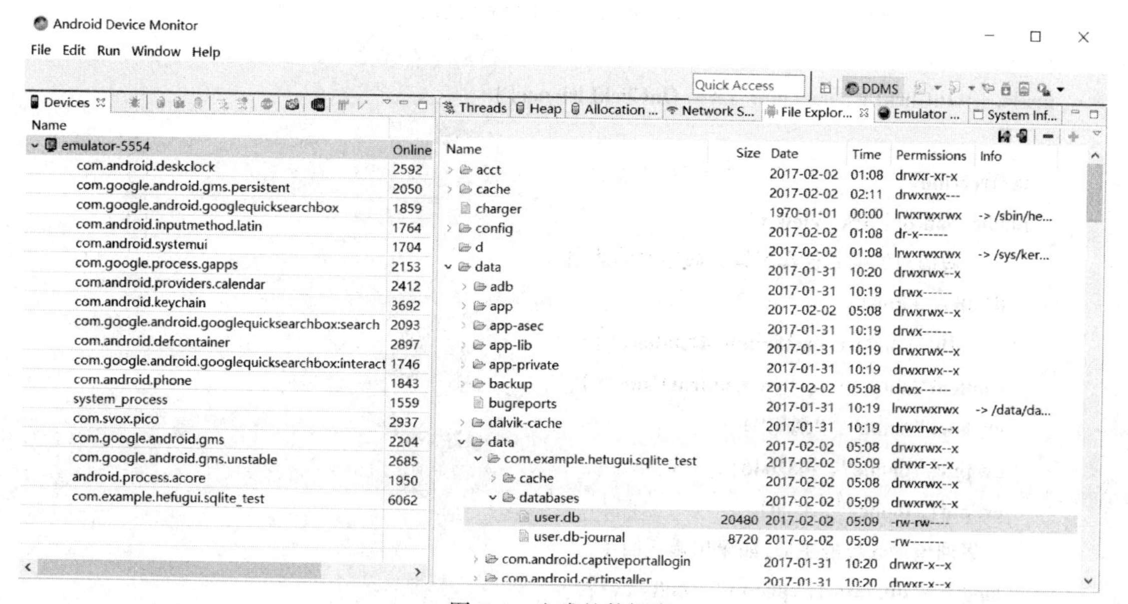

图 6-4 生成的数据库

6.2.3 操作数据库

SQLiteDataBase 操作数据库有如下两种方法。

(1) 执行 SQL 语句实现增删改查，然后使用 SQLiteDataBase 的 execSQL( )进行操作。

例如：

```
db.execSQL("insert into person(name, phone, money) values(?, ?, ?);",
        new Object[]{"张三", 15987461, 75000});
Cursor cursor = db.rawQuery("select _id, name, money from person where name = ?;",
        new String [] {" 张三"});
```

(2) 利用 ContentValues 和 HashTable 操作，ContentValues 和 HashTable 都是一种存储的机制，但是两者最大的区别就在于，ContenValues 只能存储基本类型的数据，如 String、Int 等，不能存储对象；而 HashTable 可以存储对象。然后使用 SQLiteDataBase 的 Insert( )、Update( )、Delete( )、Query( )函数操作。

例如，使用 ContentValues 进行操作示例如下：

```
//以键值对的形式保存要存入数据库的数据
ContentValues cv = new ContentValues();
cv.put("name", "刘化");
cv.put("phone", 1651646);
cv.put("money", 3500);
//返回值是改行的主键,如果出错返回-1
long i = db.insert("person", null, cv)
```

实现步骤如下。

(1) 在实现 Activity 的文件 MainActivity.java 中增加操作数据库的功能，代码如下。

```
public class MainActivity extends AppCompatActivity {
......
insert.setOnClickListener(new View.OnClickListener()
{
  @Override
  public void onClick(View v) {
    //以键值对的形式保存要存入数据库的数据
    if(db = =null)
        db = helper.getWritableDatabase();
    ContentValues cv = new ContentValues();
    cv.put("name", "刘化");
    cv.put("phone", 1651646);
    cv.put("money", 3500);
    //返回值是改行的主键,如果出错返回-1
    long i = db.insert("person", null, cv);
  }
```

```
    });
    delete.setOnClickListener(new View.OnClickListener()
    {
        @Override
        public void onClick(View v) {
          if(db = =null)
                db = helper.getWritableDatabase();
    int i = db.delete("person", "_id = ? and name = ?", new String[]{"1", "张三"});
    update.setOnClickListener(new View.OnClickListener()
    {
      @Override
      public void onClick(View v) {
          if(db = =null)
             db = helper.getWritableDatabase();
          ContentValues cv = new ContentValues();
          cv.put("money", 25000);
          int i = db.update("person", cv, "name = ?", new String[]{"赵四"});
      }
    });
    query.setOnClickListener(new View.OnClickListener()
    {
        @Override
        public void onClick(View v) {
          if(db = =null)
               db = helper.getWritableDatabase();
          Cursor c = db.query("person", null, null, null, null, null, null);
        //cursor.getCount()是记录条数
        //Toast.makeText(MainActivity.this,"当前共有" +c.getCount() +"条记录",Toast.LENGTH_
SHORT).show();
        //循环显示
          for(c.moveToFirst();! c.isAfterLast();c.moveToNext()){
          Toast.makeText(MainActivity.this,"第" + c.getInt(0)    +"条数据,姓名是" +c.getString(1) +",
电话是" +c.getString(2) +",金额是" +c.getInt(3) +"\n",Toast.LENGTH_SHORT).show();
            }
          }
        });
      }
    }
```

（2）运行项目，结果如图6-5所示，单击INSERT按钮，再单击QUERY按钮，出现查询结果。

图 6-5　数据库操作

6.3　实例：SQLite——实现会员功能

在 Android Studio 中新建项目 Member_Manager，开始本节的练习。

（1）在项目 Member_Manager 源代码目录下新建类文件 MemberInfo. java，内容为会员的信息，代码如下。

```
public class MemberInfo {
    public int _id;
    public String name;
    public int age;
    public Stringdepartment;
    public Stringtelephone;
    public MemberInfo() {}
    public MemberInfo (int _id,String name,int age,Stringdepartment,String weibo){
        this._id = _id;
        this.name = name;
        this.age = age;
        this.department = department;
```

```
        this.telephone = telephone;
    }
}
```

（2）在项目 Member_Manager 的源代码目录下新建数据库类文件 DBHelper.java，代码如下。

```
public class DBHelper extends SQLiteOpenHelper {
  public static final String DB_NAME = "Member.db";
  public static final String DB_TABLE_NAME = "info";
  private static final int DB_VERSION =1;
  public DBHelper (Context context) {
      //Context context, String name, CursorFactory factory, int version
      super (context, DB_NAME, null, DB_VERSION);
  }
  //数据第一次创建的时候会调用 onCreate
  @Override
  public void onCreate (SQLiteDatabase db) {
      //创建表
      db.execSQL ("CREATE TABLE IF NOT EXISTS info" +
      " (_id INTEGER PRIMARY KEY AUTOINCREMENT, name VARCHAR, age INTEGER, web-
site STRING, weibo STRING)");
        Log.i ("SQLite", "create table");
  }
  //数据库第一次创建时 onCreate 方法会被调用，当系统发现版本变化之后，会调用 onUpgrade
方法
  @Override
  public void onUpgrade (SQLiteDatabase db, int oldVersion, int newVersion) {
      //在表 info 中增加一列 other
      //db.execSQL ("ALTER TABLE info ADD COLUMN other STRING");
      Log.i ("WIRELESSQA", "update sqlite " +oldVersion +"---->" +newVersion);
  }
}
```

（3）在项目 Member_Manager 的源代码目录下新建数据库类文件 DBManager.java，其中封装了常用的业务方法，代码如下。

```
public class DBManager {
    private DBHelper helper;
    private SQLiteDatabase db;
    public DBManager(Context context){
        helper = new DBHelper(context);
        db = helper.getWritableDatabase();
    }
    /**向表 info 中增加成员信息*/
    public void add(List<MemberInfo> memberInfo) {
```

```
        db.beginTransaction();//开始事务
        try {
          for(MemberInfo info : memberInfo) {
          Log.i("SQLite", "------add memberInfo----------");
          Log.i("SQLite", info.name + "/" + info.age + "/" + info.department + "/" +
info.telephone);
          //向表 info 中插入数据
            db.execSQL("INSERT INTO info VALUES(null,?,?,?,?)", new Object[] {
info.name, info.age, info.department, info.telephone });
        }
        db.setTransactionSuccessful();//事务成功
      } finally {
        db.endTransaction();//结束事务
      }
    }
    public void add(int _id, String name, int age, String department, String tele-
phone) {
      Log.i (" SQLite", " ------add data----------");
        ContentValues cv = new ContentValucs();
      // cv.put (" _id", _id);
      cv.put (" name", name);
      cv.put (" age", age);
      cv.put (" website", department);
      cv.put (" weibo", telephone);
      db.insert (DBHelper.DB_TABLE_NAME, null, cv);
      Log.i (" SQLite", name + " /" + age + " /" + department + " /" + telephone);
    }
    /*通过 name 来删除数据 */
    public void delData (String name) {
        // ExecSQL (" DELETE FROM info WHERE name =" + " '" + name + " '");
      String [] args = { name };
      db.delete (DBHelper.DB_TABLE_NAME, " name = ?", args);
      Log.i (" SQLite", " delete data by " + name);
    }
    /*清空数据 */
    public void clearData() {
          ExecSQL (" DELETE FROM info");
          Log.i (" SQLite", " clear data");
      }
    /**通过名字查询信息，返回所有的数据  */
    public ArrayList<MemberInfo> searchData (final String name) {
        String sql = " SELECT * FROM info WHERE name =" + " '" + name + " '";
        return ExecSQLForMemberInfo (sql);
```

```
    }
    public ArrayList<MemberInfo> searchAllData() {
        String sql = " SELECT * FROM info";
        return ExecSQLForMemberInfo (sql);
    }
    /*通过名字来修改值 */
    public void updateData (String raw, String rawValue, String whereName) {
      String sql = " UPDATE info SET " + raw + " =" + " " + " '" + rawValue + " '" + "
WHERE name =" + " '" + whereName + " '";
        ExecSQL (sql);
        Log.i (" SQLite", sql);
    }
    /*执行 SQL 命令返回 */
    private ArrayList<MemberInfo> ExecSQLForMemberInfo (String sql) {
      ArrayList<MemberInfo> list = new ArrayList<MemberInfo>();
      Cursor c = ExecSQLForCursor (sql);
      while (c.moveToNext()) {
          MemberInfo info = new MemberInfo();
          info._id = c.getInt (c.getColumnIndex (" _id"));
          info.name = c.getString (c.getColumnIndex (" name"));
          info.age = c.getInt (c.getColumnIndex (" age"));
          info.department = c.getString (c.getColumnIndex (" department"));
          info.telephone = c.getString (c.getColumnIndex (" telephone"));
          list.add (info);
      }
      c.close();
      return list;
    }
    /*执行一个 SQL 语句 */
    private void ExecSQL (String sql) {
      try {
          db.execSQL (sql);
          Log.i (" execSql: ", sql);
        } catch (Exception e) {
          Log.e (" ExecSQL Exception", e.getMessage());
            e.printStackTrace();
        }
    }
    /*执行 SQL，返回一个游标 */
    private Cursor ExecSQLForCursor (String sql) {
        Cursor c = db.rawQuery (sql, null);
        return c;
    }
```

```
    public void closeDB() {
        db.close();
    }
}
```

(4) 在项目 Member_Manager 的源代码目录下新建数据库类文件 DisplayActivity. java，显示查询会员的结果，代码如下。

```
/* 显示结果 */
public class DisplayActivity extends MainActivity{
    private String result = null;
    private TextView display = null;
    @Override
    protected void onCreate(Bundle savedInstanceState) {
        super.onCreate(savedInstanceState);
        setContentView(R.layout.activity_display);
        Bundle extras = getIntent().getExtras();
        result = extras.getString (" searchResult");
        display = (TextView) findViewById (R.id.display_txt);
        display.setText (result);
    }
```

(5) 在项目 Member_Manager 的 layout 目录下建立两个布局文件，一个是主 Activity 对应的布局文件 activity_main. xml，另一个是会员信息显示界面 activity_dispaly. xml，如图 6-6 所示。

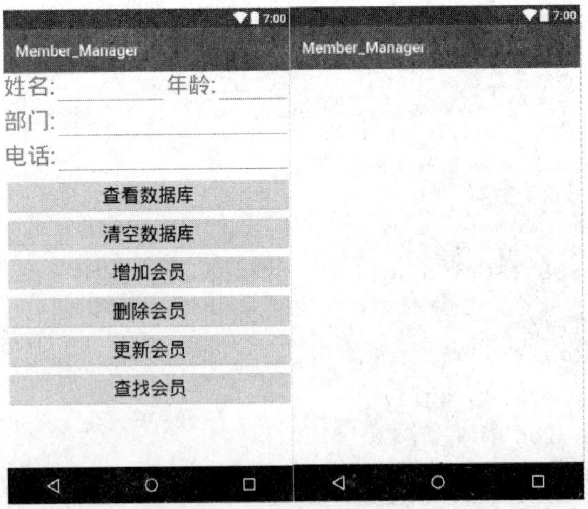

图 6-6　项目 Member_Manager 的两个布局文件

(6) 主界面的 Activity 文件 MainActivity. java 的代码如下。

```
public class MainActivity extends AppCompatActivity {
    private EditText edit_name = null;
    private EditText  edit_age = null;
    private EditText  edit_department = null;
    private EditText  edit_telephone  = null;
```

```
  private Button searchAll,clear,add,delete,update,search;
  private String    name = null;
  private int       age = 0;
  private String    department = null;
  private String    telephone = null;
  private DBManager dbManager;
  @Override
protected void onCreate(Bundle savedInstanceState) {
    super.onCreate(savedInstanceState);
    setContentView(R.layout.activity_main);
    //初始化 DBManager
    dbManager = new DBManager(this);
    edit_name = (EditText) findViewById(R.id.name_edit);
    edit_age = (EditText) findViewById(R.id.age_edit);
    edit_department = (EditText) findViewById(R.id.department_edit);
    edit_telephone = (EditText) findViewById(R.id.telephone_edit);
      add = (Button) findViewById(R.id.add);
    //监听增加会员按钮
    add.setOnClickListener(new View.OnClickListener() {
        @Override
        public void onClick(View v) {
            name = edit_name.getText().toString();
            age = Integer.valueOf(edit_age.getText().toString());
            department = edit_department.getText().toString();
            telephone = edit_telephone.getText().toString();
            ArrayList<MemberInfo> infoList = new ArrayList<MemberInfo>();
            MemberInfo m = new MemberInfo();
            m.age = age;
            m.name = name;
            m.department = department;
            m.telephone = telephone;
            infoList.add(m);
            dbManager.add(infoList);
        }
    });
    //查询数据库里的所有数据
    searchAll = (Button) findViewById(R.id.all);
    searchAll.setOnClickListener(new View.OnClickListener() {
        @Override
      public void onClick(View v) {
          ArrayList<MemberInfo> infoList = new ArrayList<MemberInfo>();
          infoList = dbManager.searchAllData();
```

```
            String result = "";
            for(MemberInfo info : infoList) {
                result = result + String.valueOf(info._id) + " |" + info.name + " |" +
String.valueOf(info.age) + " |" + info.department + " |" + info.telephone;
                result = result + "\n" + "----------------------------------------" + "\n";
            }
            Log.i("SQLite", result);
            startDisplayActivity("searchResult", result);
        }
    });
    //通过一个会员的名字来删除一个会员信息
    delete = (Button) findViewById(R.id.del);
    delete.setOnClickListener(new View.OnClickListener() {
        @Override
        public void onClick(View v) {
            name = edit_name.getText().toString();
            dbManager.delData(name);
        }
    });
    //清空会员信息
    clear = (Button) findViewById(R.id.clear);
    clear.setOnClickListener(new View.OnClickListener() {
        @Override
        public void onClick(View v) {
            dbManager.clearData();
        }
    });
    //更新会员信息
    update = (Button) findViewById(R.id.update);
    update.setOnClickListener(new View.OnClickListener() {
        @Override
        public void onClick(View v) {
            name = edit_name.getText().toString();
            age = Integer.valueOf(edit_age.getText().toString());
            department = edit_department.getText().toString();
            telephone = edit_telephone.getText().toString();
            if(name == null) {
                    Toast.makeText(getApplicationContext(), "name 不能为空",
Toast.LENGTH_LONG).show();
            } else {
                dbManager.updateData("age", String.valueOf(age), name);
                dbManager.updateData("department", department, name);
```

```
                dbManager.updateData("telephone", telephone, name);
            }
        }
    });
    //通过姓名搜索会员
    search = (Button) findViewById(R.id.search);
    search.setOnClickListener(new View.OnClickListener() {
        @Override
        public void onClick(View v) {
            name = edit_name.getText().toString();
            if(name == = null) {
                Toast.makeText (getApplicationContext ( ), " name 不能为空",
Toast.LENGTH_LONG).show();
            } else {
                ArrayList<MemberInfo> infoList = new ArrayList<MemberInfo>();
                infoList = dbManager.searchData(name);
                String result = "";
                for(MemberInfo info : infoList) {
                    result = result + String.valueOf(info._id) + " |" + info.name + "
|" + String.valueOf(info.age) + " |" + info.department + " |" + info.telephone;
                    result = result + "\n" + "----------------------------------------" + "\n";
                }
                Log.i("SQLite", result);
                startDisplayActivity("searchResult", result);
            }
        }
    });
}
private void startDisplayActivity(String intentName, String intentValue) {
    Intent intent = new Intent(MainActivity.this, DisplayActivity.class);
    intent.putExtra(intentName, intentValue);
    startActivity(intent);
}
@Override
protected void onDestroy() {
    super.onDestroy();
    dbManager.closeDB();//关闭数据库
}
}
```

（7）运行结果如图6-7所示。

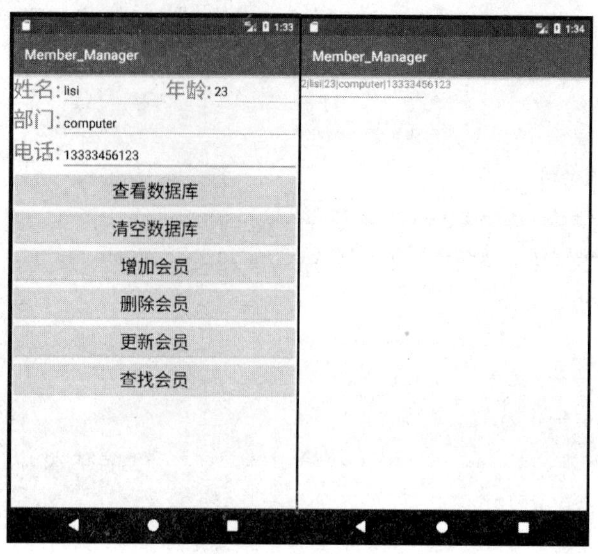

图 6-7　项目运行结果

6.4 数据共享：ContentProvider——获得联系人信息

ContentProvider 属于 Android 应用程序的组件之一，作为应用程序之间唯一的共享数据途径，ContentProvider 主要的功能就是存储并检索数据以及向其他应用程序提供访问数据的接口。

一个程序可以通过实现一个 ContentProvider 的抽象接口，将自己的数据完全暴露出去，而且 ContentProviders 是以类似数据库中表的方式将数据暴露，也就是说 ContentProvider 就像一个"数据库"。外界获取其提供的数据，应该与从数据库中获取数据的操作基本一样，只不过要采用 URI 来表示外界需要访问的"数据库"。

ContentProvider 是一个实现了一组用于提供其他应用程序存取数据的标准方法的类。应用程序可以在 ContentProvider 中执行如下操作：查询数据、修改数据、添加数据、删除数据。

ContentProvider 提供了一种多应用间数据共享的方式，Android 系统为一些常见的数据类型（如音乐、视频、图像、手机通迅录联系人信息等）内置了一系列的 ContentProvider，这些都位于 Android. provider 包下。持有特定的许可，可以在自己开发的应用程序中访问这些 ContentProvider。

ContentProvider 提供了如下方法：（1）query：查询；（2）insert：插入；（3）update：更新；（4）delete：删除；（5）getType：得到数据类型；（6）onCreate：创建数据时调用的回调函数。

ContentProvider 的使用分为如下两种方式。

（1）对于安卓系统提供的系统级的 ContentProvider，我们可以直接使用。

➢ MediaProvider：用来查询磁盘上多媒体文件。

➢ ContactsProvider：用来查询联系人信息。

➤ CalendarProvider：用来提供日历相关信息的查询。

➤ BookmarkProvider：用来提供书签信息的查询。

（2）自定义 ContentProvider 的使用方法如下。

➤ 设计数据库的储存方式。因为 ContentProvider 提供的是数据，没有数据，ContentProvider 就无法发挥作用。

➤ 定义自己的类，继承 ContentProvider 类，并实现基本的方法。重写构造方法，包括 insert、delete、getType、onCreate、query、update 等方法（根据需要来自行决定重写那些方法）。

➤ 程序添加一个 public static final Uri URI = " content：// + <包名> + URIName" 静态常量，其他程序通过这个 URI 调用此 ContentProvider 类中的数据。

➤ 在 AndroidManifest 中注册 Provider。

下面是使用 ContentProvider 的实例，读取系统的联系人信息，在 Android Studio 中新建项目 ContentProvider_CCDN1。

（1）主界面的 Activity 文件 MainActivity. java 的代码如下。

```
public class MainActivity extends AppCompatActivity {
  ListView listView;
  List <String> list;
  ArrayAdapter <String> arrayAdapter;
  @Override
protected void onCreate(Bundle savedInstanceState) {
    super.onCreate(savedInstanceState);
    setContentView(R.layout.activity_main);
      initView();
}
private void initView() {
    listView = (ListView) findViewById (R.id.mylistview);
    list = new ArrayList <String> ();
  arrayAdapter = new ArrayAdapter <String> (MainActivity.this, android.R.layout.simple_list_item_1, list);
    listView.setAdapter (arrayAdapter);
    getContentProvider();
}
private void getContentProvider() {
Cursor cursor = getContentResolver ( ). query (ContactsContract.CommonDataKinds.Phone.CONTENT_URI, null, null, null, null);
if (cursor! =null)
{
while (cursor.moveToNext()) {
      String displayNameString = cursor.getString (cursor.getColumnIndex (ContactsContract.CommonDataKinds.Phone.DISPLAY_NAME));
      String numberString = cursor.getString (cursor.getColumnIndex (Contacts-
```

```
Contract. CommonDataKinds. Phone. NUMBER));
            list. add (displayNameString + " \n" + numberString);
          }
          cursor. close ();
        }
      }
    }
```

（2）主布局文件是一个 ListView 控件，如图 6-8 所示。

（3）运行结果如图 6-9 所示。

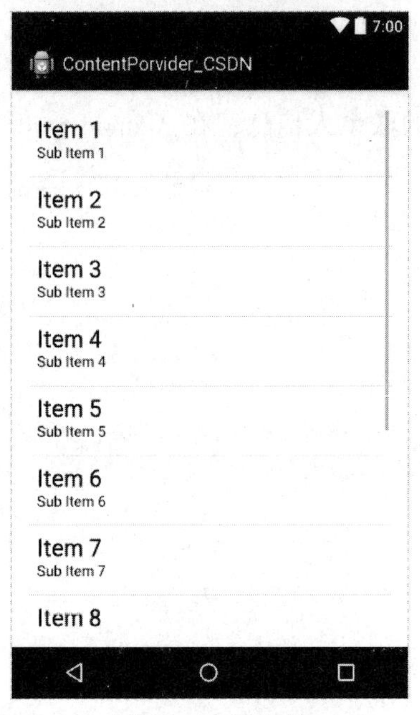

图 6-8　布局文件

图 6-9　运行结果

6.5　最新对象数据库操作——LitePal

现在的开源热潮让所有的 Android 开发者大大受益，GitHub 中含有成百上千的优秀 Android 开源项目，很多以前需要很久才能实现的功能，使用开源库可以在短短几分钟即可实现。另外，开源项目的代码都是经过时间验证的，通常比我们自己编写的代码要稳定得多。

6.5.1　LitePal 简介

LitePal 是一款开源的 Android 数据库框架，采用了对象关系映射（ORM）模式，将平时开发时最常用的一些数据库功能进行了封装，使得开发者不用编写一行 SQL 语句就可以完成各种建表、增删改查的操作。并且 LitePal 很“轻”，jar 包大小不到 100k，而且近乎零配置，这一点和 Hibernate 类的框架有很大区别。目前 LitePal 的源码已经托管到了 GitHub 上。网址：https://github. com/LitePalFramework/LitePal。

在 Android Studio 中新建项目 LitePal_Test，开始学习之旅。

6.5.2　配置 LitePal

大多开源项目将版本提交到 Jcenter 上，只需要在 app/build.gradle 文件的 dependencies 闭包添加引用就可以了，内容如下。

```
dependencies {
  compile fileTree(dir:'libs', include: ['*.jar'])
  compile 'com.android.support:appcompat-v7:25.3.1'
  testCompile 'junit:junit:4.12'
  compile 'org.litepal.android:core:1.5.1'
}
```

目前 LitePal 最新版本是 1.5.1，和前面提到的相同，因为 build.gradle 的改变，在主界面顶部会出现同步提示，如图 6-10 所示，单击 Sync Now。

Gradle files have changed since last project sync. A project sync may be necessary for the IDE to work properly.　　Sync Now

图 6-10　Sync Now 提示

同步结束，LitePal 成功引入到当前项目中了，在项目的 External Libraries 目录下可以看到导入的库 core-1.5.1，展开可以看到 LitePal，如图 6-11 所示。

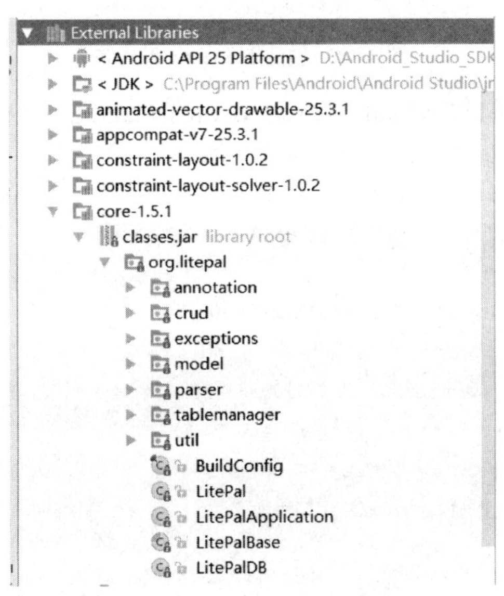

图 6-11　引入 LitePal 库到当前项目

在项目的 main 目录下新建 assets 目录，在此目录下创建一个 litepal.xml 文件，内容如下。

```
<?xml version="1.0" encoding="utf-8"?>
<litepal>
    <!--定义数据库的名称,需要以.db作为文件后缀,如果没有.db,将会默认添加-->
    <dbname value="News"></dbname>
    <!--定义数据库的版本号 -->
```

```
    <version value="1" > </version >
    <! --定义所有的映射模型,LitePal 会为每一个表创建映射类-- >
    <list >
        <mapping class="" > </mapping >
        <mapping class="" > </mapping >
    </list >
    <list >
    </list >
</litepal >
```

创建的 litepal. xml 文件如图 6-12 所示。

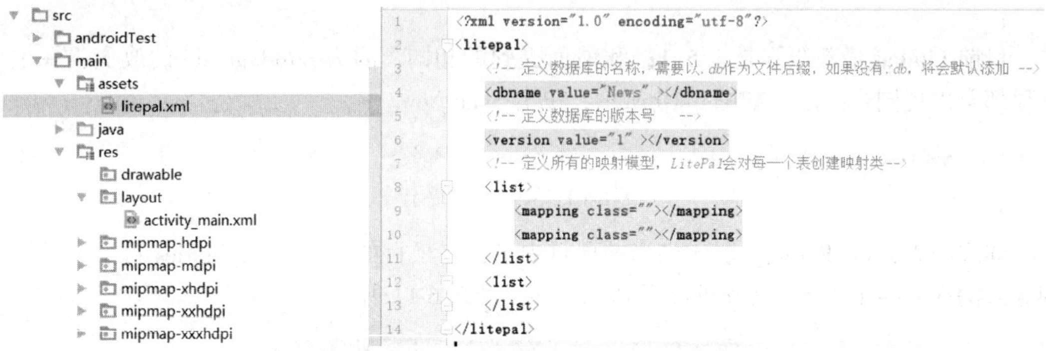

图 6-12　创建 litepal. xml 文件

最后再配置一下 LitePalApplication，修改 AndroidManifest. xml 的内容，如下所示。

```
<? xml version="1.0" encoding="utf-8"? >
<manifest xmlns:android=http://schemas.android.com/apk/res/android
package="com.example.hefugui.litepal_test" >
<application
    android: name=" org.litepal.LitePalApplication"
    android: allowBackup=" true"
    android: icon=" @mipmap/ic_launcher"
    android: label=" @string/app_name"
    android: supportsRtl=" true"
    android: theme=" @style/AppTheme" >
  ……
</application >
```

这里让项目的 application 配置为 org. litepal. LitePalApplication，这样才能让 LitePal 的所有功能都可以正常工作，现在即完成了配置。

6.5.3　数据库创建和升级

LitePal 采取的是对象关系映射（ORM）的模式，什么是对象关系映射呢？简单来说，我们使用的编程语言是面向对象语言，而使用的数据库则是关系型数据库，将面向对象的语言和面向关系的数据库之间建立一种映射关系，这就是对象关系映射。

接下来我们看一看 LitePal 中如何建表。根据对象关系映射模式的理念，每一张表都应该对应一个模型（Model），也就是说，如果我们想要建一张 News 表，就应该有一个对应的 News 模型类。在源代码目录新建一个 News 类，如下所示。

```
public class News {
    private int id;
    private String title;
    private String content;
    private Date publishDate;
    private int commentCount;
    //自动生成 get、set 方法
    ...
}
```

这是一个典型的 Java Bean，定义了 5 个字段：id、title、content、publishDate、commentCount，并生成相应的 getter 和 setter 方法，其中 id 这个字段可写可不写，因为即使不写这个字段，LitePal 也会在表中自动生成一个 id 列。这个 News 对应数据库的一张表。

LitePal 的映射规则是非常轻量级的，不像一些其他的数据库框架，需要为每个模型类单独配置一个映射关系的 XML，LitePal 的所有映射都是自动完成的。根据 LitePal 支持的数据类型，可以进行对象关系映射的数据类型一共有 8 种：int、short、long、float、double、boolean、String 和 Date。只要是声明成这 8 种数据类型的字段都会被自动映射到数据库表中，不需要进行任何额外的配置。

现在模型类已经建好了，还差最后一步，就是将它配置到映射列表当中。编辑 assets 目录下的 litepal. xml 文件，在 <list> 标签中加入 News 模型类的声明。

```
<?xml version="1.0" encoding="utf-8"?>
  <litepal>
    <dbname value="News" ></dbname>
    <!--定义所有的映射模型,LitePal 会为每一个表创建映射类-->
    <list>
    <mapping class="com.example.hefugui.litepal_test.News" ></mapping>
    </list>
</litepal>
```

注意，这里一定要填入 News 类的完整类名，包括路径。

这样，所有的工作都已完成，现在只要对数据库进行任何操作，就会自动创建 News 表。

LitePal 提供了一个来获取到 SQLiteDatabase 实例的便捷方法，如下所示。

```
SQLiteDatabase db = Connector.getDatabase();
```

调用上述代码，News 表即创建成功。

修改布局文件，添加一个按钮，如图 6-13 所示。

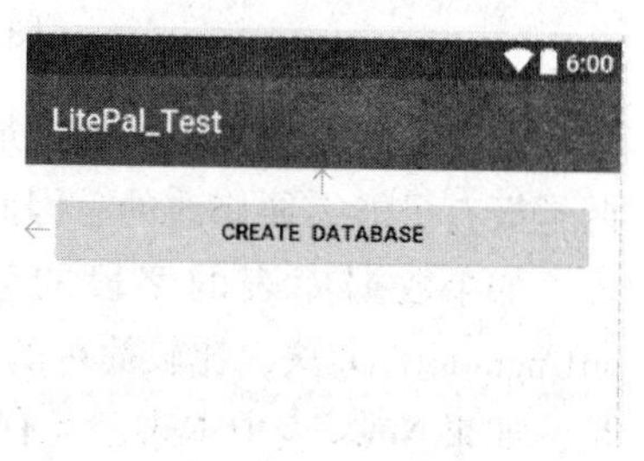

图 6-13　布局文件

修改 MainActivity 中的代码，如下所示。

```
public class MainActivity extends AppCompatActivity {
  @Override
  protected void onCreate(Bundle savedInstanceState) {
```

```
        super.onCreate(savedInstanceState);
        setContentView(R.layout.activity_main);
        Button createDatebase = (Button) findViewById (R.id.createDatebase);
        createDatebase.setOnClickListener (new View.OnClickListener()
            {
                @Override
                public void onClick (View v) {
                SQLiteDatabase db = Connector.getDatabase();
            }
        });
    }
}
```

运行项目，单击 CREATE DATABASE 按钮，在 Android Device Monitor 窗口中，可以看到新建的数据库文件“News.db”，创建的数据库位于：/data/data/包名/databases/目录中，如图 6-14 所示。

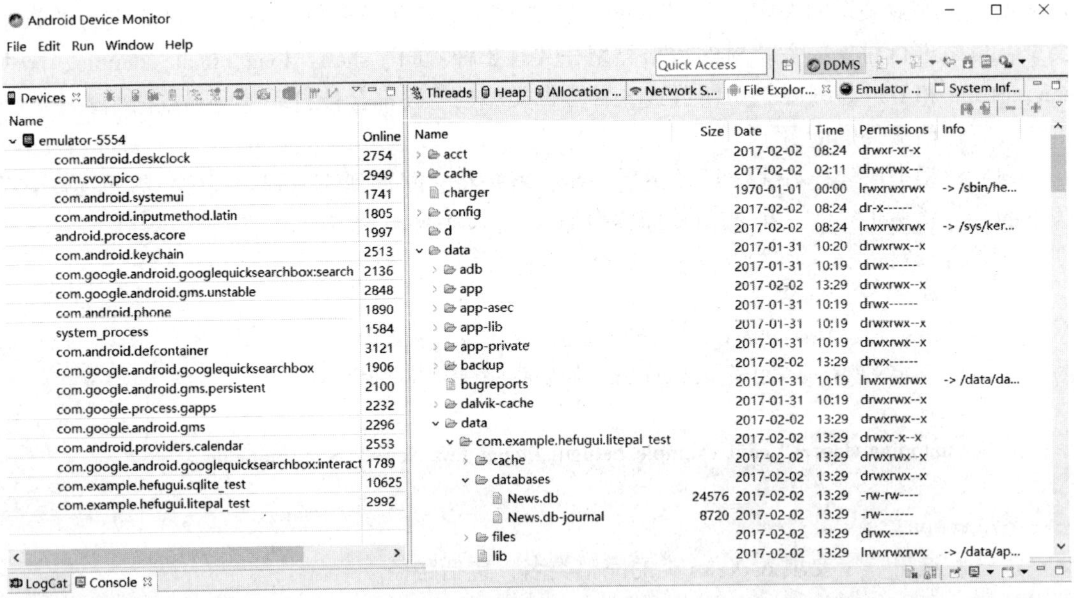

图 6-14 新建的数据库文件“News.db”

创建表只是数据库操作中最基本的一步，最初创建的表结构，随着需求的变更，到了后期是极有可能需要修改的。因此，升级表的操作对于任何一个项目都是至关重要的。

如果数据库表的字段的数量发生了变化，使用 SQLiteHelper 升级数据库时会调用 onUpgrade()方法，比较简单的方法是将数据库中现有的所有表都删除，然后重新创建。但是，如果 News 表中本来已经有数据，使用这种方式升级的话，会导致表中的数据全部丢失，所以这并不是一种值得推荐的升级方法。

下面一起来学习使用 LitePal 进行升级表操作的方法。

现在需要创建一张 Comment 表。第一步该怎么办呢？相信您已经猜到了，当然是先创建一个 Comment 类，在源代码目录新建一个 Comment 类，如下所示。

```
public class Comment {
    private int id;
    private String content;
    //自动生成 get、set 方法
    ...
}
```

Comment 类中有 id 和 content 这两个字段，也就意味着 Comment 表中会有 id 和 content 这两列。

在 assets 目录下 litepal. xml 文件的 <list> 标签中加入 Comment 模型类的声明，并将版本号加 1。

```
<?xml version = "1.0" encoding = "utf-8"? >
  <litepal>
    <dbname value = "News" > </dbname>
    <! --定义数据库的版本号 -- >
    <version value = "2" > </version>
    <! --定义所有的映射模型,LitePal 会为每一个表创建映射类-- >
  <list>
  <mapping class = "com.example.hefugui.litepal_test.News" > </mapping>
  <mapping class = " com.example.hefugui.litepal_test.Comment" > </mapping>
  </list>
</litepal>
```

现在再运行项目，单击 CREATE DATABASE 按钮，即会创建另一张表 Comment。

现在如果需要在 Comment 表中添加一个 publishdate 列，该怎么办呢？不用怀疑，相信您已经猜到应该在 Comment 类中添加字段，如下所示。

```
public class Comment {
     private int id;
     private String content;
     private Date publishDate;
       //自动生成 get、set 方法
   ...
}
```

剩下的操作就非常简单了，只需要在 litepal. xml 中对版本号加 1 即可，如下所示。

```
<litepal>
   <dbname value = "News" > </dbname>

   <version value = "3" > </version>
  ...
</litepal>
```

这样，下一次操作数据库调用 Connector. getDatabase( ) 方法时，publishDate 列会自动添加到 Comment 表中。

在 Android 中可以使用 ADB 和 SQLlite3 查看建立和修改的表。ADB 的全称为 Android Debug Bridge，就是起到调试桥的作用，该工具可以帮助管理设备或模拟器的状态。SQLite

的 sqlite3 命令被用来管理 SQLite 数据库。这两个文件在 Android SDK 的 platform-tools 文件夹下，如图 6-15 所示。

Android_Studio_SDK › sdk › platform-tools

名称	修改日期	类型	大小
api	2017/1/31 17:18	文件夹	
lib64	2017/1/31 17:18	文件夹	
systrace	2017/1/31 17:18	文件夹	
adb.exe	2017/1/10 19:56	应用程序	1,455 KB
AdbWinApi.dll	2017/1/10 19:56	应用程序扩展	96 KB
AdbWinUsbApi.dll	2017/1/10 19:56	应用程序扩展	62 KB
dmtracedump.exe	2017/1/10 19:56	应用程序	144 KB
etc1tool.exe	2017/1/10 19:56	应用程序	322 KB
fastboot.exe	2017/1/10 19:56	应用程序	788 KB
hprof-conv.exe	2017/1/10 19:56	应用程序	42 KB
NOTICE.txt	2017/1/10 19:56	文本文档	680 KB
package.xml	2017/1/10 19:56	XML 文档	18 KB
source.properties	2017/1/10 19:56	PROPERTIES 文件	1 KB
sqlite3.exe	2017/1/10 19:56	应用程序	710 KB

图 6-15 Android 的 ADB 和 SQLlite3

如果使用模拟器，则要在 Windows 的环境变量 path 中增加上述路径。

6.5.4 数据库操作

LitePal 中与存储相关的 API 其实并不多，但用法颇为丰富，而且相比于传统的 insert( ) 方法，使用 LitePal 存储数据简单到让人惊叹。下面完整地介绍 LitePal 存储数据的方法。

1. 插入记录

LitePal 要求所有的实体类都要继承自 DataSupport 这个类，因此，这里我们要把继承结构加上。修改 News 类的代码，如下所示。

```
public class News extends DataSupport{
    ......
}
public class Comment extends DataSupport {
......
}
```

继承了 DataSupport 类之后，这些实体类就拥有了进行 CRUD 操作的能力若要存储一条数据到 News 表中，可以进行如下操作。

（1）修改布局文件，增加 INSERT 按钮，如图 6-16 所示。

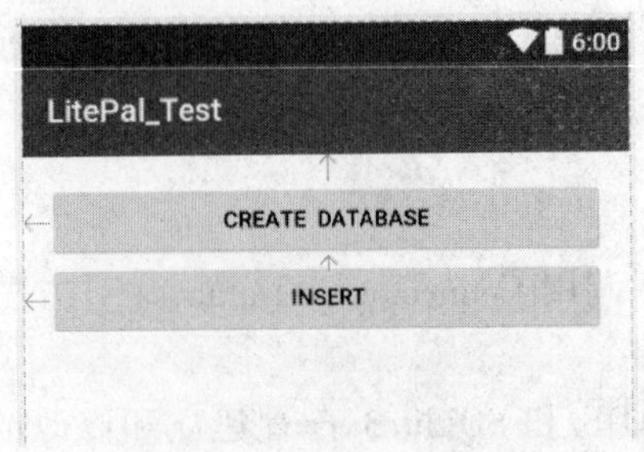

图 6-16 增加 INSERT 按钮

（2）在 INSERT 按钮的监听器中增加以下代码。

```
public class MainActivity extends AppCompatActivity {
@Override
    protected void onCreate(Bundle savedInstanceState) {
        Button insert = (Button) findViewById(R.id.insert);
        insert.setOnClickListener(new View.OnClickListener()
        {
         @Override
           public void onClick(View v) {
             News news  =  new News();
             news.setTitle("this a title");
             news.setContent("this a content");
             news.setPublishDate(new Date());
             news.save();
             }
        });
    }
}
```

这种方法非常简单，不需要 SQLiteDatabase，也不需要 ContentValues，更不需要通过列名组装数据，甚至不需要指定表名，只需要新建一个 News 对象，然后把要存储的数据通过 setter 方法传入，最后调用 save()方法即可，而这个 save()方法就是从 DataSupport 类中继承而来的，如图 6-17 所示。

```
管理员: C:\Windows\System32\cmd.exe - adb shell
sqlite> select * from news
   ...> ;
3 0 this a content 1486086764620 this a title
sqlite>
```

图 6-17　插入记录

除此之外，save()方法是有返回值的，我们可以根据返回值判断存储是否成功，代码如下。

```
if(news.save()) {
    Toast.makeText(context, "存储成功", Toast.LENGTH_SHORT).show();
} else {
    Toast.makeText (context, " 存储失败", Toast.LENGTH_SHORT).show();
}
```

2. 查询数据

（1）使用 LitePal 查询 News 表中 id 为 1 的这条记录，代码如下。

```
News news = DataSupport.find(News.class, 1);
```

（2）获取 News 表中的第一条数据，代码如下。

```
News firstNews = DataSupport.findFirst(News.class);
```

（3）获取 News 表中的最后一条数据，代码如下。

```
News lastNews = DataSupport.findLast(News.class);
```

（4）把 News 表中 id 为 1、3、5、7 的数据查找出来，代码如下。

```
List<News> newsList = DataSupport.findAll(News.class, 1, 3, 5, 7);
```

（5）把 News 表中所有数据查找出来，代码如下。

```
List<News> newsList = DataSupport.findAll(News.class);
```

（6）查询 News 表中所有评论数大于零的新闻，代码如下。

```
List < News > newsList = DataSupport.where ( " commentcount >?", " 0 "). find
(News.class);
```

首先调用了 DataSupport 的 where( )方法，在这里指定查询条件。where( )方法接收任意个字符串参数，其中第一个参数用于进行条件约束，从第二个参数开始，都用于替换第一个参数中的占位符。这个 where( )方法对应了一条 SQL 语句中的 where 部分。

（7）查询 News 表中的 title 和 content 这两列数据，也是很简单的，我们只要再增加一个连缀即可。

```
List<News> newsList = DataSupport.select("title", "content").where("comment-
count > ?", "0").find(News.class);
```

（8）若要将查询出的新闻按照发布的时间倒序排列，即最新发布的新闻放在最前面，则编写如下代码。

```
List<News> newsList = DataSupport.select("title", "content").where("comment-
count > ?", "0").order("publishdate desc").find(News.class);
```

（9）若要只查询前 10 条数据，使用连缀同样可以轻松解决这个问题，代码如下。

```
List<News> newsList = DataSupport.select("title", "content").where("comment-
count > ?", "0").order("publishdate desc").limit(10).offset(10).find(News.class);
```

3. 表关联

LitePal 的存储功能显示不仅仅只有这些用法，事实上，LitePal 在存储数据的时候默默帮我们做了很多的事情，比如多个实体类之间有关联关系的话，我们不需要考虑在存储数据的时候怎样建立数据与数据之间的关联，因为 LitePal 已经帮我们完成了，步骤如下。

（1）在 Comment 类中声明一个 News 实例，这样就清楚地表示出 News 中可以包含多个 Comment，而 Comment 中只能有一个 News，也就是多对一的关系。在 News 类增加一行代码，并自动生成 get、set 方法。

```
public class Newsextends DataSupport
{
    ...
private List<Comment> commentList = new ArrayList<Comment>();
        //自动生成 get、set 方法


}
```

（2）在 INSERT 按钮的监听器中修改代码如下。

```
public class MainActivity extends AppCompatActivity {
@Override
    protected void onCreate(Bundle savedInstanceState) {
        Button insert = (Button) findViewById(R.id.insert);
        insert.setOnClickListener(new View.OnClickListener()
```

```
        {
         @Override
          public void onClick(View v) {
             Comment comment1 = new Comment();
             comment1.setContent("好评!");
             comment1.setPublishDate(new Date());
             comment1.save();
             Comment comment2 = new Comment();
             comment2.setContent("赞一个");
             comment2.setPublishDate(new Date());
             comment2.save();
             News news = new News();
             news.getCommentList().add(comment1);
             news.getCommentList().add(comment2);
             news.setTitle("第二条新闻标题");
             news.setContent("第二条新闻内容");
             news.setPublishDate(new Date());
             news.setCommentCount(news.getCommentList().size());
             news.save();
          }
        });
    }
}
```

可以看到，这里先是存储了一条 comment1 数据，然后存储一条 comment2 数据，接着在存储 News 之前把刚才的两个 Comment 对象添加到 News 的 commentList 列表当中，这样就表示这两条 Comment 是属于这个 News 对象的，最后再把 News 存储到数据库中，这样它们之间的关联关系即自动建立。

（3）前面介绍通过 sqlite3 命令可以查看数据库，除此之外，还可以在手机上直接查看，查看软件是 Root Explorer，使用 Root Explorer 要求获取手机的 Root 权限，安装并打开软件之后，在/data/data/包名/databases 中单击 News. db 数据库，选择内置数据库查看器，然后随便单击一张表即可查看其中的数据，如图 6-18 所示。

第二条新闻已经成功存储到 News 表中，这条新闻的 id 是 2。那么，从哪里可以看出关联关系呢？多对一关联的时候，外键是存放在多方的，因此我们要到 Comment 表中查看关联关系，如图 6-19 所示。

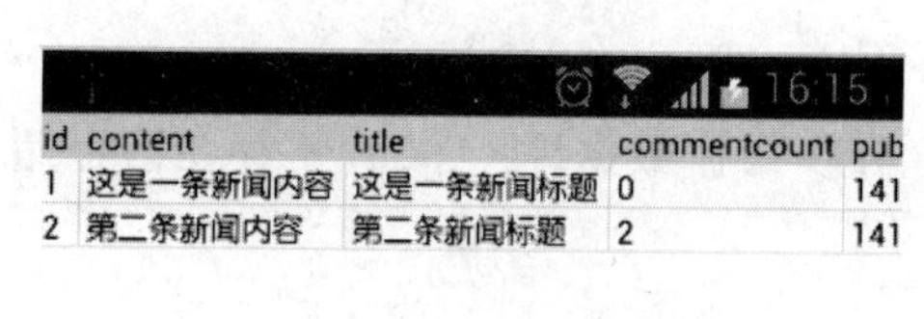

id	content	title	commentcount	pub
1	这是一条新闻内容	这是一条新闻标题	0	141
2	第二条新闻内容	第二条新闻标题	2	141

图 6-18 News 表

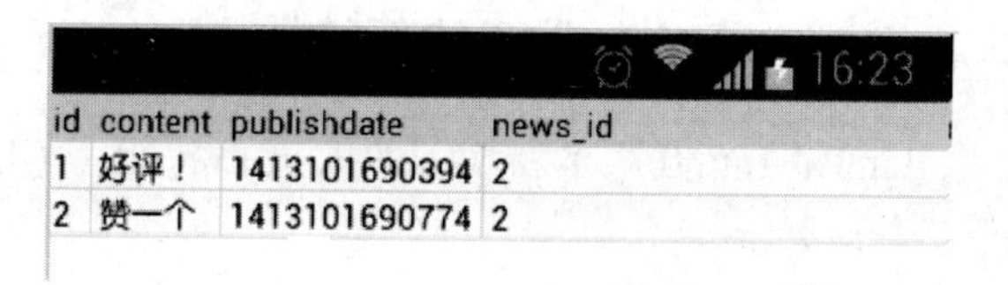

id	content	publishdate	news_id
1	好评！	1413101690394	2
2	赞一个	1413101690774	2

图 6-19 Comment 表

可以看到，两条评论都已经成功存储到 Comment 表中，并且这两条评论的 news_id 都是

2，说明它们是属于第二条新闻的。

4. 使用 LitePal 修改数据

LitePal 修改数据的 API 比较简单，并没有太多的用法，也比较容易理解，方法都是定义在 DataSupport 类中的，方法定义如下。

```
public static int update(Class<? > modelClass, ContentValues values, long id)
```

这个静态的 update( ) 方法接收三个参数，第一个参数是 Class，传入我们要修改的那个类的 Class，第二个参数是 ContentValues 对象，第三个参数是一个指定的 id，表示要修改哪一行数据。

例如，想把 News 表中 id 为 2 的记录的标题改成“今日 iPhone6 发布”，代码如下。

```
public class MainActivity extends AppCompatActivity {
@Override
protected void onCreate(Bundle savedInstanceState) {
......
Button update = (Button) findViewById(R.id.update);
update.setOnClickListener(new View.OnClickListener(){
@Override
public void onClick(View v) {
  ContentValues values = new ContentValues();
  values.put("title", "今日 iPhone6 发布");
  DataSupport.update(News.class, values, 2);
}
});
}
```

5. 使用 LitePal 删除数据

LitePal 删除数据的 API 和修改数据比较类似，但是更加简单，我们先来看一下 DataSupport 类中的方法定义，如下所示。

```
public static int delete(Class<? > modelClass, long id)
```

delete( ) 方法接收两个参数，第一个参数是 Class，传入我们要删除的那个类的 Class，第二个参数是一个指定的 id，表示我们要删除哪一行数据。

例如，想删除 News 表中 id 为 2 的记录，代码如下。

```
DataSupport.delete(News.class, 2);
```

需要注意的是，这不仅仅会将 News 表中 id 为 2 的记录删除，同时还会将其他表中以 News id 为 2 的这条记录作为外键的数据一起删除，因为外键若不存在，这条数据也没有保留的意义。

6.5.5 LitePal 1.5.0 的新特性

LitePal 1.5.0 版本新增了两大核心功能：（1）异步操作数据库；（2）不存在即存储，已存在即更新。

1. 异步操作数据库

在 1.5.0 版本之前，LitePal 操作数据库是发生在主线程中的，而 1.5.0 版本考虑到在开发过程中可能会出现对大量数据进行操作的情况，提供了异步操作数据库的 API，只需要调用对应的方法，即自动开启子线程对数据库进行一系列的增删改查操作。

由于异步操作的内部会开启线程，因此这类方法都是无返回值的，异步操作的结果只能依靠回调来完成。所以 LitePal 1.5.0 版本在每一个异步方法的后面添加了一个 listen( )方法。

异步保存的方法如下。

```
news.saveAsync().listen(new SaveCallback() {
@Override
public void onFinish(boolean success) {
      if(success) {
          Toast.makeText(context,"保存成功", Toast.LENGTH_SHORT). show();
      } else {
          Toast.makeText (context," 保存失败", Toast.LENGTH_SHORT). show();
      }
    }
});
```

异步查询的方法如下。

```
DataSupport.where("id > ?", "1")
      .select("title", "content")
      .offset(1)
      .limit(2)
      .order("id desc")
      .findAsync(News.class).listen(new FindMultiCallback(){'//此为异步查询方法'
        @Override
public <T> void onFinish(List<T> t) {
          Toast.makeText(context,"查询结束", Toast.LENGTH_SHORT). show();
      }
    });
```

2. 不存在即存储，已存在即更新

saveOrUpdate( )方法的应用示例如下。

```
News news = new News();
news.setId(1);
news.setCommentCount(2);
news.setContent("人长得漂亮? 太搞笑了。。。");
news.setTitle("我是新闻,我姓沈");
news.setPublishDate(new Date());
news.saveOrUpdate("id = ?","1");
```

saveOrUpdateAsync( )方法的应用示例如下。

```
News news = new News();
news.setId(1);
news.setCommentCount(2);
news.setContent("人长得漂亮? 太搞笑了。。。");
news.setTitle("我是新闻,我姓沈");
news.setPublishDate(new Date());
```

```
news.saveOrUpdateAsync("id = ?","1").listen(new SaveCallback() {
  @Override
  public void onFinish(boolean success) {
    if(success) {
      Toast.makeText(context, "保存成功", Toast.LENGTH_SHORT). show();
    } else {
        Toast.makeText (context, " 保存失败", Toast.LENGTH_SHORT). show();
    }
  }
});
```

1.5.0 版本中所有功能都是向下兼容的，因此升级不用付出成本。

6.6 本章小结

本章主要对常用的数据持久化方式进行了详细的讲解，包括 SharedPreferences 存储、SQLite 数据库操作和最新的 LitePal 数据库操作，SharedPreferences 存储适用于一些键值对的存储，而数据库适用于复杂数据的存储。

第七章

让界面动起来——Android动画

在日常的 Android 开发中，经常会用到动画，Android 提供了如下五种动画类型。

（1）View Animation：最简单的动画类型，只支持简单的缩放、平移、旋转、透明度等基本的动画。

（2）Drawable Animation：比较有针对性，是图片的替换动画。

（3）Property Animation：是通过动画的方式改变 View 的属性。

（4）矢量图动画。

（5）渲染动画。

本章将分别进行介绍。

7.1 绘图动画——绘制仪表盘

类 Graphics 是一个全能的绘图类，Graphics 类提供了基本的几何图形绘制方法，包括线段、矩形、圆、带颜色的图形、椭圆、圆弧、多边形、字符串等。Graphics 具有很强的绘图功能，含有很多子类。

下面以绘制仪表盘为例，实现绘图应用，在 Android 2.3 中创建应用项目：Dashboard。

（1）在源代码目录下新建源代码文件 HighlightCR. java，控制仪表盘的高亮效果的范围和颜色对象，代码如下。

```
public class HighlightCR {
    private int mStartAngle;
     private int mSweepAngle;
     private int mColor;
    public HighlightCR() {
    }
    public HighlightCR(int startAngle, int sweepAngle, int color) {
        this.mStartAngle = startAngle;
        this.mSweepAngle = sweepAngle;
        this.mColor = color;
    }
  … //Set get 方法
```

```
}
```

（2）在 res/values 目录下新建 attrs. xml，表示仪表盘的各种属性，代码如下。

```
<?xml version="1.0" encoding="utf-8"?>
<resources>
    <declare-styleable name="DashboardView">
        <attr name="radius" format="dimension"/>     <!--扇形半径-->
        <attr name="startAngle" format="integer"/> <!--起始角度-->
        <attr name="sweepAngle" format="integer"/>  <!--绘制角度-->
  <attr name="bigSliceCount" format="integer"/> <!--长刻度条数-->
<attr name="sliceCountInOneBigSlice" format="integer"/> <!--长刻度条数-->
<attr name="arcColor" format="color"/> <!--弧度颜色-->
<attr name="measureTextSize" format="dimension"/> <!--刻度字体大小-->
<attr name="textColor" format="color"/> <!--字体颜色-->
<attr name="headerTitle" format="string"/> <!--表头-->
<attr name="headerTextSize" format="string"/> <!--表头字体大小-->
<attr name="headerRadius" format="dimension"/> <!--表头半径-->
<attr name="pointerRadius" format="dimension"/> <!--指针半径-->
<attr name="circleRadius" format="dimension"/> <!--中心圆半径-->
<attr name="minValue" format="integer"/> <!--最小值-->
<attr name="maxValue" format="integer"/> <!--最大值-->
<attr name="realTimeValue" format="float"/> <!--实时值-->
<attr name="stripeWidth" format="dimension"/> <!--色带宽度-->
<attr name="stripeMode" > <!--色条显示位置-->
    <enum name="normal" value="0"/>
      <enum name="inner" value="1"/>
      <enum name="outer" value="2"/>
    </attr>
  <attr name="bgColor" format="color"/> <!--背景颜色-->
</declare-styleable>
</resources>
```

（3）在源代码目录下新建源代码文件 DashboardView. java，实现绘制仪表盘视图，这是本项目的核心文件，主要代码如下。

```
public class DashboardView extends View {
  private int mRadius; //圆弧半径
  private int mStartAngle; //起始角度
  private int mSweepAngle; //绘制角度
  private int mBigSliceCount; //大份数
  private int mSliceCountInOneBigSlice; //划分一大份长的小份数
  private int mArcColor; //弧度颜色
  private int mMeasureTextSize; //刻度字体大小
  private int mTextColor; //字体颜色
  private String mHeaderTitle = ""; //表头
  private int mHeaderTextSize; //表头字体大小
```

```
    private int mHeaderRadius; //表头半径
    private int mPointerRadius; //指针半径
    private int mCircleRadius; //中心圆半径
    private int mMinValue; //最小值
    private int mMaxValue; //最大值
    private float mRealTimeValue; //实时值
    private int mStripeWidth; //色条宽度
    private StripeMode mStripeMode = StripeMode.NORMAL;
    private int mBigSliceRadius; //较长刻度半径
    private int mSmallSliceRadius; //较短刻度半径
    private int mNumMeaRadius; //数字刻度半径
    private int mModeType;
    private List<HighlightCR> mStripeHighlight; //高亮范围颜色对象的集合
    private int mBgColor; //背景色
    private int mViewWidth; //控件宽度
    private int mViewHeight; //控件高度
    private float mCenterX;
    private float mCenterY;
    private Paint mPaintArc;
    private Paint mPaintText;
    private Paint mPaintPointer;
    private Paint mPaintValue;
    private Paint mPaintStripe;
    private RectF mRectArc;
    private RectF mRectStripe;
    private Rect mRectMeasures;
    private Rect mRectHeader;
    private Rect mRectRealText;
    private Path path;
    private int mSmallSliceCount; //短刻度个数
    private float mBigSliceAngle; //大刻度等分角度
    private float mSmallSliceAngle; //小刻度等分角度
    private String[] mGraduations; //等分的刻度值
    private float initAngle;
    private boolean textColorFlag = true; //若不单独设置文字颜色,则文字和圆弧同色
    private boolean mAnimEnable; //是否播放动画
    private MyHandler mHandler;
    private long duration = 500; //动画默认时长
    public DashboardView(Context context, AttributeSet attrs, int defStyleAttr) {
      super(context, attrs, defStyleAttr);
      TypedArray a = context.obtainStyledAttributes(attrs, R.styleable.DashboardView, def-
StyleAttr, 0);
    … //初始化各种属性
```

```
    }

private void initObjects() {
    mPaintArc = new Paint();
    mPaintArc.setAntiAlias(true);
    mPaintArc.setColor(mArcColor);
    mPaintArc.setStyle(Paint.Style.STROKE);
    mPaintArc.setStrokeCap(Paint.Cap.ROUND);
        …//初始化各种绘制对象
    }
/*绘制色带 */
  private void drawStripe(Canvas canvas) {
    if(mStripeMode != StripeMode.NORMAL && mStripeHighlight != null) {
        for(int i = 0; i < mStripeHighlight.size(); i++) {
          HighlightCR highlightCR = mStripeHighlight.get(i);
          if(highlightCR.getColor() == 0 ||highlightCR.getSweepAngle() == 0)
              continue;
}
/*绘制刻度盘 */
private void drawMeasures(Canvas canvas) {
    mPaintArc.setStrokeWidth(dpToPx(2));
      for(int i = 0; i <= mBigSliceCount; i++) {
            //绘制大刻度
      …
    }
/*绘制刻度盘的弧形 */
private void drawArc(Canvas canvas) {
    mPaintArc.setStrokeWidth(dpToPx(2));
    …
}
/*绘制圆和文字读数 */
private void drawCircleAndReadingText(Canvas canvas) {
  mPaintText.setTextSize(mHeaderTextSize);
  mPaintText.setTextAlign(Paint.Align.CENTER);
  …
}

/*依圆心坐标、半径、扇形角度,计算出扇形终射线与圆弧交叉点的 xy 坐标/
  public float[] getCoordinatePoint(int radius, float cirAngle) {
      float[] point = new float[2];
      double arcAngle = Math.toRadians(cirAngle); //将角度转换为弧度
      …
    }
```

（4）主 Activity 对应的布局文件 activity_main. xml 中包含三个刚才定义的仪表类控件，如图 7-1 所示。

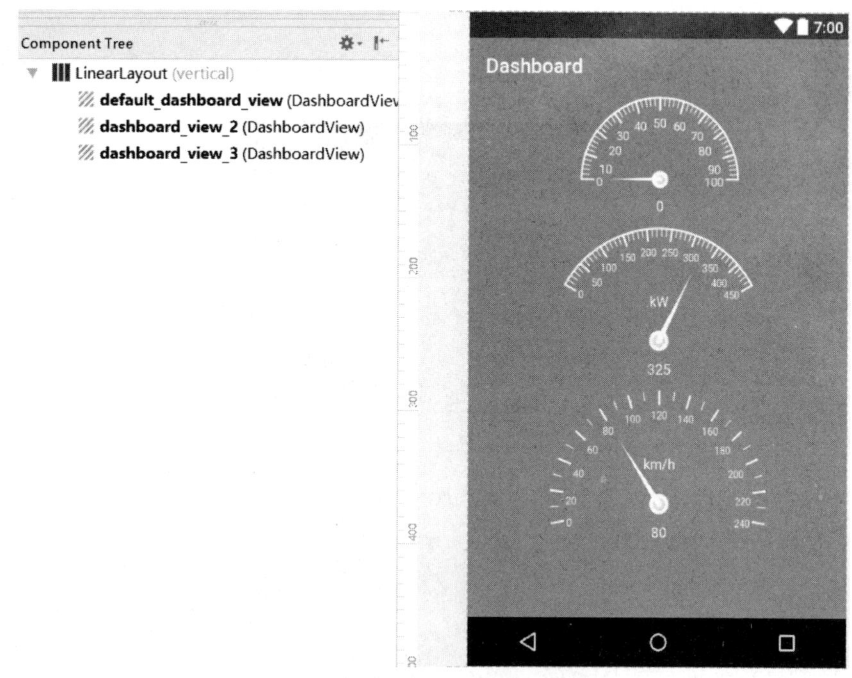

图 7-1 主布局文件

（5）主界面的 Activity 文件 MainActivity. java 的代码如下。

```
public class MainActivity extends AppCompatActivity {
  @Override
  protected void onCreate(Bundle savedInstanceState) {
    super.onCreate(savedInstanceState);
  setContentView(R.layout.activity_main);
  final DashboardView dashboardView1 = (DashboardView) findViewById (R.id.dashboard_
view_2);
  DashboardView dashboardView3 = (DashboardView) findViewById (R.id.dashboard_view_3);
dashboardView1.setOnClickListener (new View.OnClickListener() {
  @Override
    public void onClick (View v) {
      dashboardView1.setRealTimeValue (150.f, true, 100);
    }
});
List<HighlightCR> highlight1 = new ArrayList<>();
highlight1.add (new HighlightCR (210, 60, Color.parseColor ("#03A9F4")));
highlight1.add (new HighlightCR (270, 60, Color.parseColor ("#FFA000")));
dashboardView1.setStripeHighlightColorAndRange (highlight1);
List<HighlightCR> highlight2 = new ArrayList<>();
highlight2.add (new HighlightCR (170, 140, Color.parseColor ("#607D8B")));
```

```
highlight2.add (new HighlightCR (310, 60, Color.parseColor (" #795548")));
dashboardView3.setStripeHighlightColorAndRange (highlight2);
}
}
```

（6）运行结果如图 7-2 所示。

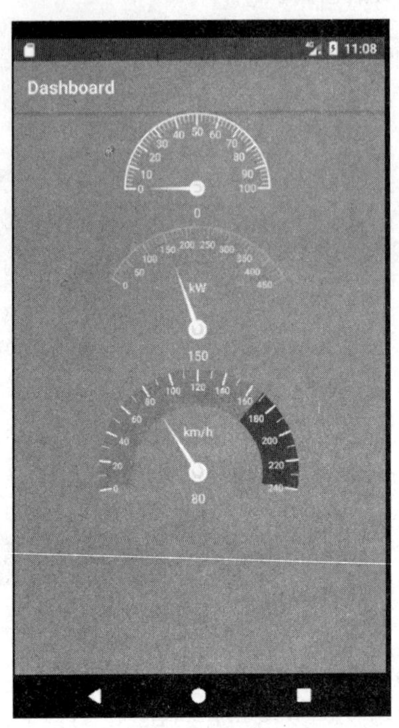

图 7-2　项目运行结果

7.2 帧动画 Drawable——模拟电扇转动

在环境控制中，要根据传感器的值打开电扇、窗户、灯光等，在移动终端界面需要同步动画，本例使用 Android 的 Drawable 实现窗户和电扇的动画。下面是具体实现过程。在 Android 2.3 中创建应用项目：eviroment_control。

（1）准备窗户动画图片，在 res/drawable-hdpi 目录下，复制窗户打开的动画图片，如图 7-3 所示。

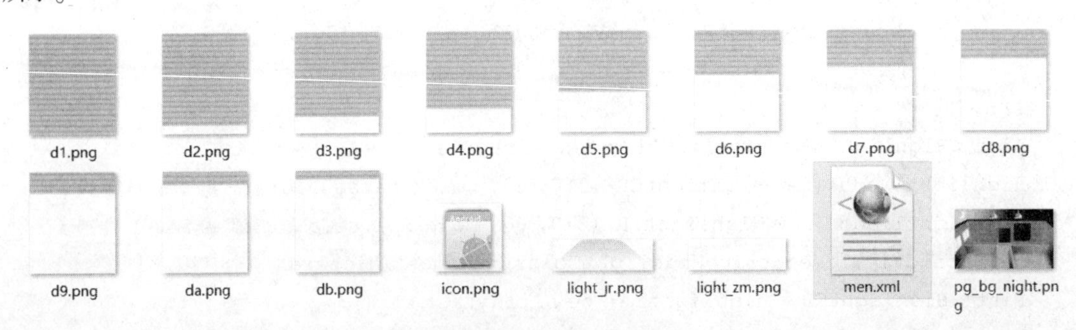

图 7-3　窗户打开的动画图片

（2）在 res/drawable-hdpi 目录下新建窗户 Drawable 动画对应的文件 chuanghu. xml，代码如下。

```
<?xml version="1.0" encoding="utf-8"?>
<animation-list xmlns:android="http://schemas.android.com/apk/res/android"
    android:oneshot="true" >
  <item android:drawable="@drawable/d1" android:duration="200"/>
  <item android:drawable="@drawable/d2" android:duration="200"/>
  <item android:drawable="@drawable/d3"  android:duration="200"/>
  <item android:drawable="@drawable/d4" android:duration="200" />
  <item android:drawable="@drawable/d5" android:duration="200"/>
  <item android:drawable="@drawable/d6"  android:duration="200"/>
  <item android:drawable="@drawable/d7"  android:duration="200"/>
  <item android:drawable="@drawable/d8"  android:duration="200"/>
  <item android:drawable="@drawable/d9"  android:duration="200"/>
  <item android:drawable="@drawable/da"  android:duration="200"/>
  <item android:drawable="@drawable/db"  android:duration="200"/>
</animation-list>
```

（3）在 res/drawable-hdpi 目录下，粘贴风扇打开的动画的图片，如图 7-4 所示。

f1.png　f2.png　f3.png　f4.png　f5.png　f6.png　f7.png　f8.png

图 7-4　风扇打开的动画图片

（4）在 res/drawable-hdpi 目录下新建风扇 Drawable 动画对应的文件 zhuan. xml，代码如下。

```
<?xml version="1.0" encoding="utf-8"?>
<animation-list xmlns:android="http://schemas.android.com/apk/res/android"
    android:oneshot="true" >
  <item android:drawable="@drawable/f1" android:duration="200"/>
    <item android:drawable="@drawable/f2" android:duration="200"/>
    <item android:drawable="@drawable/f3"  android:duration="200"/>
    <item android:drawable="@drawable/f4" android:duration="200" />
    <item android:drawable="@drawable/f5" android:duration="200"/>
    <item android:drawable="@drawable/f6"  android:duration="200"/>
    <item android:drawable="@drawable/f7"  android:duration="200"/>
    <item android:drawable="@drawable/f8"  android:duration="200"/>
</animation-list>
```

（5）编写一个布局文件 activity_main. xml，放置背景图片，放置两个 ImageView 空间，用于显示风扇和窗户的 Drawable 动画，如图 7-5 所示。

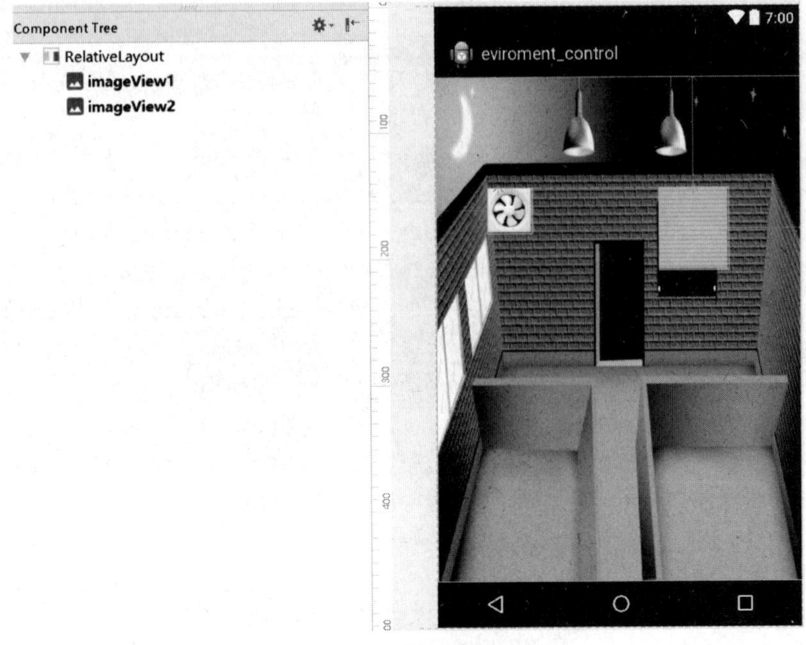

图 7-5　界面布局

(6) 编写实现代码 MainActivity. java，其功能是实现动画，具体代码如下。

```
public class MainActivity extends Activity
{
  private AnimationDrawable animationDrawable1,animationDrawable2;
ImageView chuan,fengshan;
  @Override
protected void onCreate(Bundle savedInstanceState) {
      super.onCreate(savedInstanceState);
      setContentView(R.layout.activity_main);
      chuan = (ImageView) findViewById (R.id.imageView1);
      chuan.setImageResource (R.drawable.chuanghu);
      animationDrawable1 = (AnimationDrawable) chuan.getDrawable();
       animationDrawable1.setOneShot (false);
       animationDrawable1.start();
       //animationDrawable.stop();
       fengshan = (ImageView) findViewById (R.id.imageView2);
       fengshan.setImageResource (R.drawable.zhuan);
       animationDrawable2 = (AnimationDrawable) fengshan.getDrawable();
       animationDrawable2.setOneShot (false);
       animationDrawable2.start();
    }
}
```

(7) 项目执行结果如图 7-6 所示。

图 7-6　执行结果

7.3 SurfaceView 实现下雨的天气动画效果

前面在自定义 View 中进行了绘图，但 View 的绘图机制存在如下缺陷。

（1） View 缺乏双缓冲机制。

（2） 当程序需要更新 View 中的图像时，程序必须重绘 View 上显示的整张图片。

（3） 新线程无法直接更新 View 组件。

由于 View 存在上面缺陷，在游戏开发中一般使用 SurfaceView 来进行绘制，SurfaceView 不同于 View，它可以在非 UI 线程中绘制并显示在界面上，这意味着您可以自己新开一个线程，然后把绘制渲染的代码放在该线程中。SurfaceView 一般会与 SurfaceHolder 结合使用，SurfaceHolder 用于向与之关联的 SurfaceView 上绘图，调用 SurfaceView 的 getHolder( ) 方法，即可获取 SurfaceView 关联的 SurfaceHolder。

SurfaceHolder 提供了如下方法来获取 Canvas 对象。

（1） Canvas lockCanvas( )：锁定整个 SurfaceView 对象，获取该 Surface 上的 Canvas。

（2） Canvas lockCanvas （Rect dirty）：锁定 SurfaceView 上 Rect 划分的区域，获取该 Surface 上的 Canvas。

两个方法返回的是同一个 Canvas，但是第二个方法只对圈出来的区域进行刷新，Canvas 绘图完成后通过 unlockCanvasAndPost （canvas） 方法来释放画布，提交修改。调用 Surface-

Holder 的 unlockCanvasAndPost 方法之后，该方法之前所绘制的图形还处于缓冲之中，下一次 lockCanvas( )方法锁定的区域可能会“遮挡”它。

本例使用 Android 的 SurfaceView 实现下雨的天气动画效果。下面是具体实现过程。在 Android 2.3 中创建应用项目：SurfaceView_Test。

先分析雨滴的实现方法。

➢ 每个雨滴其实是一条线，通过 canvas. drawLine( ) 绘制。

➢ 线（雨滴）的长度、宽度、下落速度、透明度以及位置是在一定范围内随机生成的。

➢ 每绘制一次后，改变雨滴的位置并重绘，即可实现雨滴的下落效果。

（1）在项目的源代码目录下包含四个类：MainActivity. java、BaseType. java、RainTypeImpl. java 和 DynamicWeatherView. java，如图 7-7 所示。

BaseType
DynamicWeatherView
MainActivity
RainTypeImpl

图 7-7　项目的类文件

其中 BaseType 实现雨的形状，RainTypeImpl 实现下雨的效果，DynamicWeatherView 类继承 SurfaceView，实现 SurfaceHolder. Callback 的接口，在接口 surfaceCreated（SurfaceHolder holder）调用线程，在线程中调用 RainTypeImpl，如图 7-8 所示。

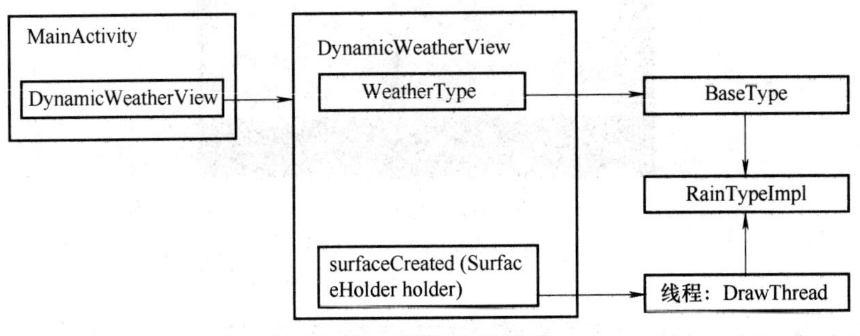

图 7-8　项目的类关系

可以通过 Android 提供给的 SurfaceHolder 接口访问下面的 Surface。可以调用 SurfaceView 的 getHolder( ) 来获取。

SurfaceView 是有生命周期的，必须在它生命周期期间执行绘制代码，所以我们需要监听 SurfaceView 的状态（例如创建以及销毁），Android 提供了 SurfaceHolder. Callback 这个接口，让我们方便地监听 SurfaceView 的状态。

下面了解一下 SurfaceHolder. Callback 接口。

```
public interface Callback {
// SurfaceView 创建时调用(SurfaceView 的窗口可见时)
public void surfaceCreated(SurfaceHolder holder);
// SurfaceView 改变时调用
public void surfaceChanged(SurfaceHolder holder, int format, int width,int height);
// SurfaceView 销毁时调用(SurfaceView 的窗口不可见时)
public void surfaceDestroyed(SurfaceHolder holder);
}
```

绘制代码需要在 surfaceCreated 和 surfaceDestroyed 之间执行，否则无效，SurfaceHolder. Callback 的回调方法是在 UI 线程中执行的，绘制线程需要我们自己手动创建。

View 适合与用户交互并且渲染时间不是很长的控件，因为 View 的绘制和用户交互都处在 UI 线程中。SurfaceView 适合迅速更新界面或者渲染时间比较长以至于影响到用户体验的场景。

（2）类 DynamicWeatherView 继承自 SurfaceView，为了监听 SurfaceView 的状态，还需要实现 SurfaceHolder. Callback 接口，主要代码如下。

```
public class DynamicWeatherView extends SurfaceView implements SurfaceHolder.Callback{
    //定义的一个接口,代表一种天气类型
    public interface WeatherType {
        void onDraw(Canvas canvas);
        void onSizeChanged(Context context, int w, int h);
    }
    public DynamicWeatherView(Context context, AttributeSet attrs, int defStyleAttr) {
        super(context, attrs, defStyleAttr);
        mContext = context;
        mHolder = getHolder();  //获得 Holder
        mHolder.addCallback(this);  //获得回调
        mHolder.setFormat(PixelFormat.TRANSPARENT);
    }
    @Override
    protected void onSizeChanged(int w, int h, int oldw, int oldh) {
        super.onSizeChanged(w, h, oldw, oldh);
        mViewWidth = w;
        mViewHeight = h;
        if(mType ! = null) {
            mType.onSizeChanged(mContext, w, h);
        }
    }
    @Override
    public void surfaceCreated(SurfaceHolder holder) {
        mDrawThread = new DrawThread();
        mDrawThread.setRunning(true);
        mDrawThread.start();
    }
    @Override
    public void surfaceChanged(SurfaceHolder holder, int format, int width, int height) {
    }
    @Override
    public void surfaceDestroyed(SurfaceHolder holder) {
        mDrawThread.setRunning(false);
```

```
    }
    /* 绘制线程* /
    private class DrawThread extends Thread {
        //用来停止线程的标记
        private boolean isRunning = false;
        public void setRunning(boolean running) {
              isRunning = running;
        }
    @Override
    public void run() {
        Canvas canvas;
        //无限循环绘制
        while(isRunning) {
            if(mType ! = null && mViewWidth ! = 0 && mViewHeight ! = 0) {
              canvas = mHolder.lockCanvas();
              if(canvas ! = null) {
                  mType.onDraw(canvas);
                    if(isRunning) {
                        mHolder.unlockCanvasAndPost(canvas);
                      } else {
                          //停止线程
                        break;
                      }
                    // sleep
                    SystemClock.sleep(1);
                  }
              }
          }
        }
    }
    }
```

(3) 类 RainTypeImpl 实现下雨效果的主要代码如下。

```
public class RainTypeImpl extends BaseType {
//背景
private Drawable mBackground;
//雨滴集合
private ArrayList<RainHolder> mRains;
//画笔
private Paint mPaint;
public RainTypeImpl(Context context, DynamicWeatherView dynamicWeatherView) {
    super(context, dynamicWeatherView);
    init();
}
```

```
    private void init() {
        mPaint = new Paint();
        mPaint.setAntiAlias(true);
        mPaint.setColor(Color.WHITE);
        //这里雨滴的宽度统一为 3
        mPaint.setStrokeWidth(3);
        mRains = new ArrayList<>();
    }
    @Override  //产生雨滴
    public void generate() {
      mBackground = getContext().getResources().getDrawable(R.drawable.rain_sky_
day));
      mBackground.setBounds (0, 0, getWidth(), getHeight());
      for (int i = 0; i < 60; i ++) {
        RainHolder rain = new RainHolder (getRandom (1, getWidth()),
getRandom (1, getHeight()), getRandom (dp2px (9), dp2px (15)),
getRandom (dp2px (5), dp2px (9)), getRandom (20, 100));
        mRains.add (rain);
      }
    }
    @Override
    public void onDraw (Canvas canvas) {
      clearCanvas (canvas);
      //画背景
      mBackground.draw (canvas);
      //画出集合中的雨点
    for (int i = 0; i < mRains.size(); i ++) {
        r = mRains.get (i);
        mPaint.setAlpha (r.a);
        canvas.drawLine (r.x, r.y, r.x, r.y+r.l, mPaint);
      }
      //将集合中的点按自己的速度偏移
      for (int i = 0; i < mRains.size(); i ++) {
        r = mRains.get (i);
        r.y += r.s;
        if (r.y > getHeight()) {
            r.y = -r.l;
        }
      }
    }
```

（4）主 Activity 对应的布局文件 activity_main.xml 的代码如下。

```
<?xml version="1.0" encoding="utf-8"?>
<RelativeLayout
```

```
    xmlns:android=http://schemas.android.com/apk/res/android
    android:layout_width=" match_parent"
  android: layout_height=" match_parent" >
  <com.example.hefugui.surfaceview_test.DynamicWeatherView
      android: id=" @ +id/dynamic_weather_view"
      android: layout_width=" match_parent"
      android: layout_height=" match_parent" />
  </RelativeLayout>
```

（5）BaseType 类是一个抽象基类，实现了 DynamicWeatherView.WeatherType 接口，内部有一些公共方法，之后要想实现不同的天气类型，只需要继承 BaseType 类重写相关方法即可。

（6）项目执行结果如图 7-9 所示。

图 7-9　项目执行结果

7.4 Android 5.0 新动画——AnimatedVectorDrawable 矢量动画

在 Android 5.0（API 级别 21）或以上的系统中，矢量图像在 Android 中被表示为 VectorDrawable 对象，AnimatedVectorDrawable 顾名思义就是针对 VectorDrawable 制作动画的类，AnimatedVectorDrawable 类可以创建一个矢量资源的动画，可以为图标制作各种动画效果。

AnimatedVectorDrawable 通过 ObjectAnimator 和 AnimatorSet 对 VectorDrawable 的属性制作

动画，从而实现各种动画效果。

AnimatedVectorDrawable 通常在三个 XML 文件中定义矢量资源的动画载体。

➢ <vector>元素的矢量资源，在 res/drawable/（文件夹）。

➢ <animated-vector>元素的矢量资源动画，在 res/drawable/（文件夹）。

➢ < objectAnimator>元素的一个或多个对象动画器，在 res/anim/（文件夹）。

矢量资源动画能创建元素属性动画。元素定义了一组路径或子组，并且元素定义了要被绘制的路径。

创建动画时，先定义矢量资源，使用 android:name 属性分配一个唯一的名字给组和路径，这样可以从动画定义中查询到。

本例使用 Android 的 AnimatedVectorDrawable 实现矢量图的笑脸效果。下面是具体实现过程。在 Android 2.3 中创建应用项目：AnimatedVectorDrawable_Demo。

（1）用一个 XML 定义一个 VectorDrawable 静态矢量图，使用<vector>、<path>和<group>标签。其中<vector>定义一个 VectorDrawable 对象，<path>定义要被绘制的路径，<group>定义一组路径或子组。使用 name 属性为<group>分配一个唯一的名字，以便做动画时使用该名字定位到需要做动画的位置。使用<vector>标签的 XML 文件应放置在 drawable 文件夹中。建立 drawable/face.xml 文件，代码如下。

```
<vector xmlns:android=http://schemas.android.com/apk/res/android
        android:height="200dp"
        android:width="200dp"
        android:viewportHeight="100"
        android:viewportWidth="100" >
    <path   android:fillColor="@color/yellow"
            android:pathData="@string/path_circle" />
    <path   android: fillColor=" @android: color/black"
            android: pathData=" @string/path_face_left_eye" />
    <path   android: fillColor=" @android: color/black"
            android: pathData=" @string/path_face_right_eye" />
    <path   android: name=" mouth"
            android: strokeColor=" @android: color/black"
            android: strokeWidth=" @integer/stroke_width"
            android: strokeLineCap=" round"
            android: pathData=" @string/path_face_mouth_sad" />
    </vector>
```

下面，简单介绍 pathData 的语法。

path 命令定义由字母后跟一个或多个数字组成的字符串，数字之间可以用“,”隔开，“,”不是必需的。字母可以是大写也可以是小写，大写代表绝对位置，小写代表相对位置。

➢ M：move to 移动绘制点。

➢ L：line to 直线。

➢ Z：close 闭合。

➢ C：cubic bezier 三次贝塞尔曲线。

➢ Q：quatratic bezier 二次贝塞尔曲线。

➤ A：ellipse 圆弧。

➤ M（x，y）：移动到（x，y）。

➤ L（x，y）：直线连到（x，y），简化命令 H（x）表示水平连接，V（y）表示垂直连接。

➤ Z：没有参数，连接起点和终点。

➤ C（x1，y1 x2，y2 x，y）：控制点（x1，y1）和（x2，y2），终点（x，y）。

➤ Q（x1，y1 x，y）：控制点（x1，y1），终点（x，y）。

➤ A（rx，ry x-axis-rotation large-arc-flag sweep-flag x，y）：画圆弧。

例如上面笑脸和悲伤的脸的嘴的两个路径数据如下所示。

```
<string name="path_face_mouth_sad" >
      M 30, 75
      Q 50, 55 70, 75
</string>
<string name=" path_face_mouth_happy" >
   M 30, 65
   Q 50, 85 70, 65
</string>
```

（2）用一个 XML 定义一个 VectorDrawable 矢量动画，使用 <animated-vector> 标签，动画的目标使用 <target> 来定义，其中属性 android：animation 为动画文件，drawable/smiling_face. xml 文件代码如下。

```
<? xml version="1.0" encoding="utf-8"? >
<animated-vector xmlns:android=http://schemas.android.com/apk/res/android
          android:drawable="@drawable/face" >
    <target  android:name="mouth"
          android:animation="@anim/smile" />
</animated-vector>
```

（3）用一个 XML 定义一个 VectorDrawable 动画过程，使用 < objectAnimator > 标签，包含动画的开始值和终值、时间间隔等，anim/smile. xml 文件代码如下。

```
<? xml version="1.0" encoding="utf-8"? >
<set xmlns:android=http://schemas.android.com/apk/res/android
   android:fillAfter="true">
<objectAnimator
  xmlns:android=http://schemas.android.com/apk/res/android
  android:duration="3000"
  android:propertyName="pathData"
android:valueFrom="@string/path_face_mouth_sad"
android: valueTo=" @string/path_face_mouth_happy"
android: valueType=" pathType"
android: interpolator=" @android: anim/accelerate_interpolator" />
</set>
```

（4）编写动画对应的界面布局文件 activity_path_morph. xml，使用的控件 ImageView，对

应的类为 PathMorphActivity，代码如下。

```
<ImageView
 xmlns:android=http://schemas.android.com/apk/res/android
 xmlns:tools=http://schemas.android.com/tools
  android:id="@+id/image"
  android:layout_width=" wrap_content"
  android: layout_height=" wrap_content"
  android: layout_gravity=" center"
  android: src=" @drawable/smiling_face"
  tools: context=" .PathMorphActivity" />
```

（5）在项目的源代码目录下建立动画类 AnimatedImageActivity，代码如下。

```
public abstract class AnimatedImageActivity extends Activity {
    private ImageView imageView;
    @Override
    protected void onCreate(Bundle savedInstanceState) {
      super.onCreate(savedInstanceState);
      setContentView(getLayoutId());
      imageView = (ImageView) findViewById(R.id.image);
      imageView.setOnClickListener(new View.OnClickListener() {
          @Override
        public void onClick(View v) {
            animate();
        }
      });
    }
  private void animate() {
    Drawable drawable = imageView.getDrawable();
      if(drawable instanceof Animatable) {
        ((Animatable) drawable).start();
      }
    }
  protected abstract int getLayoutId();
}
```

（6）在项目的源代码目录下建立类 PathMorphActivity，作为类 AnimatedImageActivity 的实现，代码如下。

```
public class PathMorphActivity extends AnimatedImageActivity {
  @Override
  protected int getLayoutId()
  {
    return R.layout.activity_path_morph;
  }
}
```

（7）主布局文件 activity_main. xml 只包含一个 Button 控件，用于启动动画界面。

（8）项目执行结果如图 7-10 所示。

图 7-10　项目执行结果

至此，一个简单的 AnimatedVectorDrawable 即制作完成。

7.5 三维动画：OpenGL ES——书本翻页动画

OpenGL 规范被广泛用于 PC 和移动设备，目前最新的 OpenGL 相关版本如下。

- 面对移动领域的 OpenGL ES（OpenGL for Embedded System）版本更新到 3.0。
- 面对桌面领域的 OpenGL 版本更新到 4.3。
- 可运用在增强现实领域的图形接口 OpenVL。

三者中，OpenGL ES 3.0 成为主角，因为它是 Android、iOS 等主流移动平台上的图形接口标准。

OpenGL ES 是 OpenGL 三维图形 API 的子集，针对手机、PDA 和游戏主机等嵌入式设备而设计。OpenGL ES 是从 OpenGL 裁剪定制而来的，去除了 glBegin/glEnd、四边形（GL_QUADS）、多边形（GL_POLYGONS）等复杂图元等许多非绝对必要的特性。

OpenGL ES 3.0 带来如下新特性。

（1）支持更多缓冲区对象。在 OpenGL ES 2.0 时中，缓冲区对象的规范有模糊之处。名字一样的缓冲区对象，在实际渲染中对表现却有细微的差别。针对这个问题，OpenGL ES 3.0 制定了更详细的格式规范。新版 OpenGL ES 还增加对 Uniform Buffer Object 的支持。

（2）新版 OpenGL ES 3.0 着色语言，支持 32 位整数和浮点数据类型以及相关操作。之前版本的着色语言只支持精度低的数据，这样虽然能够加快计算的速度，减少所需的资源，但当着色器的复杂度增加时，出错率也随之增加。同时，新版着色语言的语法更贴近日常语言习惯。

（3）支持遮挡查询（Occlusion Query）以及几何体实例化（Geometry Instancing）。通过遮挡查询，能够让 GPU 知道 3D 场景中，哪些物体被其他物体完全遮挡，GPU 不会去渲染这些完全被遮挡的物体。几何体实例化是通过对具有相同顶点数据的几何体赋予不同的空间位置、颜色或纹理等特征的技术实现的。这两个特性都能够节省硬件资源，提高 3D 图形渲染的性能。

（4）增加多个纹理的支持。包括浮点纹理、深度纹理、顶点纹理等。

（5）多重渲染目标（Multiple Render Targets），让 GPU 一次性渲染多个纹理。

（6）多重采样抗锯齿（MSAA Render To Texture），让 3D 物体边缘不出现毛刺，可提升图像效果。

（7）使用统一的纹理压缩格式 ETC：多年来阻碍 OpenGL 发展的一大瓶颈，就是没有统一的纹理压缩格式，包括 S3TC、GPUs、PVPRTC、ETC 等。因为没有统一标准，开发者不得不根据不同的硬件环境将纹理重复压缩多次，对于 Android 开发者而言，这个过程苦不堪言。显然，统一纹理压缩格式，能够提高开发者的开发效率。

Android 提供了两个基本的类供我们使用 OpenGL ES API 来创建和操纵图形：GLSurfaceView 和 GLSurfaceView. Renderer。

（1）GLSurfaceView

这是一个视图类，可以调用 OpenGL API 绘制图形和操纵物体，功能和 SurfaceView 相似。可以创建一个 GLSurfaceView 类的实例，并添加自己的渲染器。如果要自己实现一些触摸屏的操作，必须扩展这个类来实现触摸监听器。

SurfaceView 是基于 View 视图进行拓展的视图类，更适合 2D 游戏的开发，是 View 的子类，类似于使用双缓机制，在新的线程中更新画面，所以刷新界面速度比 View 快。GLSurfaceView 是基于 SurfaceView 视图再次进行拓展的视图类，是专用于 3D 游戏开发的视图，也是 SurfaceView 的子类，它内嵌的 Surface 专门负责 OpenGL 渲染，供 OpenGL 专用。

GLSurfaceView 提供了下列特性。

- 管理一个 Surface，这个 Surface 就是一块特殊的内存，能直接排版到 Android 的视图 View 上。
- 管理一个 EGL display，它能让 OpenGL 把内容渲染到上述的 Surface 上。
- 用户自定义渲染器（render）。
- 让渲染器在独立的线程里运作，和 UI 线程分离。
- 支持按需渲染（on-demand）和连续渲染（continuous）。

（2）GLSurfaceView. Renderer

这个接口定义了在 OpenGL 的 GLSurfaceView 中绘制图形所需要的方法。必须在一个单独的类中为这些接口提供实现，并使用 GLSurfaceView. setRenderer( ) 方法将它依附到 GLSurfaceView 实例对象上。

需要实现 GLSurfaceView. Renderer 的以下方法。

- onSurfaceCreated()：系统在创建 GLSurfaceView 时调用这个方法一次。我们可以使用它来设置 OpenGL 的环境变量，或初始化 OpenGL 的图形物体。
- onDrawFrame()：系统在每次重绘 GLSurfaceView 时调用这个方法。这个方法主要完成绘制图形的操作。
- onSurfaceChanged()：系统在 GLSurfaceView 的几何属性发生改变时调用该方法，包括大小或设备屏幕的方向发生变化。例如，系统在屏幕从直立变为水平时调用此方法。这个方法主要用于对 GLSurfaceView 容器的变化进行响应。

使用步骤如下。

（1）创建自定义类 1，继承自 GLSurfaceView，并创建构造器。

（2）创建自定义类 2，实现 GLSurfaceView. Renderer 接口。重写 onDrawFrame（GL10 gl）、onSurfaceChanged（GL10 gl，int width，int height）、onSurfaceCreated（GL10 gl，EGLConfig config）方法。

（3）在自定义 1 中定义自定义类 2，并在自定义类 1 的构造方法中调用 setRenderer（renderer）方法进行渲染器设置。

（4）在主 Activity 中创建自定义类 1，并将其设置为视图，为了使 GLSurfaceView 能够与 Activity 同步，要重写 Activity 的 onPause 和 onResume 方法，并分别在相应的方法中调用 GLSurfaceView 的 onPause 和 onResume 方法。

使用 GLSurfaceView 的框架代码如下。

（1）创建一个 GLSurfaceView()。

（2）setRenderer（自定义 renderer）。

（3）setContentView（GLSurfaceView 对象）。

（4）自定义的 Renderer 要实现 GLSurfaceView. Renderer 接口。

例如在主 Activity 中使用 GLSurfaceView，然后设置一个实现 Renderer 接口类的对象，这个类是 OpenGLERender，实现 Renderer 方法。

```
public class MainActivity extends Activity {
    @Override
    public void onCreate(Bundle savedInstanceState) {
     super.onCreate(savedInstanceState);
     //full screen
    requestWindowFeature(Window.FEATURE_NO_TITLE);
     getWindow().setFlags (WindowManager.LayoutParams.FLAG_FULLSCREEN, WindowManager.LayoutParams.FLAG_FULLSCREEN);
    GLSurfaceView glSurfaceView=new GLSurfaceView (this);
    glSurfaceView.setRenderer (new OpenGLESRenderer());
    setContentView (glSurfaceView);
  }
}
public class OpenGLESRender implements Renderer {
    @Override
  public void onSurfaceCreated (GL10 gl, EGLConfig config) {
```

```
    }
    @Override
    public void onDrawFrame (GL10 gl) {
    }
    @Override
    public void onSurfaceChanged (GL10 gl, int width, int height) {

    }
}
```

本例使用 OpenGL ES 的 Android 的 GLSurfaceView 实现书本翻页效果。下面是具体实现过程。在 Android 2.3 中创建应用项目：OPENGL_PageCurl。

（1）在项目的源代码目录下包含五个类：MainActivity. java、CurlView. java、CurlRender. java、CurlPage. java 和 CurlMesh. java，如图 7-11 所示。

CurlMesh. java 实现卷页曲线的计算及曲线封闭图形的绘制，CurlPage. java 设置前景和背景的位图选择设置，CurlRender. java 实现 GLSurfaceView. Renderer 接口，实现对卷页效果的渲染，CurlView. java 继承 GLSurfaceView 类，实现卷页的整体效果。

CurlMesh
CurlPage
CurlRenderer
CurlView
MainActivity

图 7-11　项目的源代码文件

（2）类 CurlMesh. java 实现卷页曲线的计算及曲线封闭图形的绘制，主要代码如下。

```
public class CurlMesh {
    /*计算卷页曲线的最大点数,点越多,曲线越平滑*/
    public CurlMesh(int maxCurlSplits) {
      mMaxCurlSplits = maxCurlSplits < 1 ? 1 : maxCurlSplits;
      mArrScanLines = new Array<Double>(maxCurlSplits+2);
      mArrOutputVertices = new Array<Vertex>(7);
      mArrRotatedVertices = new Array<Vertex>(4);
      mArrIntersections = new Array<Vertex>(2);
      mArrTempVertices = new Array<Vertex>(7+4);
      ...
  }
  /*由曲线的中心点、方向和半径计算卷页曲线*/
  public synchronized void curl(PointF curlPos, PointF curlDir, double radius) {
    if(DRAW_CURL_POSITION) {
      mBufCurlPositionLines.position (0);
      mBufCurlPositionLines.put (curlPos.x);
      mBufCurlPositionLines.put (curlPos.y-1.0f);
      mBufCurlPositionLines.put (curlPos.x);
      mBufCurlPositionLines.put (curlPos.y+1.0f);
      mBufCurlPositionLines.put (curlPos.x-1.0f);
      ...
  }
  /*计算给定扫描线的交叉点*/
```

```
    private Array <Vertex > getIntersections (Array <Vertex > vertices, int [] []
lineIndices, double scanX) {
        mArrIntersections.clear();
        for (int j = 0; j < lineIndices.length; j ++) {
            Vertex v1 = vertices.get (lineIndices [j] [0]);
            Vertex v2 = vertices.get (lineIndices [j] [1]);
            ...
        }
        ...
    }
    /*渲染曲线网格*/
    public synchronized void onDrawFrame (GL10 gl) {
        if (DRAW_TEXTURE && mTextureIds = = null) {
            mTextureIds = new int [2];
            gl.glGenTextures (2, mTextureIds, 0);
            ...
            }
            ...
    }
    /*翻转纹理*/
    public synchronized void setFlipTexture (boolean flipTexture) {
  mFlipTexture = flipTexture;
  if (flipTexture) {
      setTexCoords (1f, 0f, 0f, 1f);
  } else {
      setTexCoords (0f, 0f, 1f, 1f);
  }
}
    /*更新边界*/
    public void setRect (RectF r) {
      mRectangle [0] .mPosX = r.left;
      mRectangle [0] .mPosY = r.top;
      mRectangle [1] .mPosX = r.left;
      mRectangle [1] .mPosY = r.bottom;
      mRectangle [2] .mPosX = r.right;
      mRectangle [2] .mPosY = r.top;
      mRectangle [3] .mPosX = r.right;
      mRectangle [3] .mPosY = r.bottom;
    }
    /*设置矩形的坐标*/
    private synchronized void setTexCoords (float left, float top, float right,
float bottom) {
      mRectangle [0] .mTexX = left;
```

```
  mRectangle [0] .mTexY = top;
  mRectangle [1] .mTexX = left;
  mRectangle [1] .mTexY = bottom;
  mRectangle [2] .mTexX = right;
  mRectangle [2] .mTexY = top;
  mRectangle [3] .mTexX = right;
  mRectangle [3] .mTexY = bottom;
}
…
}
```

（3）类 CurlPage. java 的主要代码如下。

```
public class CurlPage {
…
  /*使显示的图像扩大2倍*/
  private Bitmap getTexture(Bitmap bitmap, RectF textureRect) {
    int w = bitmap.getWidth();
    int h = bitmap.getHeight();
    int newW = getNextHighestPO2(w);
    int newH = getNextHighestPO2(h);
    Bitmap bitmapTex = Bitmap.createBitmap(newW, newH, bitmap.getConfig());
    Canvas c = new Canvas(bitmapTex);
    c.drawBitmap(bitmap, 0, 0, null);
    // Calculate final texture coordinates.
    float texX = (float) w / newW;
    float texY = (float) h / newH;
    textureRect.set(0f, 0f, texX, texY);
    return bitmapTex;
}
/*前景和背景纹理图像的选择*/
public Bitmap getTexture(RectF textureRect, int side) {
  switch(side) {
      case SIDE_FRONT:
        return getTexture (mTextureFront, textureRect);
      default:
        return getTexture (mTextureBack, textureRect);
  }
}
/*纹理图像的设置*/
public void setTexture (Bitmap texture, int side) {
  if (texture == null) {
      texture = Bitmap.createBitmap (1, 1, Bitmap.Config.RGB_565);
      if (side == SIDE_BACK) {
        texture.eraseColor (mColorBack);
```

```
    } else {
      texture.eraseColor (mColorFront);
  }
}
switch (side) {
  case SIDE_FRONT:
    if (mTextureFront ! = null)
      mTextureFront.recycle();
    mTextureFront = texture;
    break;
  case SIDE_BACK:
    if (mTextureBack ! = null)
        mTextureBack.recycle();
    mTextureBack = texture;
    break;
case SIDE_BOTH:
    if (mTextureFront ! = null)
      mTextureFront.recycle();
    if (mTextureBack ! = null)
      mTextureBack.recycle();
    mTextureFront = mTextureBack = texture;
    break;
  }
  mTexturesChanged = true;
  }
  …
}
```

(4) 类 CurlRender.java 的主要代码如下。

```
public class CurlRenderer implements GLSurfaceView.Renderer {
  /*构造函数*/
public CurlRenderer(CurlRenderer.Observer observer) {
  mObserver = observer;
  mCurlMeshes = new Vector<CurlMesh>();
   mPageRectLeft = new RectF();
   mPageRectRight = new RectF();
  }
  /*将 CurlMesh 增加到渲染*/
public synchronized void addCurlMesh(CurlMesh mesh) {
    removeCurlMesh(mesh);
    mCurlMeshes.add(mesh);
}
/*选择已保存的左页或右页的矩形*/
public RectF getPageRect(int page) {
```

```
        if(page == PAGE_LEFT) {
            return mPageRectLeft;
        } else if (page == PAGE_RIGHT) {
          return mPageRectRight;
      }
        return null;
      }
      //所有的绘图操作都在此方法中执行
      @Override
      public synchronized void onDrawFrame (GL10 gl) {
          mObserver.onDrawFrame();
          gl.glClearColor (Color.red (mBackgroundColor) / 255f,
          Color.green (mBackgroundColor) / 255f,
          Color.blue (mBackgroundColor) / 255f,
          Color.alpha (mBackgroundColor) / 255f);
          gl.glClear (GL10.GL_COLOR_BUFFER_BIT);
          gl.glLoadIdentity();
          if (USE_PERSPECTIVE_PROJECTION) {
          gl.glTranslatef (0, 0, -6f);
        }
        for (int i = 0; i < mCurlMeshes.size(); ++i) {
            mCurlMeshes.get (i). onDrawFrame (gl);
        }
      }
      //当窗口大小改变时调用
      @Override
      public void onSurfaceChanged (GL10 gl, int width, int height) {
        gl.glViewport (0, 0, width, height);
        mViewportWidth = width;
        mViewportHeight = height;
        float ratio = (float) width / height;
        mViewRect.top = 1.0f;
        mViewRect.bottom = -1.0f;
        mViewRect.left = -ratio;
        mViewRect.right = ratio;
        updatePageRects();
        gl.glMatrixMode (GL10.GL_PROJECTION);
        gl.glLoadIdentity();
        if (USE_PERSPECTIVE_PROJECTION) {
            GLU.gluPerspective (gl, 20f, (float) width / height, .1f, 100f);
        } else {
      GLU.gluOrtho2D (gl, mViewRect.left, mViewRect.right, mViewRect.bottom, mVie-
wRect.top);
```

```
    }
    gl.glMatrixMode (GL10.GL_MODELVIEW);
    gl.glLoadIdentity();
  }
  @Override
  public void onSurfaceCreated (GL10 gl, EGLConfig config) {
    gl.glClearColor (0f, 0f, 0f, 1f);
    gl.glShadeModel (GL10.GL_SMOOTH);
    gl.glHint (GL10.GL_PERSPECTIVE_CORRECTION_HINT, GL10.GL_NICEST);
    gl.glHint (GL10.GL_LINE_SMOOTH_HINT, GL10.GL_NICEST);
    gl.glHint (GL10.GL_POLYGON_SMOOTH_HINT, GL10.GL_NICEST);
    gl.glEnable (GL10.GL_LINE_SMOOTH);
    gl.glDisable (GL10.GL_DEPTH_TEST);
    gl.glDisable (GL10.GL_CULL_FACE);
    mObserver.onSurfaceCreated();
  }
  …
  }
```

（5）类 CurlView.java 继承 GLSurfaceView 类，实现卷页的整体效果，主要代码如下。

```
public class CurlView extends GLSurfaceView implements View.OnTouchListener,
      CurlRenderer.Observer
{
/*构造函数*/
public CurlView(Context ctx, AttributeSet attrs) {
  super(ctx, attrs);
  init(ctx);
}
/*构造函数*/
private void init(Context ctx) {
  mRenderer = new CurlRenderer(this); //使用前面实现的类,新建渲染对象
  setRenderer(mRenderer);  //设置渲染对象
    setRenderMode(GLSurfaceView.RENDERMODE_WHEN_DIRTY);
    setOnTouchListener (this);   //设置触屏事件处理方法
    mPageLeft = new CurlMesh (10);
    mPageRight = new CurlMesh (10);
    mPageCurl = new CurlMesh (10);
    mPageLeft.setFlipTexture (true);
    mPageRight.setFlipTexture (false);
}
/*绘图操作*/
@Override
public void onDrawFrame() {
…
```

```
    }
//当窗口被创建时调用
@Override
public void onSurfaceCreated() {
  mPageLeft.resetTexture();
  mPageRight.resetTexture();
  mPageCurl.resetTexture();
}
/*触屏事件处理函数*/
@Override
public boolean onTouch (View view, MotionEvent me) {
…
}
/*界面显示一个图片还是两个图片*/
public void setViewMode (int viewMode) {
  switch (viewMode) {
  case SHOW_ONE_PAGE:
      mViewMode = viewMode;
      mPageLeft.setFlipTexture (true);
      mRenderer.setViewMode (CurlRenderer.SHOW_ONE_PAGE);
      break;
  case SHOW_TWO_PAGES:
      mViewMode = viewMode;
      mPageLeft.setFlipTexture (false);
      mRenderer.setViewMode (CurlRenderer.SHOW_TWO_PAGES);
      break;
  }
}
/*根据页装入新位图*/
private void startCurl (int page) {
…
}
/*更新卷页的位置*/

private void updateCurlPos (PointerPosition pointerPos) {
…
}
/*更新页的位图*/
private void updatePages() {
…
}
/*提供位图图片的接口，在调用此类时，实现此接口设置翻页的位图及其数量*/
public interface PageProvider {
```

```
...
}
```

（6）项目的主 Activity 类 MainActivity. java 的主要代码如下。

```
public class MainActivity extends Activity {
@Override
public void onCreate(Bundle savedInstanceState) {
    super.onCreate(savedInstanceState);
    setContentView(R.layout.activity_main);
    int index = 0;
    if (getLastNonConfigurationInstance() != null) {
         index = (Integer) getLastNonConfigurationInstance();
    }
    mCurlView = (CurlView) findViewById (R.id.curl);
    mCurlView.setPageProvider (new MainActivity.PageProvider());
    mCurlView.setSizeChangedObserver (new MainActivity.SizeChangedObserver());
    mCurlView.setCurrentIndex (index);
    mCurlView.setBackgroundColor (0xFF202830);
}
private class PageProvider implements CurlView.PageProvider {
  //位图资源文件
  private int [] mBitmapIds = { R.drawable.m01, R.drawable.m02,
           R.drawable.m03, R.drawable.m04, R.drawable.m05, R.drawable.m06  };
      //页数
      @Override
      public int getPageCount() {
          return 6;
  }
  ...
}
```

（7）在资源目录/res/drawable/下准备 6 张翻页的图片，如图 7-12 所示。

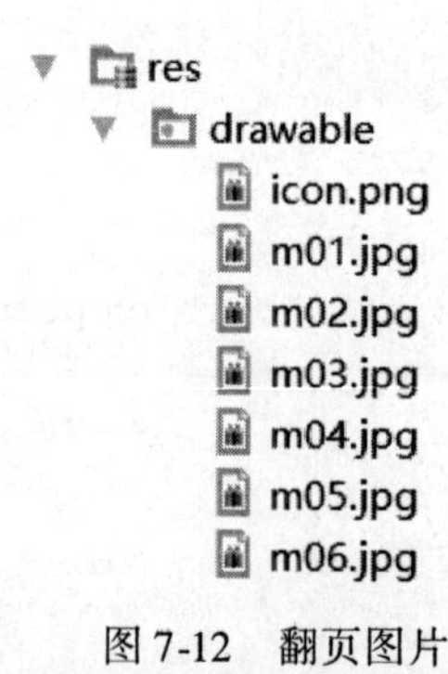

图 7-12　翻页图片

（8）主布局文件 activity_main. xml 中只有一个自定义的 CurlView 类，如图 7-13 所示。
（9）项目运行结果如图 7-14 所示。

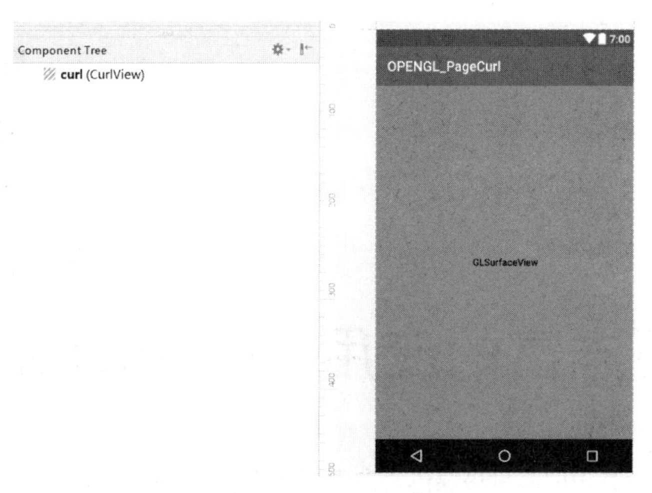

图 7-13　主布局文件

图 7-14　项目运行结果

7.6 本章小结

Android 动画在相关应用场景和游戏中有广泛的应用，是 Android 应用的重要组成部分。本章首先介绍了 Android 的有关动画，包括绘图动画、Drawable 动画、矢量动画等基本的图形类及二位动画，在此基础上，进一步深入介绍了 OpenGL ES、Android 的三维动画，进一步丰富了 Android 的应用效果。

第八章

更丰富的应用——Android多媒体

很多人对移动设备中的海量多媒体资源很感兴趣，多媒体资源一般包括视频、音频和图片等。Android 多媒体是 Android 的重要组成部分，本章主要讲解 Android 开发中访问和操作音频与视频的方法。

8.1 视频播放器1——MediaController + VideoView 播放视频

VideoView 用于播放一段视频媒体，它继承了 SurfaceView，是一个视频控件，包的位置：android. widget. VideoView。

播放一段视频，不可避免地要涉及到开始、暂停、停止等操作，VideoView 也为开发人员提供了对应的方法，下面是一些常用的方法。

- int getCurrentPosition()：获取当前播放的位置。
- int getDuration()：获取当前播放视频的总长度。
- isPlaying()：当前 VideoView 是否在播放视频。
- void pause()：暂停。
- void seekTo (int msec)：从第几毫秒开始播放。
- void resume()：重新播放。
- void setVideoPath (String path)：以文件路径的方式设置 VideoView 播放的视频源。
- void setVideoURI (Uri uri)：以 URI 的方式设置 VideoView 播放的视频源，可以是网络 URI，也可以是本地 URI。
- void start()：开始播放。
- void stopPlayback()：停止播放。
- setMediaController (MediaController controller)：设置 MediaController 控制器。
- setOnCompletionListener (MediaPlayer. onCompletionListener l)：监听播放完成的事件。
- setOnErrorListener (MediaPlayer. OnErrorListener l)：监听播放发生错误时的事件。
- setOnPreparedListener (MediaPlayer. OnPreparedListener l)：：监听视频装载完成的事件。

与 MediaPlayer 配合 SurfaceView 播放视频不同，VideoView 播放之前无需编码装载视频，它会在 start()开始播放的时候自动装载视频。并且，VideoView 在使用完成后，无需编码回收资源。

提到 VideoView，不得不介绍 MediaController。虽然 VideoView 提供了方便的 API 用于播

放、暂停、停止等操作，但还是需要编码完成，如果使用 MediaController，这些操作都可以省去。

MediaController 可以配合 VideoView 播放一段视频，它为 VideoView 提供悬浮的操作栏，在操作栏中可以对 VideoView 播放的视频进行控制，默认情况下，会悬浮显示三秒。它通过 MediaController. setMediaPlayer( )方法指定需要控制的 VideoView，但是仅仅这样是不够的，MediaController 的控制类似于双向控制，MediaController 指定控制的 VideoView，VideoView 还需要指定由哪个 MediaController 来控制，这需要使用 VideoView. setMediaController( )方法。

下面是 MediaController 的一些常用方法。

- boolean isShowing( )：当前悬浮控制栏是否显示。
- void setMediaPlayer（MediaController. MediaPlayerControl player)：设置控制的组件。
- void setPrevNextListeners（View. OnClickListener next，View. OnClickListener prev)：设置上一个视频、下一个视频的切换事件。

通过上面的方法可以看出，setMediaPlayer( )指定的并不是一个 VideoView，而是一个 MediaPlayerControl 接口，MediaPlayerControl 接口内部定义了一些播放相关的播放、暂停、停止等操作，而 VideoView 实现了 MediaPlayerControl。

下面是 MediaController + VideoView 播放视频的实例。在 Android 2. 3 中创建应用项目：VideoView_Play。

（1）在主布局文件 activity_main. xml 中放置一个 VideoView 控件，代码如下。

```
<LinearLayout xmlns:android=http://schemas. android. com/apk/res/android
    android:orientation="vertical"
    android:layout_width=" fill_parent"
    android: layout_height=" fill_parent"
    android: background=" #ffc" >
    <VideoView
        android: id=" @ +id/VideoView01"
        android: layout_width=" match_parent"
        android: layout_height=" wrap_content" />
</LinearLayout>
```

（2）主 Activity 处理文件 MainActivity. java 的代码如下。

```
public class MainActivity extends AppCompatActivity {
private VideoView videoView; //声明
@Override
protected void onCreate(Bundle savedInstanceState) {
    super. onCreate(savedInstanceState);
    setContentView(R. layout. activity_main);
    //网络视频
    String videoUrl2 = Utils. videoUrl;
    Uri uri = Uri. parse (videoUrl2);
    videoView = (VideoView) this. findViewById (R. id. VideoView01);
    //设置视频控制器
    videoView. setMediaController (new MediaController (this));
```

```
        //播放完成回调
        videoView.setOnCompletionListener (new MyPlayerOnCompletionListener());
        //设置视频路径
        videoView.setVideoURI (uri);
        //开始播放视频
        videoView.start();
        }
        class MyPlayerOnCompletionListener implements MediaPlayer.OnCompletionListener
{
        @Override
        public void onCompletion (MediaPlayer mp) {
          Toast.makeText (MainActivity.this, " 播放完成了", Toast.LENGTH_SHORT).show();
        }
      }
    }
```

(3) 在源代码目录下新建文件 Utils.java，设置网络视频的地址，代码如下。

```
public class Utils {
public static final String videoUrl = "http://clips.vorwaerts-gmbh.de/big_buck_
bunny.mp4";
}
```

(4) 在配置文件 AndroidManifest.xml 中增加访问权限，代码如下。

```
<uses-permission android:name="android.permission.INTERNET" />
<uses-permission android:name="android.permission.WRITE_EXTERNAL_STORAGE" />
<uses-permission android: name=" android.permission.READ_EXTERNAL_STORAGE" />
```

(5) 项目运行结果如图 8-1 所示。

图 8-1　项目运行结果

8.2 视频播放器2——MediaPlayer + SurfaceView播放视频

前面介绍了使用VideoView播放视频的方法，使用VideoView播放视频方法简单、方便，这个类其实也是继承了SurfaceView类，并且实现了MediaController。

在VideoView上有一个用于对媒体播放进行控制的面板，即MediaPlayer，其中包括快进、快退、播放、暂停按钮以及一个进度条。MediaPlayer具有如下：

优点：比较简单，可以直接进行使用；

缺点：灵活性不高；

第二种方式是使用MediaPlayer和SurfaceView来播放视频，通过MediaPlayer来控制视频的播放、暂停、进度等。但是MediaPlayer主要用于播放音频，没有提供输出图像的输出界面，这时就要用到SurfaceView控件，将它与MediaPlayer结合起来，就能实现视频的输出。

通过SurfaceView显示视频内容的特点如下。

优点：灵活性高，可以进行自定义。

缺点：难度比较大。

MediaPlayer类是Android的SDK中实现多媒体支持的非常重要的一部分，内嵌了支持的格式。MediaPlayer类包含了7种设定数据源的方法，具体如下。

（1）void setDataSource（String path）：设定使用的数据源（文件路径或http/rtsp地址）。

（2）void setDataSource（FileDescriptorfd，long offset，long length）：设定使用的数据源(filedescriptor)。

（3）void setDataSource（FileDescriptor fd）：设定使用的数据源（filedescriptor）。

（4）void setDataSource（Context context，Uri uri）：设定一个如URI内容的数据源。

（5）static MediaPlayercreate（Context context，Uri uri）：根据给定的URI方便地创建MediaPlayer对象的方法。

（6）static MediaPlayercreate（Context context，int resid）：根据给定的资源id方便地创建MediaPlayer对象的方法。

（7）static MediaPlayercreate（Context context，Uri uri，SurfaceHolder holder）：根据给定的URI方便地创建MediaPlayer对象的方法。

SurfaceView类的主要方法如表8-1所示。

表8-1 SurfaceView类主要方法

种 类	方法名称	描 述
构造方法	public SurfaceView（Context context）	通过Context创建SurfaceView对象
	public SurfaceView（Context context，AttributeSet attrs）	通过Context对象和AttributeSet创建SurfaceView对象
	public SurfaceView（Context context，AttributeSet attrs，int defStyle）	通过Context对象和AttributeSet创建SurfaceView对象并可以指定样式
常用方法	public SurfaceHolder getHolder()	得到SurfaceHolder对象，用于管理SurfaceView
	public void setVisibility（int visibility）	设置是否可见，其值可以是VISIBLE、INVISIBLE或GONE

SurfaceView 是视图类 View 的子类，其中内嵌了一个专门用于绘制的 Surface，SurfaceView 可以控制这个 Surface 的格式和尺寸，以及 Surface 的绘制位置。可以这样理解，Surface 就是管理数据的地方，SurfaceView 就是展示数据的地方。

SurfaceHolder 是一个接口，类似于一个 Surface 的监听器。通过三个回调方法监听 Surface 的创建、销毁或者改变。

使用 MediaPlayer + SurfaceView 播放视频的步骤如下。

(1) 创建 MediaPlayer 对象，并设置加载的视频文件。

(2) 在界面布局文件中定义 SurfaceView 控件。

(3) 通过 MediaPlayer. setDisplay (SurfaceHolder sh) 指定视频画面输出到 SurfaceView 之上，SurfaceHolder 可以通过 Surfaceview 的 getHolder( ) 方法获得。

(4) 将 MediaPlayer 的其他一些方法用于播放视频。例如，调用 MediaPlayer. prepare( ) 进行准备，调用 MediaPlayer. start( ) 播放视频。

视频播放时，先确定视频的格式，这和解码相关，不同的格式视频编码不同，通过编码格式进行解码，最后得到一帧一帧的图像，并把这些图像快速显示在界面上，即为播放一段视频。SurfaceView 在 Android 中正是完成这个功能的。SurfaceView 是配合 MediaPlayer 使用的，MediaPlayer 提供了相应的方法设置 SurfaceView 显示图片，只需要为 MediaPlayer 指定 SurfaceView 显示图像即可。它的完整签名为 void setDisplay (SurfaceHolder sh)。它需要传递一个 SurfaceHolder 对象，SurfaceHolder 可以理解为 SurfaceView 装载需要显示的一帧帧图像的容器，它可以通过 SurfaceHolder. getHolder( ) 方法获得。使用 MediaPlayer 配合 SurfaceView 播放视频的步骤与使用 MediaPlayer 播放 MP3 大体一致，只需要额外设置显示的 SurfaceView 即可。

准备完成 SurfaceHolder 后需要给 SurfaceHolder 设置一个 Callback，调用 addCallback( ) 方法。Callback 有如下三个回调函数。

```
surfaceView.getHolder().addCallback(new SurfaceHolder.Callback() {
        @Override
          public void surfaceCreated(SurfaceHolder holder) {
          }
          @Override
          public void surfaceChanged(SurfaceHolder holder, int format, int width,
int height) {
          }
          @Override
          public void surfaceDestroyed(SurfaceHolder holder) {
          }
     }
```

surfaceCreated( ) 会在 SurfaceHolder 被创建的时候回调，在这里可以进行一些初始化的操作，surfaceDestroyed( ) 会在 SurfaceHolder 被销毁的时候回调，在这里可以进行一些释放资源的操作，防止内存泄漏。

一般会在 surfaceCreated 中为 MediaPlayer 设置 surfaceHolder，例如：

```
        @ Override
        public void surfaceCreated(SurfaceHolder holder) {
            player.setDisplay(holder);
        }
```

下面是 MediaPlayer + SurfaceView 播放视频的实例。在 Android 2.3 中创建应用项目：SurfaceView_Player。

（1）在主布局文件 activity_main.xml 中放置一个 SurfaceView 控件和 3 个 Button，如图 8-2 所示。

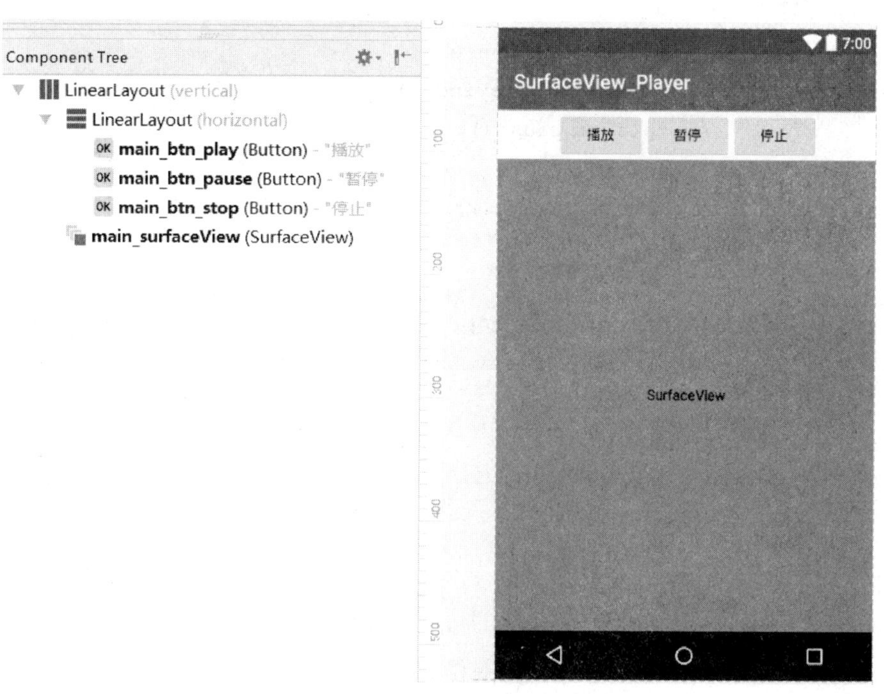

图 8-2　项目布局

（2）主 Activity 文件 MainActivity.java 的代码如下。

```
public class MainActivity extends Activity {
    private SurfaceView surfaceView;
    private MediaPlayer mplayer;
    //记录当前视频的播放位置
  private int position;
  @ Override
  protected void onCreate(Bundle savedInstanceState) {
      super.onCreate(savedInstanceState);
      setContentView(R.layout.activity_main);
        //创建 MediaPlayer 对象
      mplayer = new MediaPlayer();
      //获取 SurfaceView 组件实例
      surfaceView = (SurfaceView) findViewById (R.id.main_surfaceView);
      //设置播放时打开屏幕
```

```
        surfaceView.getHolder().setKeepScreenOn(true);
        surfaceView.getHolder().addCallback(new SurfaceListener());
    }
    public void click(View v) {
        try {
            switch (v.getId()) {
                case R.id.main_btn_play:
                    play();
                    break;
                case R.id.main_btn_pause:
                    if (mplayer.isPlaying()) {
                        mplayer.pause();
                    } else {
                        mplayer.start();
                    }
                    break;
                case R.id.main_btn_stop:
                    if (mplayer.isPlaying())
                        mplayer.stop();
                    break;
                }
            } catch (Exception e) {
                e.printStackTrace();
            }
        }
    private void play() throws IOException {
        mplayer.reset();
        //设置需要播放的视频
        mplayer.setDataSource("http://clips.vorwaerts-gmbh.de/big_buck_
bunny.mp4");
        //把视频画面输出到 surfaceView
        mplayer.setDisplay(surfaceView.getHolder());
        mplayer.prepare();
        mplayer.start();
    }
    private class SurfaceListener implements SurfaceHolder.Callback {
        @Override
        public void surfaceChanged(SurfaceHolder holder, int format, int width, int
height)
        {
        }
        @Override
```

```
    public void surfaceCreated (SurfaceHolder holder) {
        if (position > 0) {
          try {
              //开始播放
              play();
              //直接从指定位置开始播放
              mplayer.seekTo (position);
              position = 0;
          } catch (IOException e) {
              e.printStackTrace();
          }
        }
    }
    @Override
    public void surfaceDestroyed (SurfaceHolder holder) {
      }
  }
@Override
protected void onPause() {
    super.onPause();
    if (mplayer.isPlaying()) {
        //保存当前的播放位置
        position = mplayer.getCurrentPosition();
        mplayer.stop();
    }
}
@Override
protected void onDestroy() {
    super.onDestroy();
    //停止播放
    if (mplayer.isPlaying())
        mplayer.stop();
    //释放资源
    mplayer.release();
}
}
```

(3) 在配置文件 AndroidManifest. xml 中增加访问权限代码如下。

```
<uses-permission android:name="android.permission.INTERNET" />
<uses-permission android:name="android.permission.WRITE_EXTERNAL_STORAGE" />
<uses-permission android: name=" android.permission.READ_EXTERNAL_STORAGE" />
```

(4) 项目运行结果如图 8-3 所示。

图 8-3　项目运行结果

8.3 实现按住说话录音

MediaRecorder 与 MediaPlayer 类似，用于录像录音。MediaRecorder 在录像录音时必须按照 API 说明的调用顺序依次调用，否则会报错，可能会出现无法调用 start( )方法或者调用 start( )后闪退的情况。

要想使用 MediaRecorder，需要在配置文件 AndroidManifest. xml 中增加访问权限。

```
<uses-permission android:name="android.permission.RECORD_AUDIO" />
<user-permission android: name=" android.permission.WRITE_EXTERNAL_STORAGE" />
<user-permission android: name=" android.permission.CAMERA" />
<user-permission android: name=" android.permission.FLASHLIGHT" />
<user-permission android: name=" android.permission.MOUNT_UNMOUNT_FILRSYSTEMS" />
<user-permission android: name=" android.hardware.camera" />
<user-permission android: name=" android.hardware.camera.autofocus" />
```

（1）下面是使用 MediaRecorder 录音的过程。

```
MediaRecorder recorder=new MediaRecorder();
recorder.setAudioSource(MediaRecorder.AudioSource.MIC);
recorder.setOutputFormat(MediaRecorder.OutputFormat.THREE_GPP);
//音频编码格式：default, AAC_ELD, AMR_NB, AMR_WB,
recorder.setAudioEncoder (MediaRecorder.AudioEncoder.AMR_NB);
```

```
recorder.setOutputFile (PATH_NAME);
recorder.prepare();
recorder.start(); // Recording is now started
...
recorder.stop();
recorder.reset(); //You can reuse the object by going back to setAudioSource
() step
recorder.release();
```

（2）下面是使用 MediaRecorder 录像的过程。

```
//设置调用的摄像头
MediaReorder mediavecovder = new MediaRecovder();
mediarecorder.setCamera(Camera);
//指定 Audio、Video 来源
mediarecorder.setAudioSource(MediaRecorder.AudioSource.CAMCORDER);
mediarecorder.setVideoSource(MediaRecorder.VideoSource.CAMERA);
//指定 CamcorderProfile(需要 API Level 8 以上版本)
mediarecorder.setProfile(CamcorderProfile.get(CamcorderProfile.QUALITY_HIGH));
//设置输出格式和编码格式 (针对低于 API Level 8 版本)
mediarecorder.setOutputFormat (MediaRecorder.OutputFormat.THREE_GPP); //设置输
出格式，.THREE_GPP 为 3gp，.MPEG_4 为 mp4
mediarecorder.setAudioEncoder (MediaRecorder.AudioEncoder.DEFAULT); //设置声
音编码类型 mic
mediarecorder.setVideoEncoder (MediaRecorder.VideoEncoder.H264); //设置视频编
码类型，一般为 H263、H264
mediarecorder.setOutputFile (" /sdcard/myVideo.3gp");
mediarecorder.setVideoSize (640, 480); //设置视频分辨率，若设置错误，调用 start()
时会报错，可注释掉在运行程序测试，有时注释掉可以运行
mediarecorder.setVideoFrameRate (24); //设置视频帧率，可省略
mediarecorder.setVideoEncodingBitRate (10* 1024* 1024);
mediarecorder.setPreviewDisplay (surfaceHolder.getSurface()); //设置视频预览
try {
    //准备录制
    mediarecorder.prepare();
    //开始录制
    mediarecorder.start();
} catch (IllegalStateException e) {
    e.printStackTrace();
} catch (IOException e) {
    e.printStackTrace();
}
//停止录制
mediarecorder.stop(); //先停止
mediarecorder.reset(); //再重置 mediarecorder
```

```
//释放资源
mediarecorder.release(); //释放 mediarecorder
mediarecorder = null;
if (mCamera ! = null) {
    mCamera.release(); //释放摄像头
    mCamera = null;
    }
}
```

下面是使用 MediaRecorder 录音和使用 MediaPlayer 播放录音的实例。在 Android 2.3 中创建应用项目：Record_Sound。

（1）在主布局文件 activity_main.xml 中放置两个按钮，一个为“录音”，另一个为“播放”，再放置一个文本控件 TextView 以显示录音文件，如图 8-4 所示。布局文件 layout_microphone.xml 为按住录音的弹出窗口，如图 8-5 所示。

图 8-4　主布局文件

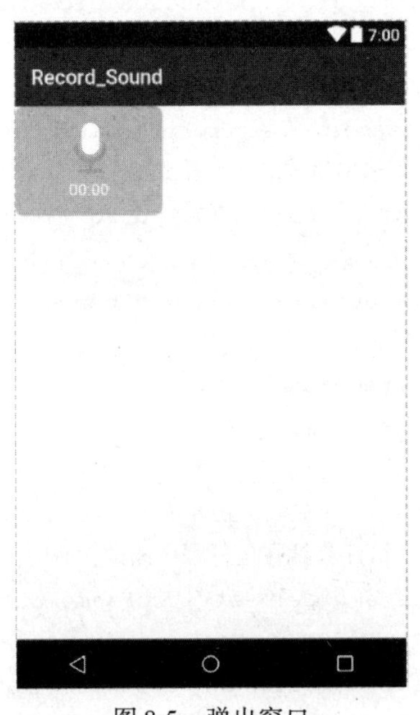

图 8-5　弹出窗口

（2）在项目的源代码目录下包含 4 个类：MainActivity.java、AudioRecordUtils.java、PopWindowFactory.java 和 TimeUtil.java，如图 8-6 所示。

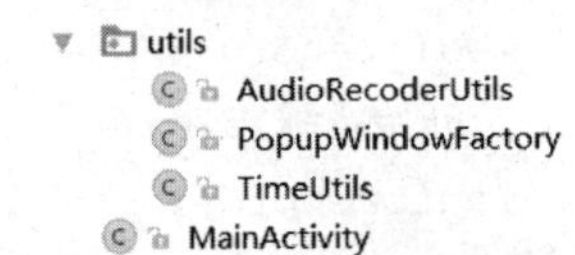

图 8-6　项目的源代码文件

AudioRecordUtils.java 实现录音，PopWindowFactory.java 实现弹出窗口，TimeUtil.java 实现系统时间获取和格式转换，MainActivity.java 实现总体调用。

（3）AudioRecordUtils.java 实现录音，其代码如下。

```
public class AudioRecoderUtils {
  private String filePath; //文件路径
  private String FolderPath;
  private MediaRecorder mMediaRecorder;
```

```
private final String TAG = "fan";
public static final int MAX_LENGTH = 1000 * 60 * 10; //最大录音时长1000*60*10;
private OnAudioStatusUpdateListener audioStatusUpdateListener;
/*文件存储默认 sdcard/record*/
public AudioRecoderUtils() {
  //默认保存路径为/sdcard/record/下
  this (Environment.getExternalStorageDirectory() +" /record/");
}
public AudioRecoderUtils (String filePath) {
  File path = new File (filePath);
    if (! path.exists())
      path.mkdirs();
    this.FolderPath = filePath;
}
private long startTime;
private long endTime;
/*开始录音，使用 MP3 格式*/
public void startRecord() {
  //开始录音
  /*实例化 MediaRecorder 对象 */
  if (mMediaRecorder == null)
    mMediaRecorder = new MediaRecorder();
  try {
  /* setAudioSource/setVedioSource */
  mMediaRecorder.setAudioSource (MediaRecorder.AudioSource.MIC); //设置麦克风
  /*设置音频文件的编码：AAC/AMR_NB/AMR_MB/Default 声音的采样 */
  mMediaRecorder.setOutputFormat (MediaRecorder.OutputFormat.THREE_GPP);
  /*设置输出文件的格式：THREE_GPP/MPEG-4/RAW_AMR/Default THREE_GPP (3gp 格式，
H263 视频/ARM 音频编码)、MPEG-4、RAW_AMR (只支持音频且音频编码要求为 AMR_NB) */
  mMediaRecorder.setAudioEncoder (MediaRecorder.AudioEncoder.AMR_NB);
  filePath = FolderPath + TimeUtils.getCurrentTime() +" .mp3";
  /*准备 */
  mMediaRecorder.setOutputFile (filePath);
  mMediaRecorder.setMaxDuration (MAX_LENGTH);
  mMediaRecorder.prepare();
  /*开始 */
  mMediaRecorder.start();
  /*获取开始时间*/
  startTime = System.currentTimeMillis();
  updateMicStatus();
  Log.e (" fan", " startTime" +startTime);
} catch (IllegalStateException e) {
  Log.i (TAG, " call startAmr (File mRecAudioFile) failed!" +e.getMessage());
} catch (IOException e) {
```

```
        Log.i (TAG, " call startAmr (File mRecAudioFile) failed!" +e.getMessage());
    }
    }
    /*停止录音*/
    public long stopRecord() {
      if (mMediaRecorder == null)
          return 0L;
      endTime = System.currentTimeMillis();
      try {
          mMediaRecorder.stop();
          mMediaRecorder.reset();
          mMediaRecorder.release();
          mMediaRecorder = null;
          audioStatusUpdateListener.onStop (filePath);
          filePath = "";
      } catch (RuntimeException e) {
          mMediaRecorder.reset();
          mMediaRecorder.release();
          mMediaRecorder = null;
          File file = new File (filePath);
          if (file.exists())
              file.delete();
          filePath = "";
          }
        return endTime - startTime;
      }
      /*取消录音*/
    public void cancelRecord() {
      try {
          mMediaRecorder.stop();
          mMediaRecorder.reset();
          mMediaRecorder.release();
          mMediaRecorder = null;
      } catch (RuntimeException e) {
          mMediaRecorder.reset();
          mMediaRecorder.release();
          mMediaRecorder = null;
      }
      File file = new File (filePath);
      if (file.exists())
        file.delete();
      filePath = "";
      }
      private final Handler mHandler = new Handler();
```

```
    private Runnable mUpdateMicStatusTimer = new Runnable() {
      public void run() {
          updateMicStatus();
        }
    };
    private int BASE = 1;
    private int SPACE = 100; //间隔取样时间
    public void setOnAudioStatusUpdateListener (OnAudioStatusUpdateListener au-
dioStatusUpdateListener) {
    this.audioStatusUpdateListener = audioStatusUpdateListener;
    }
    /*更新麦克状态*/
  private void updateMicStatus() {
    if (mMediaRecorder != null) {
      double ratio = (double) mMediaRecorder.getMaxAmplitude() / BASE;
      double db = 0; //分贝
      if (ratio > 1) {
        db = 20 * Math.log10 (ratio);
        if (null != audioStatusUpdateListener) {
        audioStatusUpdateListener.onUpdate (db, System.currentTimeMillis()-startTime);
        }
      }
        mHandler.postDelayed (mUpdateMicStatusTimer, SPACE);
      }
  }
  public interface OnAudioStatusUpdateListener {
    /*录音中...@param db 当前声音分贝 @param time 录音时长*/
    public void onUpdate (double db, long time);
    /*停止录音* @param filePath 保存路径*/
    public void onStop (String filePath);
  }
  }
```

(4) 主 Activity 文件 MainActivity.java 的代码如下。

```
  public class MainActivity extends AppCompatActivity {
    static final int VOICE_REQUEST_CODE = 66;
    private Button mButton, pButton;
    private ImageView mImageView;
    private TextView mTextView;
    private AudioRecoderUtils mAudioRecoderUtils;
    private Context context;
    private PopupWindowFactory mPop;
    private RelativeLayout rl;
    private TextView txt;
    @Override
```

```
    protected void onCreate (Bundle savedInstanceState) {
        super.onCreate (savedInstanceState);
        setContentView (R.layout.activity_main);
        context = this;
        rl = (RelativeLayout) findViewById (R.id.rl);
        mButton = (Button) findViewById (R.id.button);
        pButton = (Button) findViewById (R.id.play);
        txt = (TextView) findViewById (R.id.show_sound);
        pButton.setOnClickListener (new View.OnClickListener()
         {
          @Override
            public void onClick (View v) {
              MediaPlayer player = new MediaPlayer();
              try {
                player.setDataSource (txt.getText().toString());
                player.prepare();
                player.start();
              } catch (Exception e) {
                e.printStackTrace();
              }
            }
        });
        //PopupWindow的布局文件
      final View view = View.inflate (this, R.layout.layout_microphone, null);
      mPop = new PopupWindowFactory (this, view);
      //PopupWindow布局文件里面的控件
      mImageView = (ImageView) view.findViewById (R.id.iv_recording_icon);
      mTextView = (TextView) view.findViewById (R.id.tv_recording_time);
      mAudioRecoderUtils = new AudioRecoderUtils();
      //录音回调
        mAudioRecoderUtils.setOnAudioStatusUpdateListener (new AudioRecoderU-
tils.OnAudioStatusUpdateListener() {
      //录音中....db为声音分贝, time为录音时长
      @Override
      public void onUpdate (double db, long time) {
         mImageView.getDrawable().setLevel ((int) (3000 + 6000 * db / 100));
         mTextView.setText (TimeUtils.long2String (time));
      }
      //录音结束, filePath为保存路径
        @Override
      public void onStop (String filePath) {
        Toast.makeText (MainActivity.this, "录音保存在:" + filePath, Toast.LENGTH_
SHORT).show();
        txt.setText (filePath);
```

```
        mTextView.setText (TimeUtils.long2String (0));
      }
    });
    //6.0 以上需要权限申请
    requestPermissions();
  }
  /*开启扫描之前判断权限是否打开*/
  private void requestPermissions() {
      //判断是否开启摄像头权限
        if ((ContextCompat.checkSelfPermission (context,
Manifest.permission.WRITE_EXTERNAL_STORAGE) ==
PackageManager.PERMISSION_GRANTED) &&
(ContextCompat.checkSelfPermission (context,
Manifest.permission.RECORD_AUDIO) == PackageManager.PERMISSION_GRANTED)
      ) {
          StartListener();
          //判断是否开启语音权限
      } else {
          //请求获取摄像头权限
      ActivityCompat.requestPermissions ((Activity) context, new String [] {Mani-
fest.permission.WRITE_EXTERNAL_STORAGE, Manifest.permission.RECORD_AUDIO}, VOICE_
REQUEST_CODE);
    }
    }
  /*请求权限回调*/
  @Override
  public void onRequestPermissionsResult (int requestCode, String [] permis-
sions, int [] grantResults) {
      super.onRequestPermissionsResult (requestCode, permissions, grantResults);
      if (requestCode == VOICE_REQUEST_CODE) {
          if ((grantResults [0] == PackageManager.PERMISSION_GRANTED) &&
(grantResults [1] == PackageManager.PERMISSION_GRANTED)) {
              StartListener();
        } else {
          Toast.makeText (context, "已拒绝权限!", Toast.LENGTH_SHORT).show();
        }
      }
  }
  public void StartListener() {
    //Button的touch监听
      mButton.setOnTouchListener (new View.OnTouchListener() {
        @Override
        public boolean onTouch (View v, MotionEvent event) {
        switch (event.getAction()) {
```

```
            case MotionEvent.ACTION_DOWN:
                mPop.showAtLocation (rl, Gravity.CENTER, 0, 0);
                mButton.setText (" 松开保存");
                mAudioRecoderUtils.startRecord();
                break;
            case MotionEvent.ACTION_UP:
                mAudioRecoderUtils.stopRecord();    //结束录音（保存录音文件）
                mAudioRecoderUtils.cancelRecord(); //取消录音（不保存录音文件）
                mPop.dismiss();
                mButton.setText (" 按住说话");
                break;
              }
              return true;
            }
        });
      }
    }
```

（5）在配置文件 AndroidManifest.xml 中增加访问权限，代码如下。

```
<uses-permission android:name = "android.permission.WRITE_EXTERNAL_STORAGE" >
</uses-permission>
<uses-permission android: name = " android.permission.RECORD_AUDIO" > </uses-permission>
<uses-permission android: name =" android.permission.MOUNT_UNMOUNT_FILESYSTEMS" />
```

（6）项目运行结果如图 8-7 所示。

图 8-7 项目运行结果

8.4 实现二维码识别

二维码（Two-dimentional code）是用某种特定的几何图形按一定规律在平面（二维方向）分布的黑白相间的图形，是记录数据符号信息的方式。现实生活中，二维码的应用已经非常普遍，例如产品防伪/溯源、广告推送、网站链接、数据下载、商品交易、定位/导航、电子凭证、车辆管理、信息传递、名片交流、WiFi 共享等。

二维码在代码编制上巧妙地利用构成计算机内部逻辑基础的 0、1 比特流的概念，使用若干个与二进制相对应的几何形体来表示文字数值信息，通过图像输入设备或光电扫描设备自动识读，以实现信息自动处理。二维条码有许多种类，常用的码制有 Data Matrix、Maxi-Code、Aztec、QR Code、Vericode、PDF417、Ultracode、Code 49、Code 16K 等。每种码制有其特定的字符集，每个字符占有一定的宽度，具有一定的校验功能等，还具有对不同行的信息自动识别功能及处理图形旋转变化等特点。二维码是一种比一维码更高级的条码格式。一维码只能在一个方向（一般是水平方向）上表达信息，而二维码在水平和垂直方向都可以存储信息。一维码只能由数字和字母组成，而二维码能存储汉字、数字和图片等信息，因此，二维码的应用领域要广得多。常见的二维码如图 8-8 所示。

图 8-8　常见的二维码

二维码可以大致分为矩阵式和行排式两种。

（1）矩阵式

矩形式二维码在一个矩形空间通过黑、白像素在矩阵中的不同分布进行编码。

在矩阵元素位置上，出现方点、圆点或其他形状点表示二进制 1，不出现点表示二进制的 0，点的排列组合确定了矩阵式二维码所代表的意义。矩阵式二维码是建立在计算机图像处理技术、组合编码原理等基础上的一种新型图形符号自动识读处理码制。具有代表性的矩阵式二维码有 Code One、Maxi Code、QR Code、Data Matrix 等。

在 21 * 21 的矩阵中，黑白的区域在 QR 码规范中被指定为固定的位置，称为寻像图形（finder pattern）和定位图形（timingpattern）。寻像图形和定位图形用来帮助解码程序确定图形中具体符号的坐标。黄色的区域用来保存被编码的数据内容以及纠错信息码。蓝色的区域，用来标识纠错的级别（也就是 Level L 到 Level H）和所谓的 Mask pattern，这个区域被称为“格式化信息”（format information）。

（2）行排式

行排式二维码（又称堆积式二维码或层排式二维码）的编码原理是：在一维码基础之上，按需要堆积成二行或多行。它在编码设计、校验原理、识读方式等方面继承了一维码的一些特点，识读设备与条码印刷与一维码技术兼容。但由于行数的增加，需要对行进行判定、其译码算法与软件也不完全与一维码相同。有代表性的行排式二维码有 CODE49、CODE 16K、PDF417 等。

通过图像的采集设备，得到含有条码的图像，此后经过条码定位、分割和解码三个步骤实现条码的识别。

条码的定位采用以下步骤。

（1）利用点运算的阈值理论将采集到的图像变为二值图像，即对图像进行二值化处理。

（2）得到二值化图像后，对其进行膨胀运算。

（3）对膨胀后的图像进行边缘检测得到条码区域的轮廓。

经过上述处理后得到一系列图像。

现在很多 App 都集成了“扫一扫”功能，如微信、QQ、手机助手等。二维码使生活变得更加简洁，扫一扫订餐、扫一扫下载等。说到二维码，不得不提 Google 一个开源的扫码框架：ZXing（下载地址：http://code. google. com/p/zxing/）。

ZXing 是基于多种 1D/2D 条码处理的开源库，是一个完整的项目。它可以通过手机摄像头实现条码的扫描以及解码，功能极其强大。如果要实现二维码的扫描以及解码，我们需要在该开源项目的基础上进行简化和修改。

本例中二维码扫描的技术采用的是 Google 提供的 ZXing 开源项目。扫描条形码就是直接读取条形码的内容，扫描二维码是按照自己指定的二维码格式进行编码和解码。可以到 http://code. google. com/p/zxing/下载 ZXing 项目的源码，然后按照官方文档进行开发。ZXing 支持的二维码格式如表 8-2 所示。

表 8-2　ZXing 支持的二维码格式

一维产品码	一维工业码	二　维　码
UPC-A	Code 39	QR Code
UPC-E	Code 93	Data Matrix
EAN-8	Code 128	Aztec（beta）
EAN-13	Codabar	PDF 417（beta）
	ITF	MaxiCode
	RSS-14	
	RSS-Expanded	

分析项目结构，明确扫描框架需求。在 ZXing 中，有很多其他的功能，项目结构比较复杂二维码 QRCode 扫描需要如下几个包。

- com. google. zxing. client. Android. Camera：基于 Camera 调用以及参数配置，属于核心包。
- DecodeFormatManager、DecodeThread、DecodeHandler：基于解码格式、解码线程、解码结果处理的解码类。

➢ ViewfinderView、ViewfinderResultPointCallBack：基于取景框视图定义的 View 类。
➢ CaptureActivity、CaptureActivityHandler：基于扫描 Activity 以及扫描结果处理的 Capture 类。
➢ InactivityTimer、BeepManager、FinishListener：基于休眠、声音、退出的辅助管理类。
➢ Intents、IntentSource、PrefrencesActivity：基于常量存储的常量类。

下面是使用 ZXing 库识别 QR 码的实例。在 Android 2.3 中创建应用项目：ZxingDemo1。

（1）在项目配置文件 AndroidManifest.xml 中增加访问权限，代码如下。

```
<uses-permission android:name="android.permission.CAMERA"/>
<uses-permission android:name="android.permission.INTERNET" />
<uses-permission android:name="android.permission.VIBRATE" />
<uses-permission android:name="android.permission.FLASHLIGHT" />
```

（2）添加 core-3.0.0.jar 文件到 app/libs 目录下，如图 8-9 所示。

图 8-9　添加 ZXing 库

（3）在 res/layout 目录下新建布局文件 capture.xml，代码如下。

```
<merge xmlns:android="http://schemas.android.com/apk/res/android"
  xmlns:tools="http://schemas.android.com/tools" >
    <!--整体透明画布 -->
  <SurfaceView android:id="@+id/preview_view"
       android:layout_width="fill_parent"
       android:layout_height="fill_parent" />
    <!--扫描取景框 -->
    <com.karics.library.zxing.view.ViewfinderView
       android:id="@+id/viewfinder_view"
       android:layout_width="fill_parent"
       android:layout_height="fill_parent" />
<!--标题栏 -->
    <RelativeLayout android:layout_width="fill_parent"
       android:layout_height="50dp"
       android:layout_gravity="top"
       android:background="#99000000" >
      <ImageButton android:id="@+id/capture_imageview_back"
         android:layout_width="42dp"
         android:layout_height="42dp"
```

```
        android: layout_centerVertical = " true"
        android: background = " @drawable/selector_capture_back" />
    <TextView android: layout_width = " wrap_content"
        android: layout_height = " wrap_content"
        android: layout_centerInParent = " true"
        android: textColor = " #ffffffff"
        android: textSize = " 20sp"
        android: text = " 扫一扫" />
  </RelativeLayout>
</merge>
```

(4) 源代码 captureActivity. java 用于实现 SurfaceHolder. Callback 接口，对应的函数有 onCreate( )、onPause( )、onResume( )、onDestroy( )，涉及到 Camera 的初始化或销毁，主要代码如下。

```
public final class CaptureActivity extends Activity implements
  SurfaceHolder. Callback
{
@Override
  public void onCreate(Bundle icicle) {
      super. onCreate(icicle);
      //保持 Activity 处于唤醒状态
      Window window = getWindow();
      window. addFlags(WindowManager. LayoutParams. FLAG_KEEP_SCREEN_ON);
      setContentView (R. layout. capture);
      hasSurface = false;
      inactivityTimer = new InactivityTimer (this);
      beepManager = new BeepManager (this);
      imageButton_back = (ImageButton) findViewById (R. id. capture_imageview_back);
      imageButton_back. setOnClickListener (new View. OnClickListener() {
  @Override
        public void onClick (View v) {
          finish();
        }
        });
        }
        @Override
        protected void onResume() {
            super. onResume();
          // CameraManager 必须在这里初始化，而不是在 onCreate()中初始化
          cameraManager = new CameraManager (getApplication());
          viewfinderView = (ViewfinderView) findViewById (R. id. viewfinder_view);
          viewfinderView. setCameraManager (cameraManager);
          handler = null;
```

```
        SurfaceView surfaceView = (SurfaceView) findViewById (R.id.preview_view);
        SurfaceHolder surfaceHolder = surfaceView.getHolder();
        if (hasSurface) {
              //初始化 camera
            initCamera (surfaceHolder);
        } else {
           //重置 callback,等待 surfaceCreated()初始化 camera
           surfaceHolder.addCallback (this);
        }
        beepManager.updatePrefs();
        inactivityTimer.onResume();
        source = IntentSource.NONE;
        decodeFormats = null;
        characterSet = null;
      }
      @Override
      protected void onPause() {
          if (handler != null) {
              handler.quitSynchronously();
              handler = null;
          }
          inactivityTimer.onPause();
          beepManager.close();
          cameraManager.closeDriver();
          if (! hasSurface) {
            SurfaceView surfaceView = (SurfaceView) findViewById (R.id.preview_view);
            SurfaceHolder surfaceHolder = surfaceView.getHolder();
            surfaceHolder.removeCallback (this);
          }
          super.onPause();
      }
    @Override
    protected void onDestroy() {
      inactivityTimer.shutdown();
      super.onDestroy();
  }
```

Surfaceview 是基于 Camera 实现的，SurfaceView 的使用需要实现 SurfaceHolder. Callback 接口，在开启屏幕 SurfaceView 时初始化 Camera。

```
@Override
public void surfaceCreated(SurfaceHolder holder) {
      if(! hasSurface) {
        hasSurface = true;
```

```
            initCamera(holder);
        }
    }
    @Override
    public void surfaceDestroyed(SurfaceHolder holder) {
            hasSurface = false;
    }
    @Override
    public void surfaceChanged(SurfaceHolder holder, int format, int width,
            int height) {
    }
```

接下来要初始化 Camera，代码简化之后如下。

```
private void initCamera(SurfaceHolder surfaceHolder) {
      if(surfaceHolder == null) {
          throw new IllegalStateException("No SurfaceHolder provided");
      }
      if(cameraManager.isOpen()) {
           return;
      }
      try {
        //打开 Camera 硬件设备
        cameraManager.openDriver(surfaceHolder);
          //创建一个 handler 打开预览,并抛出一个运行时异常
          if(handler == null) {
              handler = new CaptureActivityHandler(this, decodeFormats,
                      decodeHints, characterSet, cameraManager);
          }
        } catch(IOException ioe) {
            Log.w(TAG, ioe);
            displayFrameworkBugMessageAndExit();
        } catch(RuntimeException e) {
            Log.w(TAG, "Unexpected error initializing camera", e);
            displayFrameworkBugMessageAndExit();
        }
}
```

在 CaptureActivity 中，有一个核心方法，用来返回并处理解码结果，即扫描结果。handleDecode( )将解码的 bitmap 以及内容回传到开启扫描的 Activity 进行处理。

```
public void handleDecode(Result rawResult, Bitmap barcode, float scaleFactor) {
    inactivityTimer.onActivity();
    boolean fromLiveScan = barcode != null;
    //这里处理解码完成后的结果,将参数回传到 Activity 处理
      if(fromLiveScan) {
```

```
        beepManager.playBeepSoundAndVibrate();
        Toast.makeText(this, "扫描成功", Toast.LENGTH_SHORT).show();
        Intent intent = getIntent();
        intent.putExtra (" codedContent", rawResult.getText());
        intent.putExtra (" codedBitmap", barcode);
        setResult (RESULT_OK, intent);
            finish();
        }
    }
```

（5）源代码 CodeCreator. java，用于生成 Url 生成二维码，代码如下。

```
public static Bitmap createQRCode(String url) throws WriterException {
        if(url == null ||url.equals("")) {
            return null;
        }
    //生成二维矩阵,编码时指定大小
      BitMatrix matrix = new MultiFormatWriter().encode(url,
              BarcodeFormat.QR_CODE, 300, 300);
         int width = matrix.getWidth();
        int height = matrix.getHeight();
        //二维矩阵转为一维像素数组
        int [] pixels = new int [width * height];
        for (int y = 0; y < height; y ++) {
            for (int x = 0; x < width; x ++) {
                if (matrix.get (x, y)) {
                    pixels [y * width + x] = 0xff000000;
                }
            }
        }
        Bitmap bitmap = Bitmap.createBitmap (width, height,
            Bitmap.Config.ARGB_8888);
        bitmap.setPixels (pixels, 0, width, 0, 0, width, height);
        return bitmap;
    }
```

通过以上的操作过程，ZXing 项目的简化工作基本完成，二维码扫描的整体构架主要包含如下三部分。

- 定义取景框，即扫描的 View，通过 SurfaceView 进行绘制。
- Camera，扫描的核心在于 Camera 的配置使用，包括预览、自动聚焦、打开设备等处理。
- Decode 解码，扫描完成后整个工程的核心。

除了以上三个模块之外，需要明确的是 CaptureActivitiy 中 handleDeCode( )方法要进行自定义处理。

（6）项目运行结果如图 8-10 所示。

图 8-10　项目运行结果

8.5 Android TTS 文字识别——实现文字朗读

Text-To-Speech 简称 TTS，是 Android 1.6 之后版本中重要的功能，能够将指定的文本转成不同语言音频输出。TTS 可以方便地嵌入到游戏和应用程序中，增加用户体验。Android. speech. tts. TextToSpeench 库如图 8-11 所示。

要使用 Android 的 TTS，需要获取第三方提供的 TTS 支持，因此需要安装一款合适的第三方 TTS 应用，在系统中进行设置即可，例如“讯飞语音 +”，语音流畅度较好，可选语速，而且该软件还在不断更新中。安装最新版的“讯飞语音 +”，完成设置即可，如图 8-12 所示。

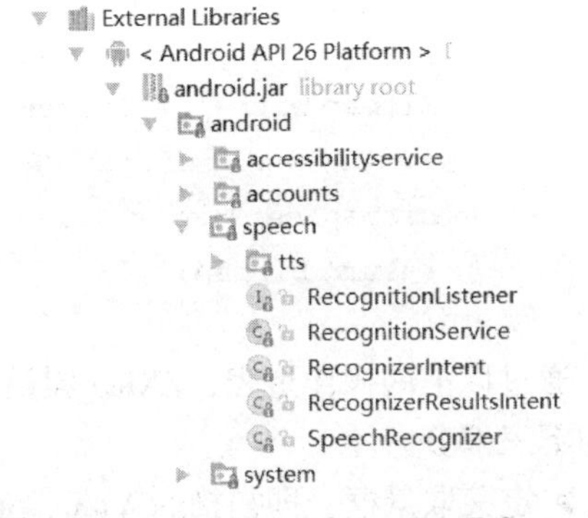

图 8-11　Android. speech. tts. TextToSpeench

开源项目 eyes-free（http://code. google. com/p/eyes-free/，Android 上的 TTS 功能应该也是基于这个开源项目提供的）除了提供 Pico 外，还把支持其他更多语言语音合成的另一个 TTS 引擎 eSpeak 也移植到了 Android 平台，并支持中文的语音合成。

在安装了 eyes-free 提供的 TTS Service Extended 的 apk 后，在程序中使用 eyes-free 提供的 TTS library，并把 TTS Engine 设置为 eSpeak，从而实现朗读中文。不过，经过测试，实际的效果比较差。

图 8-12　“讯飞语音 +”安装与设置

8. 5. 1　Text-To-Speech 开发流程

Android 的自动朗读支持主要是通过 TextToSpeech 完成的，该类提供了如下 TextToSpeech 构造器。

在创建 TextToSpeech 对象时，必须先提供一个 OnInitListener 监听器，该监听器负责监听 TextToSpeech 的初始化结果。

TextToSpeech 最常用的两个方法如下。

（1）speak（String text，int queueMode，HashMap < String，String > params）

（2）synthesizeToFile（String text，HashMap < String，String > params，String filename）

上面两个方法都用于把 text 文字内容转换为音频，区别在于 speak 方法播放转换的音频，而 synthesizeToFile 把转换得到的音频保存成声音文件。

上面两个方法中的 params 都用于指定声音转换时的参数，speak 方法中的 queueMode 参数指定 TTS 的发音队列模式，该参数支持如下两个常量。

（1）TextToSpeech. QUEUE_FLUSH：如果指定该模式，当 TTS 调用 speak 方法时，会中断当前实例正在运行的任务（也可以理解为清除当前语音任务，转而执行新的语音任务）。

（2）TextToSpeech. QUEUE_ADD：如果指定为该模式，当 TTS 调用 speak 方法时，会把新的发音任务添加到当前发音任务队列之后，也就是等任务队列中的发音任务执行完成后再执行 speak 方法指定的发音任务。

当程序用完 TextToSpeech 对象之后，可以在 Activity 的 OnDestroy( ) 方法中调用它的 shutdown( ) 来关闭 TextToSpeech，释放它所占用的资源。

从 Android 5. 0 开始，上述两个方法已被弃用，改用下面的格式。

(1) speak (CharSequence text, int queueMode, Bundle params, String utteranceId)

(2) synthesizeToFile (CharSequence text, Bundle params, File file, String utteranceId)

参数 CharSequence text 为转换的文本；参数 int queueMode 与前面介绍的 int queueMode 意义相同；参数 utteranceId 为一个请求的唯一标识符；参数 Bundle params 为请求的参数，使用的参数名有三个：KEY_PARAM_STREAM、KEY_PARAM_VOLUME、KEY_PARAM_PAN。

- KEY_PARAM_STREAM：指定声音流的类型，它的定义 AudioManager 的格式 STREAM_constants 的其中一个。
- KEY_PARAM_VOLUME：语音的音量，值的范围为从 0 到 1，0 是静音，1 是最大，缺省是 1。
- KEY_PARAM_PAN：指定从左到右如何发音文本，值的范围为从-1 到 1。

使用 TextToSpeech 的步骤如下。

(1) 创建 TextToSpeech 对象，创建时传入 OnInitListener 监听器，监听创建是否成功。

(2) 设置 TextToSpeech 所使用语言、国家选项，通过返回值判断 TTS 是否支持该语言、国家选项。

(3) 调用 speak( )或 synthesizeToFile( )方法。

(4) 关闭 TTS，回收资源。

8.5.2 Text-To-Speech 实现文字朗读

下面是 Android 使用 TTS 实现朗读文字并保存声音的实例，在 Android 2.3 中创建应用项目：TTS。

(1) 在主布局文件 activity_main.xml 中放置两个按钮 Button，一个为“朗读”，一个为“记录声音”，并添加一个文本控件 EditText，用于显示朗读文本，如图 8-13 所示。

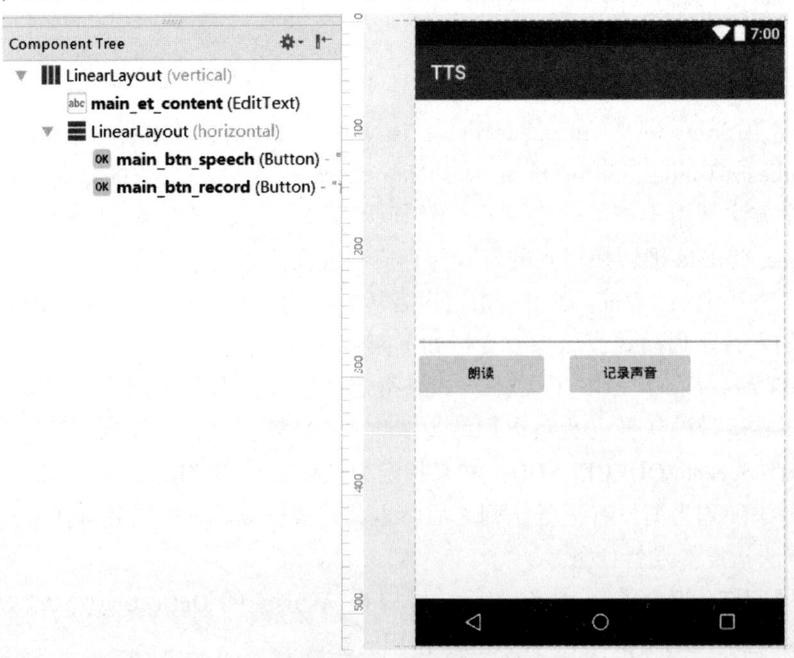

图 8-13　主布局文件

（2）主 Activity 文件 MainActivity. java 的代码如下。

```
public class MainActivity extends Activity {
  private TextToSpeech tts;
  private EditText editText;
  @Override
  protected void onCreate(Bundle savedInstanceState) {
      super.onCreate(savedInstanceState);
      setContentView(R.layout.activity_main);
    //初始化 TextToSpeech 对象
    tts = new TextToSpeech (this, new TextToSpeech.OnInitListener() {
          @Override
          public void onInit (int status) {
            //如果装载 TTS 引擎成功
            if (status == TextToSpeech.SUCCESS) {
                //设置使用中文朗读
                int result = tts.setLanguage (Locale.CHINA);
                //如果不支持所设置的语言
                if (result != TextToSpeech.LANG_COUNTRY_AVAILABLE
                         && result != TextToSpeech.LANG_AVAILABLE) {
                  Toast.makeText (MainActivity.this,
                 " TTS 暂时不支持这种语言的朗读!", Toast.LENGTH_LONG).show();
                }
            }
          }
      });
      editText = (EditText) findViewById (R.id.main_et_content);
  }
  @Override
  protected void onDestroy() {
      super.onDestroy();
    //关闭 TextToSpeech 对象
      if (tts != null) {
          tts.shutdown();
      }
  }
  public void click (View v) {
    switch (v.getId()) {
      case R.id.main_btn_record:
      //将朗读文本的音频记录到指定文件
          File f1 =new File (" /mnt/sdcard/sound.wav");
          //Android 5.0 及以上
          tts.synthesizeToFile (editText.getText(), null, f1," 111");
        //Android 5.0 以下
```

```
        //tts.synthesizeToFile (editText.getText().toString(), null, " /mnt/sd-
card/sound.wav");
            Toast.makeText (MainActivity.this, " 声音记录成功!",
                    Toast.LENGTH_SHORT).show();
                break;
        case R.id.main_btn_speech:
          //执行朗读
          //Android 5.0 及以上
          tts.speak (editText.getText().toString(), TextToSpeech.QUEUE_ADD, null,
" 111");
          //Android 5.0 以下
          // tts.speak (editText.getText().toString(), TextToSpeech.QUEUE_ADD, null);
            break;
        }
    }
}
```

(3) 项目运行结果如图 8-14 所示。

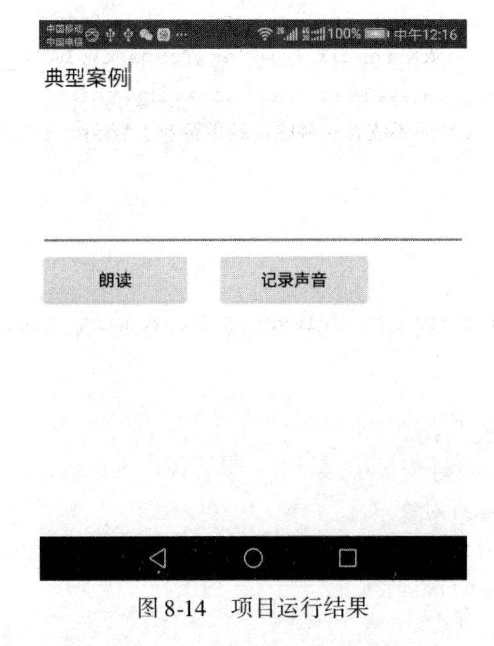

图 8-14　项目运行结果

8.6 Android 语音识别——多种语言语音识别

在 Android 中使用 Google 的 Voice Recognition 的方法极其简单，通过一个 Intent 的 Action 动作完成的方法，主要有以下两种模式。

(1) ACTION_RECOGNIZE_SPEECH：一般语音识别，在这种模式下可以捕捉到语音的处理后的文字列。

(2) ACTION_WEB_SEARCH：网络搜索。

下面实例中采用 ACTION_RECOGNIZE_SPEECH 模式，需要实现 onActivityResult 方法，这是在语音识别结束之后的回调函数，下面是实现代码。

```
public class VoiceRecognition extends Activity implements OnClickListener {
    private static final int VOICE_RECOGNITION_REQUEST_CODE = 1234;
    private ListView mList;
        /* * 呼叫与活动首先建立 * /
        @Override
        public void onCreate (Bundle savedInstanceState)
         {
          super.onCreate (savedInstanceState);
          setContentView (R.layout.voice_recognition);
          Button speakButton = (Button) findViewById (R.id.btn_speak);
          mList = (ListView) findViewById (R.id.list);
          // Check to see if a recognition activity is present
          PackageManager pm = getPackageManager();
           List < ResolveInfo > activities  = pm.queryIntentActivities  (new Intent
(RecognizerIntent.ACTION_RECOGNIZE_SPEECH), 0);
           if (activities.size() ! = 0)
           {
              speakButton.setOnClickListener (this);
          }
          else
           {
              speakButton.setEnabled (false);
              speakButton.setText (" Recognizer not present");
          }
    }
    public void onClick (View v)
    {
          if (v.getId() = = R.id.btn_speak)
           {
              startVoiceRecognitionActivity();
          }
    }
    private void startVoiceRecognitionActivity()
    {
      //通过 Intent 传递语音识别的模式
      Intent intent = new Intent (RecognizerIntent.ACTION_RECOGNIZE _SPEECH);
      //语言模式和自由形式的语音识别
       intent.putExtra  ( RecognizerIntent.EXTRA _ LANGUAGE _ MODEL,  RecognizerIn-
tent.LANGUAGE_MODEL_FREE_FORM);
      //提示语音开始
      intent.putExtra (RecognizerIntent.EXTRA_PROMPT, " Speech recognition demo");
```

```
    //开始执行我们的 Intent、语音识别
  startActivityForResult (intent, VOICE_RECOGNITION_REQUEST_CODE);
    }
    //当语音结束时的回调函数 onActivityResult
    @Override
  protected void onActivityResult (int requestCode, int resultCode, Intent data)
     {
    if (requestCode == VOICE_RECOGNITION_REQUEST_CODE && resultCode == RESULT_
OK)
  {
    //取得语音的字符
    ArrayList<String> matches = data.getStringArrayListExtra (RecognizerIn-
tent.EXTRA_RESULTS);
    mList.setAdapter (new ArrayAdapter<String> (this, android.R.layout.simple_
list_item_1, matches));
      }
      super.onActivityResult (requestCode, resultCode, data);
    }
  }
```

本例要求 Android 系统中安装了支持 RecognizerIntent. ACTION_RECOGNIZE_SPEECH 的应用，例如 Google 的 Voice Search 应用。

下面是 Android 使用 Voice Recognition 实现语音识别并显示对应文字的实例，在 Android 2.3 中创建应用项目：Voice_Recognition。

（1）在主布局文件 voice_recognition.xml 中放置两个按钮 Button 和 ListView 控件，一个实现按住发声，一个实现显示发声的文本，如图 8-15 所示。

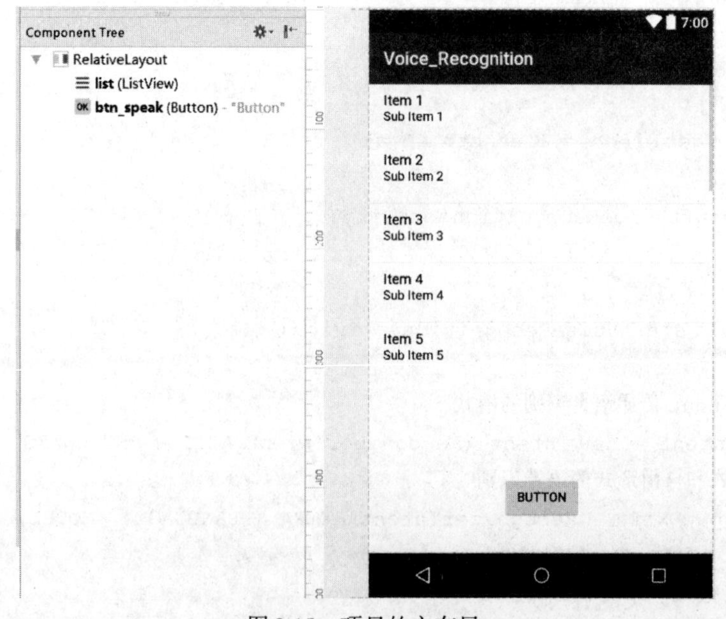

图 8-15　项目的主布局

（2）主 Activity 文件 MainActivity. java 的代码如下。

```
public class MainActivity extends AppCompatActivity implements View.OnClickListener {
      private static final int VOICE_RECOGNITION_REQUEST_CODE = 1234;
      private ListView mList;
    @Override
    public void onCreate (Bundle savedInstanceState)
    {
      super.onCreate (savedInstanceState);
      setContentView (R.layout.voice_recognition);
      Button speakButton = (Button) findViewById (R.id.btn_speak);
      mList = (ListView) findViewById (R.id.list);
      // Check to see if a recognition activity is present
      PackageManager pm = getPackageManager();
      List <ResolveInfo> activities = pm.queryIntentActivities (new Intent (Rec-
ognizerIntent.ACTION_RECOGNIZE_SPEECH), 0);
      if (activities.size() != 0)
       {
          speakButton.setOnClickListener (MainActivity.this);
      }
      Else
       {
          speakButton.setEnabled (false);
          speakButton.setText (" Recognizer not present");
      }
    }
    public void onClick (View v)
    {
      if (v.getId() == R.id.btn_speak)
       {
          startVoiceRecognitionActivity();
      }
    }
    private void startVoiceRecognitionActivity()
    {
      //通过 Intent 传递语音识别的模式
      Intent intent = new Intent (RecognizerIntent.ACTION_RECOGNIZE_SPEECH);
      //语言模式和自由形式的语音识别
       intent.putExtra (RecognizerIntent.EXTRA_LANGUAGE_MODEL, RecognizerIn-
tent.LANGUAGE_MODEL_FREE_FORM);
      //提示语音开始
      intent.putExtra (RecognizerIntent.EXTRA_PROMPT, " Speech recognition demo");
      //开始执行我们的 Intent、语音识别
      startActivityForResult (intent, VOICE_RECOGNITION_REQUEST_CODE);
```

```
    }
    //当语音结束时的回调函数 onActivityResult
    @Override
    protected void onActivityResult (int requestCode, int resultCode, Intent data)
    {
      if (requestCode == VOICE_RECOGNITION_REQUEST_CODE && resultCode == RESULT_
OK)
       {
          //取得语音的字符
    ArrayList<String> matches =
       data.getStringArrayListExtra (RecognizerIntent.EXTRA_RESULTS);
            mList.setAdapter ( new  ArrayAdapter  <  String  >  ( this,
android.R.layout.simple_list_item_1, matches));
       }
       super.onActivityResult (requestCode, resultCode, data);
     }
    }
```

(3) 项目运行结果如图 8-16 所示。

图 8-16　项目运行结果

8.7 基于 Ijkplayer 的视频播放器

Ijkplayer 是 Bilibili 基于 ffmpeg 开发并开源的轻量级视频播放器，支持播放本地网络视频，也支持流媒体播放，这应该是目前 github 最火的开源视频播放器了。Ijkplayer 源码官方下载地址：https://github.com/Bilibili/ijkplayer。Ijkplayer 是一个基于 ffplay 的轻量级 Android/iOS 视频播放器，实现了跨平台功能，API 易于集成，编译配置可裁剪，方便控制安装包大小，并支持硬件加速解码，更加省电。Ijkplayer 还提供 Android 平台下应用弹幕集成的解决方案，此方案已用于美拍和斗鱼 App，目前 Ijkplayer 最新的版本是 0.7.7。

FFplay 是一个使用了 FFmpeg 和 SDL 库的、简单可移植的媒体播放器，FFmpeg 是全球领先的多媒体框架，能够解码、编码、转码、复用、解复用、流、过滤器和播放大部分的视频格式。它提供了录制、转换以及流化音视频的完整解决方案，包含了非常先进的音频/视频编解码库 libavcodec。为了保证较高的可移植性和编解码质量，libavcodec 中很多 code 都是从头开发的。

下载 Ijkplayer 源码后，需要进行编译，Ijkplayer 的编译是在 Ubuntu 下实现的，具体实现过程如下。

（1）需要为 Ubuntu 安装 homebrew、Git、yasm。

```
# install homebrew, git, yasm
ruby -e "$(curl -fsSL https://raw.githubusercontent.com/Homebrew/install/master/install)"
brew install git
brew install yasm
# add these lines to your ~/.bash_profile or ~/.profile
# export ANDROID_SDK=<your sdk path>
# export ANDROID_NDK=<your ndk path>
# on Cygwin (unmaintained)
# install git, make, yasm
```

（2）开始编译。

```
git clone https://github.com/Bilibili/ijkplayer.git ijkplayer-android
cd ijkplayer-android
git checkout -B latest k0.6.1
./init-android.sh
cd android/contrib
./compile-ffmpeg.sh clean
./compile-ffmpeg.sh all
cd..
./compile-ijk.sh all
# Android Studio:
#    Open an existing Android Studio project
#    Select android/ijkplayer/ and import
#    define ext block in your root build.gradle
#    ext {
#      compileSdkVersion = 23       // depending on your sdk version
#      buildToolsVersion = "23.0.0" // depending on your build tools version
#      targetSdkVersion = 23        // depending on your sdk version
```

```
# }
# Eclipse:(obselete)
#    File -> New -> Project -> Android Project from Existing Code
#    Select android/ and import all project
#    Import appcompat-v7
#    Import preference-v7
# Gradle
#    cd ijkplayer
#    gradle
```

Ijkplayer 编译后的结果如图 8-17 所示。

› ijkplayer

名称	修改日期	类型
gradle	2017/7/1 9:25	文件夹
ijkplayer-arm64	2017/7/1 9:25	文件夹
ijkplayer-armv5	2017/7/1 9:25	文件夹
ijkplayer-armv7a	2017/7/1 9:25	文件夹
ijkplayer-example	2017/7/1 9:25	文件夹
ijkplayer-exo	2017/7/1 9:25	文件夹
ijkplayer-java	2017/7/1 9:25	文件夹
ijkplayer-x86	2017/7/1 9:25	文件夹
ijkplayer-x86_64	2017/7/1 9:25	文件夹
tools	2017/7/1 9:25	文件夹
.gitignore	2016/7/18 16:41	文本文档
build.gradle	2016/7/18 16:41	GRADLE 文件
gradle.properties	2016/7/18 16:41	PROPERTIES 文件
gradlew	2016/7/18 16:41	文件
gradlew.bat	2016/7/18 16:41	Windows 批处理...
settings.gradle	2016/7/18 16:41	GRADLE 文件

图 8-17　Ijkplayer 编译结果

各目录的含义如下。

（1）ijkplayer-java：Ijkplayer 的一些操作封装及定义。这里面是通用的 API 接口，其中最主要的是 IMediaPlayer，用于渲染显示多媒体。

（2）ijkplayer-exo：Google 的一个新的开源播放器 ExoPlayer，在 Demo 中和 Ijkplayer 对比使用。通过安装 Ijkplayer 可以发现 setting 里可以选择不同 player 来渲染多媒体显示，该模块下面就是一个 MediaPlayer。

（3）ijkplayer-example：测试程序。

（4）ijkplayer-{arch}：编译出来的各个版本的 .so 文件。

首先需要的是 ijkplayer-{arch}、ijkplayer-Java 两个库。ijkplayer-exo 是 Google 提供的新的播放器，这里不需要使用。

下面是 Android 使用 Ijkplayer 实现播放器的实例，在 Android 2.3 中创建应用项目：Ijkplayer_Example。

（1）把 ijkplayer-armv7a/src/main/libs 中的文件复制到新工程 app 目录的 libs 中。

（2）把 ijkplayer-java/build/outputs/aar/ijkplayer-java-release.aar 复制到新工程 app 目录的 libs 中。

(3) 修改 App 中的 build. gradle，主要设置 . so 及 . aar 的位置。

```
apply plugin:'com.android.application'
android {
  compileSdkVersion 26
  buildToolsVersion "26.0.0"
...
sourceSets {
  main {
      jniLibs.srcDirs = ['libs']  /** 在 libs 文件夹中找 so 文件 */
    }
  }
}
repositories {
  mavenCentral()
  flatDir {
  dirs 'libs'/** 在 libs 文件夹中找 aar 文件 */
}
}
dependencies {
...
compile(name:'ijkplayer-java-release', ext:'aar') /* 编译 ijkplayer-java-release.aar 文件 */
}
```

(4) 复制 ijkplayer-example 下面的 tv. danmaku. ijk. media. example. widget. media 到项目的源代码目录中，如图 8-18 所示。

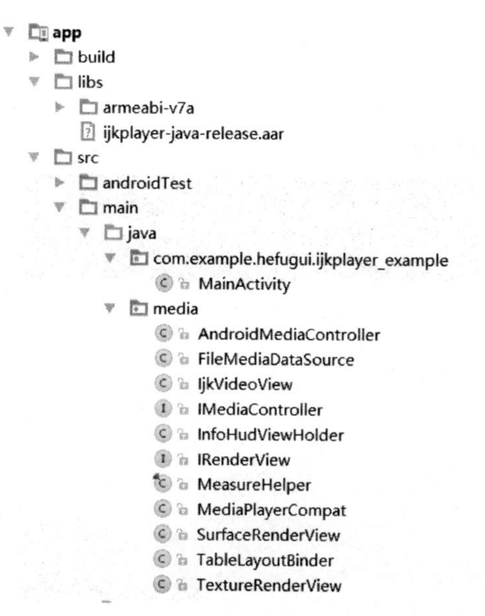

图 8-18　拷贝 ijkplayer-example 的源代码到项目

(5) 在 AndroidManifest. xml 中增加网络权限，代码如下。

```
<uses-permission android:name = "android. permission. INTERNET"/ >
```

（6）主布局文件 activity_main. xml 的代码如下。

```
<? xml version = "1.0" encoding = "utf-8"? >
<RelativeLayout xmlns:android = http://schemas. android. com/apk/res/android
  xmlns:tools = http://schemas. android. com/tools
  android:layout_width = " match_parent"
  android: layout_height = " match_parent"
  tools: context = " . MainActivity" >
   <media. IjkVideoView
    android: id = " @ + id/video_view"
    android: layout_width = " match_parent"
    android: layout_height = " match_parent" / >
</RelativeLayout >
```

（7）主 Activity 文件 MainActivity. java 用于实现对 IjkVideoView 的调用，代码如下。

```
public class MainActivity extends AppCompatActivity {
    private IjkVideoView videoView;
    @ Override
    protected void onCreate (Bundle savedInstanceState) {
        super. onCreate (savedInstanceState);
        setContentView (R. layout. activity_main);
        videoView = (IjkVideoView) findViewById (R. id. video_view);
        videoView. setAspectRatio (IRenderView. AR_ASPECT_FIT_PARENT);
  videoView. setVideoURI (Uri. parse (" http: //clips. vorwaerts-gmbh. de/big_buck_
bunny. mp4"));
        videoView. start ();
    }
}
```

（8）项目运行结果如图 8-19 所示 .

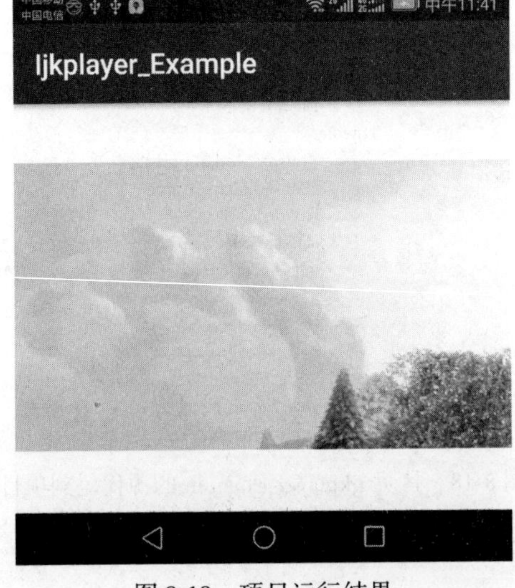

图 8-19　项目运行结果

上面展示的是 Ijkplayer 的最基本应用方法，Ijkplayer 有丰富的播放器功能，可以基于 Ijkplayer 定制更高级的播放器。

8.8 本章小结

本章介绍了 Android 音视频的操作，主要使用 Android 的系统类来实现，最后介绍了基于 Ijkplayer 的视频播放器，Ijkplayer 是 Bilibili 基于 ffmpeg 开发并开源的轻量级视频播放器，被 Android 开发者广泛应用。

另外，Android 7.0 添加了新的 VR 模式的平台支持和优化，帮助开发者为用户打造高质量移动 VR 体验，并增加了一些性能增强特性，包括允许 VR 应用访问某个专属的 CPU 核心。在您的应用中，可以充分利用专为 VR 设计的智能头部跟踪和立体声通知功能。最重要的是，Android 7.0 的图形延时非常低。

第九章

连接到远方——Android网络开发

如果说21世纪进入了互联网时代，那么21世纪的第二个十年，则是移动互联网时代的真正来临，这是因为支持移动互联网普及的关键设备——智能手机在2010年代后得到了迅速普及。

Android开发中最重要的组成部分就是通过网络与服务器端的交互操作，以获取数据。目前移动终端运用非常广泛的有Wifi、NFC、蓝牙等。本章介绍Android基本网络技术和常用传输数据格式的使用方法。

9.1 Android应用程序的权限

Android是一个多进程系统，每一个应用程序（和系统的组成部分）都运行在自己的进程中。通过进程ID，系统可以区分不同的应用程序和系统组件，并赋予不同的权限。更细粒度的安全特性则通过“许可”机制来提供，该机制能够对一个进程可执行的操作进行约束。

9.1.1 Android权限机制详解

权限是一种安全机制，Android权限主要用于限制应用程序内部某些具有限制性特性的功能使用以及应用程序之间的组件访问。

Android安全机制中的一个重要特点是在默认情况下应用程序没有权限执行对其他应用程序、操作系统或用户有害的操作。这些操作包括读/写用户的隐私数据（例如联系方式或E-mail）、读/写其他应用程序的文件、访问网络、阻止设备休眠等。应用程序的进程是一个安全的Sandbox，它的资源不能被外界访问，也不能访问其他应用程序。为了在应用之间共享资源，必须显式地声明，而在声明自己提供的资源的同时，可以要求使用方具备相应的权限。应用程序在manifest中声明自己需要拥有的权限，在安装时向系统请求，安装程序通过用户的反馈和验证应用程序的作者予以确认。

在Android应用程序运行时，若使用的权限没有申请或将设置里的权限手动关闭，那么应用程序将直接崩溃。

一个新建的Android应用默认是没有权限的，这意味着它不能执行任何可能对用户体验有不利影响的操作或者访问设备数据。为了使用受保护的功能，必须在应用程序的Android-

Manifest. xml 文件中添加一个或多个 <uses-permission> 标签。

在 AndroidManifest. xml 文件的 <manifest> 标签内使用权限标签声明使用某一个权限，可输入权限标签，Android Studido 会出现提示，主要权限标签有 <permission> 和 <uses-permission>。

1. <uses-permission>——系统权限标签

如果 Android 应用程序需要访问系统一个受保护的操作，例如访问网络、写外部存储器、定位等官方定义的权限，则要在 Android 应用程序的配置文件 AndroidManifest. xml 中添加相应的 <uses-permission>，否则应用程序在运行时会发生崩溃。

在 Android Studio 中添加权限，在 AndroidManifest. xml 文件中，选择 <uses-permission> 标签，Android Studio 会出现提示，从权限列表中进行选择，如图 9-1 所示。

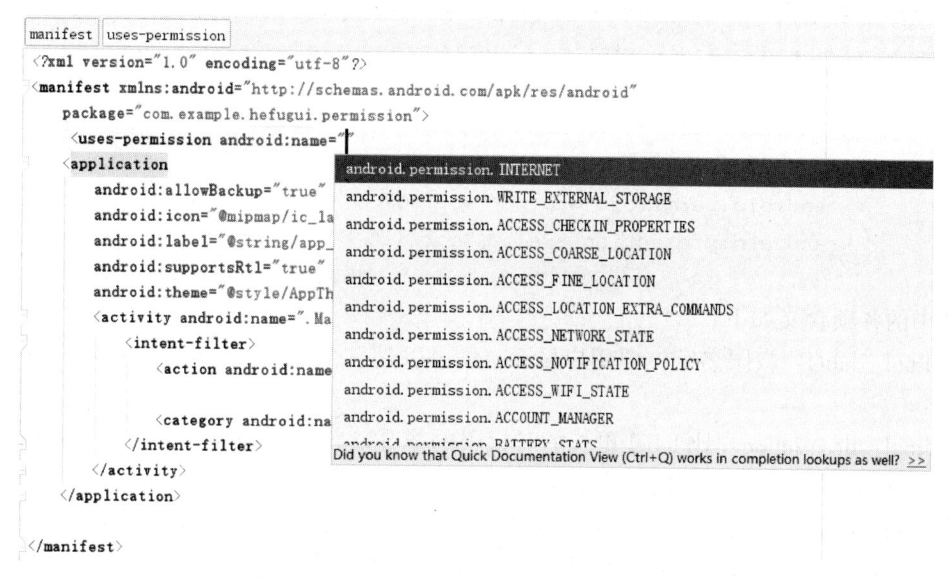

图 9-1　<uses-permission> 标签的权限列表

例如，若要访问网络，则增加如下代码。

```
<? xml version = "1.0" encoding = "utf-8"? >
  <manifest xmlns:android = http://schemas.android.com/apk/res/android
   package = "com.example.hefugui.okhttp" >
   <uses-permission android:name = "android.permission.INTERNET"/ >
   <application >
   ……
   </application >
  </manifest >
```

使用蓝牙的权限代码如下。

```
<uses-permission android:name = "android.permission.BLUETOOTH"/ >
```

允许程序读取用户联系人数据的代码如下。

```
<uses-permission android:name = "android.permission.READ_CONTACTS " / >
```

允许程序读取短信息的代码如下。

```
<uses-permission android: name = " android.permission.READ_SMS" / >
```

允许程序从非系统拨号器里输入电话号码的代码如下。

```
<uses-permission android: name =" android. permission. CALL_PHONE " />
```

允许程序拨打电话并替换系统的拨号器界面的代码如下。

```
<uses-permission android: name =" android. permission. CALL_PRIVILEGED " />
```

允许访问摄像头进行拍照的代码如下。

```
<uses-permission android: name =" android. permission. CAMERA " />
```

2. <permission>——自定义权限标签

应用程序能用 permission 保护自己的组件，可以使用 Android 系统定义的或者其他应用定义或者自身应用定义的 permission。

自定义权限的使用分两步，第一步是权限定义，第二步应用程序权限声明。

（1）如果要定义一个新的 permission，可以用 <permission> 节点来定义，格式如下。

```
<permission android:description = "string resource"
          android:icon = "drawable resource"
          android:label = "string resource"
          android:name = "string"
          android:permissionGroup = "string"
          android:protectionLevel = ["normal" |"dangerous" |
                                    "signature" |"signatureOrSystem"] />
```

其中的各项含义如下。

android：label：权限名字，向用户显示，值可以是一个 string 数据，例如这里的“自定义权限”。

android：description：比 label 更长的对权限的描述。值是在 resource 文件中获取的，不能直接写 string 值，例如这里的“@ string/hello”。

android：name：权限名字，如果其他 App 引用该权限，需要填写这个名字。

android：protectionLevel：权限级别，分为如下 4 个级别。

- normal：低风险权限，在安装的时候，系统会自动授予权限给 application。
- dangerous：高风险权限，系统不会自动授予权限给 App，在用到的时候，会向用户提示。
- signature：签名权限，在其他 App 引用声明的权限时，需要保证两个 App 的签名一致。这样系统会自动授予权限给第三方 App，而不提示给用户。
- signatureOrSystem：引用该权限的 App 需要有和系统同样的签名才能授予权限，一般不推荐使用。

android：permissionGroup：将本权限归为某个权限组。属性值是组的名称。

（2）在应用程序中使用前面定义的 <permission> name 属性声明。

<permission> 和 <uses-permission> 两者之间不同之处如下

- <uses-permission> 是 Android 预定义的权限，<permission> 是自己定义的权限，<permission> 相对来说使用较少。
- <uses-permission> 是官方定义的权限，是调用其他应用时自己需要声明的权限，<permission> 是自己定义的权限，别人调用这个程序时需要用 <uses-permission> 来声明。

在一般情况下，不需要为自己的应用程序声明某个权限，除非提供了供其他应用程序调用的代码或者数据，在这个时候才需要使用 < permission > 这个标签，很显然，这个标签可以让我们声明自己程序的权限。

9. 1. 2　Android 6. 0 网络权限管理

从 Android 6. 0 开始，权限分成两种，第一种是普通权限，不涉及用户隐私，只需要在 Manifest 中声明即可，比如网络、蓝牙、NFC 等；第二种是危险权限（即运行时权限，下面统称为运行时权限），涉及到用户隐私信息，需要用户授权后才可使用，比如 SD 卡读写、联系人、短信读写等。

（1）普通权限：如果应用程序在 manifest 中声明了普通权限，系统会自动授予这些权限。

（2）运行时权限：如果应用程序在 manifest 中声明了运行时权限（也就是说，这些权限可能会影响用户隐私和设备的普通操作），如表 9-1 所示，应用程序在运行时会明确地让用户决定是否授予这些权限。系统请求用户授予这些权限的方式是由当前应用运行的系统版本来决定的。安装器会决定是否授予它所声明的权限，这有时候会询问用户。只有权限被授予，这个应用才能使用受保护的特性，否则，访问失败并且不会通知用户。

表 9-1　危险权限列表

Permission Group（权限组）	Permissions（权限）
Android. permission-group. CALENDAR	Android. permission. READ_CALENDAR Android. permission. WRIT_CALENDAR
Android. permission-group. CAMERA	Android. permission. CAMERA
Android. permission-group. CONTATCS	Android. permission. READ_CONTATCS Android. permission. WRITE_CONTATCS Android. permission. GET_CONTATCS
Android. permission-group. LOCATION	Android. permission. ACCESS_FINE_LOCATION Android. permission. ACESS_COARSE_LOCATION
Android. permission-group. MICRIPHONE	Android. permission. RECORD_AUDIO
Android. permission-group. PHONE	Android. permission. READ_PHONE_STATE Android. permission. CALL_PHONE Android. permission. READ_CALL_LOG Android. permission. WRITE_CALL_LOG Android. permission. USE_SIP Android. permission. PROCESS_OUTGOING_CALLS
Android. permission-group. SENSORS	Android. permission. BODY_SENSORS
Android. permission-group. SMS	Android. permission. SEND_SMS Android. permission. RECEIVE_SMS Android. permission. READ_SMS Android. permission. RECEIVE_WAP_PUSH Android. permission. RECEIVE_SMS Android. permission. READ_CELL_BROADCASTS
Android. permission-group. STORAGE	Android. permission. READ_EXTERNAL_STORAGE Android. permission. WRITE_EXTERNAL_STORAGE

这里以 CALL_PHONE 这个权限作为实例进行说明。

(1) 在布局文件 activity_main. xml 中增加一个按钮，如图 9-2 所示。

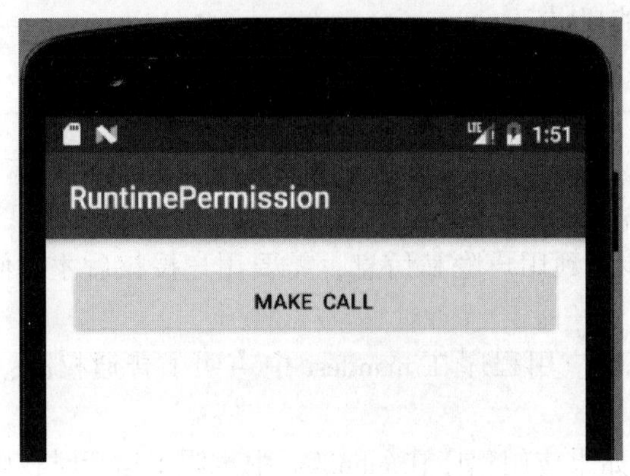

图 9-2　RuntimePermission 项目布局

(2) 在 AndroidManifest. xml 中添加权限，代码如下。

```
<? xml version="1.0" encoding="utf-8"? >
<manifest xmlns:android=http://schemas. android. com/apk/res/android
    package="com. example. hefugui. okhttp" >
<uses-permission android:name="android. permission. CALL_PHONE" / >
    <application >
    ……
    </application >
</manifest >
```

(3) 在 MainActivity 的 Button 监听事件中增加拨打电话的代码。

```
public class MainActivity extends AppCompatActivity {
    @Override
    protected void onCreate(Bundle savedInstanceState) {
        super. onCreate(savedInstanceState);
        setContentView(R. layout. activity_main);
        Button makeCall = (Button) findViewById (R. id. make_call);
        makecall. setOnClickListener (new View. OnClickListener()
         {
            @Override
            public void onClick (View v) {
                try
                 {
                    Intent intent = new Intent (Intent. ACTION_CALL);
                    intent. setData (Uri. parse (" tel: 10086"));
                    startActivity (intent);
                }
                catch (Exception e)
                 {
```

```
                    e.printStackTrace();
                }
            }
        });
    }
}
```

（4）运行程序，在低于 Android 6.0 系统的手机上运行是正常的，但是如果在 6.0 以上高版本中运行，单击 Make Call 按钮将没有任何效果，这时观察 logcat 的打印信息，会看到如图 9-3 所示的信息。

```
.err: java.lang.SecurityException: Permission Denial: starting Intent { act=android.intent.action.CALL dat=tel:xxxxx c
m.example.hefugui.runtimepermission/u0a76} (pid=5613, uid=10076) with revoked permission android.permission.CALL_PHONE
.err:     at android.os.Parcel.readException(Parcel.java:1683)
.err:     at android.os.Parcel.readException(Parcel.java:1636)
.err:     at android.app.ActivityManagerProxy.startActivity(ActivityManagerNative.java:3071)
```

图 9-3　项目运行日志信息

信息中提示 Permission Denial，这是由于权限禁止造成的，因为在 6.0 以上版本的系统中，使用危险权限必须在运行时做出处理。

在 Android 6.0 系统及以上的源代码编辑中，Android Studio 环境也会出现权限的提示，如图 9-4 所示。

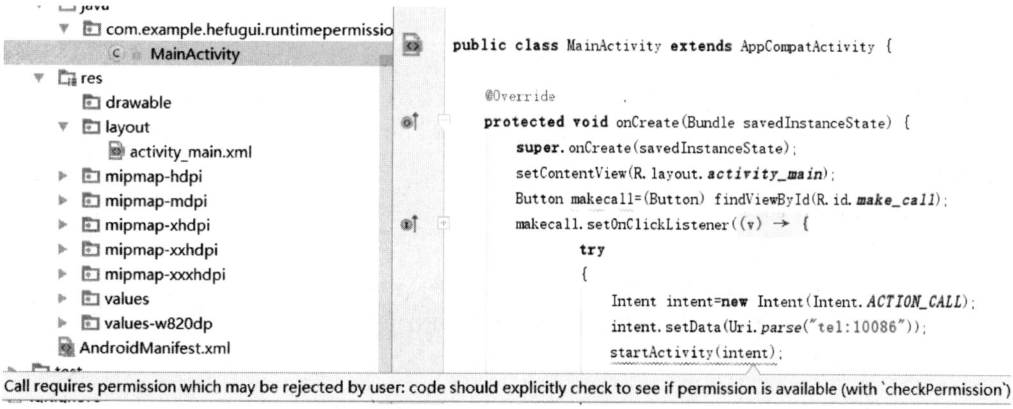

```
com.example.hefugui.runtimepermissio
    MainActivity
res
    drawable
    layout
        activity_main.xml
    mipmap-hdpi
    mipmap-mdpi
    mipmap-xhdpi
    mipmap-xxhdpi
    mipmap-xxxhdpi
    values
    values-w820dp
AndroidManifest.xml

public class MainActivity extends AppCompatActivity {

    @Override
    protected void onCreate(Bundle savedInstanceState) {
        super.onCreate(savedInstanceState);
        setContentView(R.layout.activity_main);
        Button makecall=(Button) findViewById(R.id.make_call);
        makecall.setOnClickListener((v) -> {
            try
            {
                Intent intent=new Intent(Intent.ACTION_CALL);
                intent.setData(Uri.parse("tel:10086"));
                startActivity(intent);

Call requires permission which may be rejected by user: code should explicitly check to see if permission is available (with `checkPermission`)
```

图 9-4　权限代码提示

下面修复这个问题，修改 MainActivity 的代码如下。

```
public class MainActivity extends AppCompatActivity {
    @Override
    protected void onCreate(Bundle savedInstanceState) {
        super.onCreate(savedInstanceState);
        setContentView(R.layout.activity_main);
        Button makeCall = (Button) findViewById(R.id.make_call);
        makeCall.setOnClickListener(new View.OnClickListener() {
            @Override
            public void onClick(View v) {
                if (ContextCompat.checkSelfPermission(MainActivity.this,
```

```
Manifest. permission. CALL_PHONE) ! =PackageManager. PERMISSION_GRANTED)
    {
      ActivityCompat. requestPermissions(MainActivity. this,
              new String[]{Manifest. permission. CALL_PHONE}, 1);
        }
        else{
             call();
        }
        }
      });
    }
    private void call() {
      try {
          Intent intent = new Intent(Intent. ACTION_CALL);
          intent. setData(Uri. parse("tel:10086"));
            startActivity(intent);
        } catch(SecurityException e) {
              e. printStackTrace();
          }
        }

    @Override
    public void onRequestPermissionsResult(int requestCode, String[]
    permissions, int[] grantResults) {
        switch(requestCode) {
          case 1:
            if(grantResults. length > 0 && grantResults[0] ==
                PackageManager. PERMISSION_GRANTED) {
                call();
            } else {
                Toast. makeText(this, "You denied the permission",
                  Toast. LENGTH_SHORT). show();
            }
              break;
            default:
        }
    }}
```

如果执行的操作需要一个 dangerous permission，那么每次在执行操作的地方都必须检查是否有这个 permission，因为用户可以在应用设置里随意地更改授权情况，所以必须每次在使用前都检查是否有权限。

检查权限的方法为 ContextCompat. checkSelfPermission()，其两个参数分别是 Context 和权限名，返回值是 PERMISSION_GRANTED 或 PERMISSION_DENIED。

如果已经授权，将直接拨打电话，如果没有，则调用 ActivityCompat. requestPermissions()申请授权，requestPermissions() 方法 3 个参数，Activity、permission 名字的数组和一个整型的 request code。

调用 requestPermissions() 方法，会弹出一个对话框，用户可选择同意或拒绝权限申请，不论哪种结果，授权的结果会封装在 grantResults 参数中，如果用户同意将调用 call()方法，如果不同意，则放弃操作。

现在重新运行程序，单击 Make Call 按钮，结果如图 9-5 所示，如果用户同意授权，结果如图 9-6 所示。

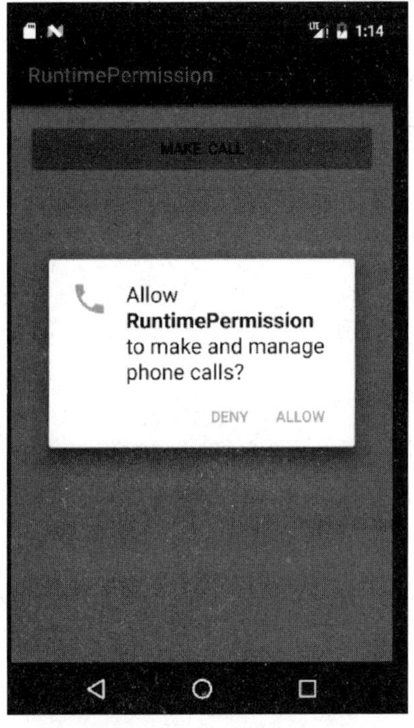

图 9-5　申请电话权限对话框

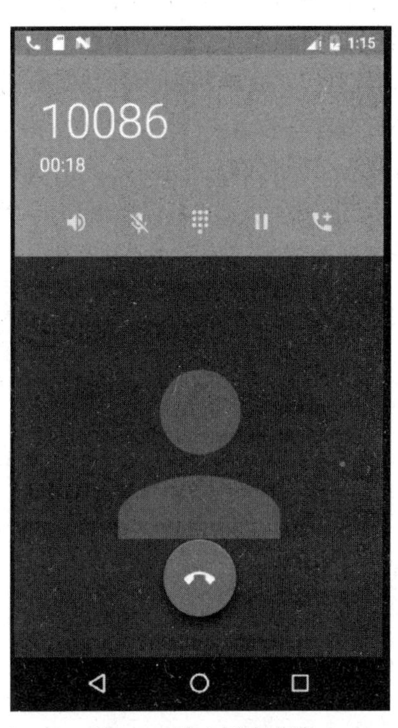

图 9-6　用户同意申请

9.2 解析 JSON 格式数据

JSON（JavaScript Object Notation）是一种轻量级的数据交换格式。JSON 采用完全独立于语言的文本格式，这些特性使 JSON 成为理想的数据交换语言，易于阅读和编写，同时也易于机器解析和生成。

JSON 建构于如下两种结构。

（1）“名称/值”对的集合：不同的语言中，它被理解为对象（object）、记录（record）、结构（struct）、字典（dictionary）、哈希表（hash table）、有键列表（keyed list），或者关联数组（associative array）。

例如：{ " firstName" : " Brett" , " lastName" :" McLaughlin" , " email" : " aaaa" }

（2）值的有序列表：在大部分语言中，它被理解为数组（array）。例如，表示人名的列表如下。

```
{ "people": [
    { "firstName": "Brett", "lastName":"McLaughlin", "email": "aaaa" },
    { "firstName": "Jason", "lastName":"Hunter", "email": "bbbb"},
    { "firstName": "Elliotte", "lastName":"Harold", "email": "cccc" }
]}
```

9.2.1 使用 JSONObject

本节介绍 JSONObject 的使用方法。

1. 将数据封装成为 JSON 格式

JSON 可以把各种数据，包括对象数据，装成为 JSON 格式，封装方法如下。

JSONObject jsonObject = new JSONObject( )：定义 JSON 对象。

jsonObject. put （key，value）：放入值。

return jsonObject. toString( )：变为字符串。

例子：现在要创建如下 JSON 文本。

```
{
    "phone" : ["12345678", "87654321"], // 数组
    "name" : "yuanzhifei89", // 字符串
    "age" : 100, // 数值
    "address" : { "country" : "china", "province" : "jiangsu" }, // 对象
    "married" : false //布尔值
}
```

创建代码如下。

```
try {
    //首先创建一个对象
    JSONObject person = new JSONObject();
    //第一个键 phone 的值是数组,所以需要创建数组对象
    JSONArray phone = new JSONArray();
    phone.put("12345678").put("87654321");
    person.put("phone", phone);
    person.put("name", "yuanzhifei89");
    person.put("age", 100);
    //键 address 的值是对象,所以又要创建一个对象
    JSONObject address = new JSONObject();
    address.put("country", "china");
    address.put("province", "jiangsu");
    person.put("address", address);
    person.put("married", false);
} catch(JSONException ex) {
        throw new RuntimeException(ex);
}
```

2. JSON 对象解析

JSON 可以把接收到的数据解析为封装 JSON 格式的数据，解析方法如下。

➢ 从网络获得字符串数据，例如名字为 retSrc。

➢ JSONObject result =new JSONObject（retSrc）：字符串构造 JSON 对象。

➢ result. get( )：取出数据。

JSONObject 对象有各种 get( )方法，例如 getInt( )用于取出整数；getBoolean( )用于取出布尔型；getJSONObjec( )用于取出 JSONObject 对象，如图 9-7 所示。

```
JSONObject result = new JSONObject( person.toString());
result.
getBoolean(String name)                      boolean
get(String name)                             Object
getJSONObject(String name)                   JSONObject
getInt(String name)                          int
getString(String name)                       String
accumulate(String name, Object value)        JSONObject
getDouble(String name)                       double
getJSONArray(String name)                    JSONArray
getLong(String name)                         long
```

图 9-7　JSONObject 对象的各种 get( )方法

对于前面封装的 JSON 对象，其解析方法如下。

```
JSONObject result = new JSONObject(person.toString());
String name = result.getString("name");
int age =  result.getInt("age");
JSONObject address_resut =result.getJSONObject (" address");
boolean married =result.getBoolean (" address");
```

9.2.2　使用 GSON

GSON 为 Google 的一个开源 JSON 解析工具包，Gighub 网址：https://github.com/google/gson。使用 GSON 解析 JSON 数据，可以大大简化 JSON 数据解析过程，并避免参数缺少或对应不上等问题。

在使用 GSON 之前，在内层的 build.gradle 文件，也就是 app/build.gradle 文件的 dependencies 中增加如下内容。

```
dependencies {
    compile fileTree(dir: 'libs', include: ['* .jar'])
    compile 'com.android.support:appcompat-v7:25.1.0'
    compile 'com.google.code.gson:gson:2.8.0'
    testCompile 'junit:junit:4.12'
}
```

添加以上内容，应用会自动下载 GSON 库，目前最新的版本是 2.8.0，如图 9-8 所示

在 GSON 的 API 中，提供了两个重要的方法：toJson( )和 fromJson( )方法。其中，toJson( )方法用于实现将 Java 对象转换为相应的 JSON 数据，fromJson( )方法则用于实现将 JSON 数据转换为相应的 Java 对象。

1. toJson( )方法

toJson( )方法用于将 Java 对象转换为相应的 JSON 数据，主要有以下几种形式。

- String toJson（JsonElement jsonElement）
- String toJson（Object src）
- String toJson（Object src，Type typeOfSrc）

其中，第一个方法用于将 JsonElement 对象（可以是 JsonObject、JsonArray 等）转换成 JSON 数据；第二个方法用于将指定的 Object 对象序列化成相应的 JSON 数据；第三个方法用于将指定的 Object 对象（可以包括泛型类型）序列化成相应的 JSON 数据。

图 9-8　下载的 GSON 库

2. fromJson( )方法

fromJson( )方法用于将 JSON 数据转换为相应的 Java 对象，主要有以下几种形式。

- <T> T fromJson（JsonElement json，Class <T> classOfT）
- <T> T fromJson（JsonElement json，Type typeOfT）
- <T> T fromJson（JsonReader reader，Type typeOfT）
- <T> T fromJson（Reader reader，Class <T> classOfT）
- <T> T fromJson（Reader reader，Type typeOfT）
- <T> T fromJson（String json，Class <T> classOfT）
- <T> T fromJson（String json，Type typeOfT）

以上的方法用于将不同形式的 JSON 数据解析成 Java 对象。

新建项目 GSON_Test，MainActivity 的代码如下。

```
public class MainActivity extends AppCompatActivity {
  Button btn;
  EditText et1,et2,et3,et4;
    @Override
  protected void onCreate(Bundle savedInstanceState) {
      super.onCreate(savedInstanceState);
      setContentView(R.layout.activity_main);
      btn = (Button) findViewById (R.id.btn);
      et1 = (EditText) findViewById (R.id.et1);
      et2 = (EditText) findViewById (R.id.et2);
      et3 = (EditText) findViewById (R.id.et3);
      et4 = (EditText) findViewById (R.id.et4);
      btn.setOnClickListener (new View.OnClickListener()
       {
         @Override
        public void onClick (View v) {
          Gson gson = new Gson();
          //第 1 种类对象
            Student student = new Student();
```

```
                student. setName (" xuanyouwu");
                student. setAge (26);
                String jsonStr = gson. toJson (student);
                et1. setText (jsonStr);
                //第 2 种列表对象
                List < String > list = Arrays. asList (" 1", " a", " 3", " rt", " 5");
                et2. setText (gson. toJson (list));
                //第 3 种映射对象
                Map < String, Object > content = new HashMap < String, Object > ();
                content. put (" name", " xiaoming");
                content. put (" age", " 23");
                et3. setText (gson. toJson (content));
                //使用 GSON 的 fromJson 方法
                Type type = new TypeToken < ArrayList < String > > () { } . getType();
                ArrayList < String > sList = gson. fromJson (list. toString(), type);
                et4. setText (sList. toString());
            }
        });
    }
    //定义的类
    public static class Student {
        private int age;
        private String name;
        public String getName() { return name; }
        public int getAge() {  return age;   }
        public void setAge (int age) {  this. age = age; }
        public void setName (String name) { this. name = name; }
    }
}
```

运行结果如图 9-9 所示。

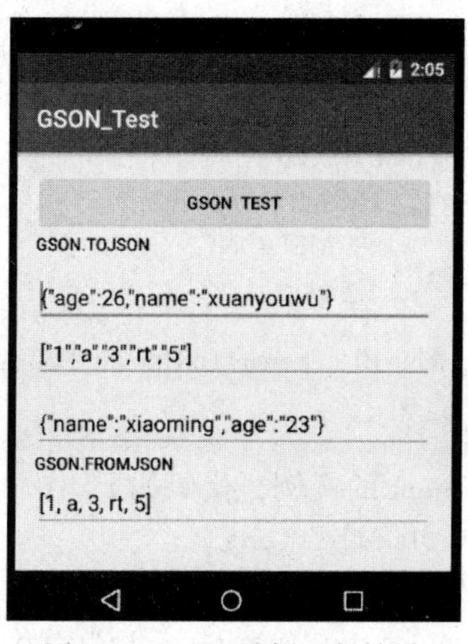

图 9-9 GSON 示例运行结果

9.3 使用 OkHttp3 请求天气预报

OkHttp 是由大名鼎鼎的 Square 公司开发的，除了 OkHttp 之外，该公司还开发了 Picasso、Retrofit 等著名的开源项目。OkHttp 不仅在封装上做得简单容易，连底层实现也独具特色，和 HttpURLConnection 相比，更加出色，OkHttp 官网地址：http://square.github.io/okhttp/，OkHttp GitHub 地址：https://github.com/square/okhttp。

OkHttp 是一个 Http 请求框架，相当于 Android 原生的 HttpChent 和 httpURLConnectiond 的封装，写法更加简单，可以处理更加复杂的网络请求。

在使用 OkHttp 之前，要在内层的 build.gradle 文件，即 app/build.gradle 文件的 dependencies 中增加如下内容。

```
dependencies {
  compile fileTree(dir:'libs', include: ['*.jar'])
  androidTestCompile('com.android.support.test.espresso:espresso-core:2.2.2', {
        exclude group:'com.android.support', module:'support-annotations'
    })
   compile'com.android.support:appcompat-v7:26.+'
  compile 'com.squareup.okhttp3:okhttp:3.8.1'
   testCompile'junit:junit:4.12'
}
```

添加以上内容，程序会自动下载两个库，一个是 OkHttp，另一个是 Okio 库，后者是前者通信的基础。目前 OkHttp 最新的版本是 3.8.1，可以从 OkHttp 的主页查看版本。下载后，可在项目的 External Libraries 项看到下载的库，如图 9-10 所示。

External Libraries
< Android API 26 Platform > D:\Android_
< JDK > C:\Program Files\Android\Androi
animated-vector-drawable-26.0.0-alpha1
okhttp-3.8.1
okhttp-3.8.1.jar library root
META-INF
okhttp3
publicsuffixes.gz
okio-1.13.0
okio-1.13.0.jar library root
META-INF
okio

图 9-10　下载的 OkHttp 和 Okio 库

下面是 OkHttp 的具体用法。

（1）首先创建一个 OkHttpClient 的实例，代码如下。

```
OkHttpClient client =new OkHttpClient();
```

（2）创建一个 Request 对象，设置目标地址，代码如下。

```
Request  request=new Request.Builder().url("http://www.baidu.com").build();
```

（3）调用 OkHttpClient 的 newCall( ) 方法，创建 Call 对象，并调用它的 execute( ) 方法，发送请求并获取服务器返回的数据。

```
Response response = client.newCall(request).execute();
```

（4）其中，Response 对象就是服务器返回的数据，获取返回的内容。

```
String responseData = response.body().string();
```

（5）如果发起的是 POST 请求，则会比 GET 请求复杂一些，先构造出一个 RequestBody 对象以存放参数，如下所示。

```
RequestBody requestBody = new FormBody.Builder()
    .add("username","admin")
    .add("password","123456")
    .build();
```

然后在 Request.Builder() 调用 post() 方法，并将 RequestBody 对象传入：

```
Request  request = new Request.Builder()
    .url("http://www.baidu.com")
    .post(requestBody)
    .build();
```

接下来的操作和前面相同。

本项目为天气预报，首要的问题是如何获得天气信息，本项目从中国天气网获得这些信息，天气网的网址：http://www.weather.com.cn/。

如何获取全国所有省份的信息呢？我们只要访问以下网址 http://www.weather.com.cn/data/list3/city.xml，即可返回中国所有省份的名称和代号，如图 9-11 所示。

```
01|北京,02|上海,03|天津,04|重庆,05|黑龙江,06|吉林,07|辽宁,08|内蒙古,09|河北,10|山西,11|陕西,12|山东,13|新疆,14|西藏,15|青海,16|
甘肃,17|宁夏,18|河南,19|江苏,20|湖北,21|浙江,22|安徽,23|福建,24|江西,25|湖南,26|贵州,27|四川,28|广东,29|云南,30|广西,31|海
南,32|香港,33|澳门,34|台湾| Shanghai 01 | Beijing, 02, 03 | tianjin, chongqing, 04 | | heilongjiang 05, 06 | jilin, 07 |
liaoning, 08 | in Inner Mongolia, 09 | hebei, 10 | shanxi, 11 | shanxi, 12 | | xinjiang, 13, shandong | Tibetan 14, 15 |
qinghai, 16 | gansu, 17 | ningxia, 18 | henan, 19 | jiangsu, 20 | hubei, 21 | zhejiang, 22 | anhui, 23 | fujian | jiangxi 24,
25 | hunan, 26 | guizhou, sichuan, 27 | | guangdong 28, 29 | yunnan, 30 | guangxi, 31 | hainan, 32 | Hong Kong, macau, 33 |, 34
| Taiwan
```

图 9-11　http://www.weather.com.cn/data/list3/city.xml 中的信息

返回的值：01 | 北京，02 | 上海，03 | 天津，21 | 浙江等，可以看到，城市与其代号之间通过“ | ”相隔开，省份与省份之间用逗号隔开，要记住这个结构，之后会用到这种表达式截取信息。

如何查看浙江省内的城市的信息呢？其实非常简单，只需要访问以下网址 http://www.weather.com.cn/data/list3/city21.xml，也就是将省级代号添加至 city 后面即可，服务器将返回数据 2101 | 杭州，2102 | 湖州，2103 | 嘉兴等，如图 9-12 所示。

```
2101|杭州,2102|湖州,2103|嘉兴,2104|宁波,2105|绍兴,2106|台州,2107|温州,2108|丽水,2109|金华,2110|衢州,2111|舟山2101 | 2101 |
| 2101, hangzhou, huzhou, jiaxing, ningbo, 2104 | 2105 | shaoxing, 2106 | taizhou, 2107 | wenzhou, 2108 | lishui, 2109 |
jinhua, 2110 | quzhou, 2111 | zhoushan
```

图 9-12　查看省内城市信息

采用同样的方法，访问杭州以下的县市的信息，只需要在 city 后添加 2101 即可：http://www.weather.com.cn/data/list3/city2101.xml，如图 9-13 所示。

```
210101|杭州,210102|萧山,210103|桐庐,210104|淳安,210105|建德,210106|昌化,210107|临安,210108|富阳,210109|余杭Xiaoshan,
hangzhou, 210101 | 210102 | 210103 | tonglu, 210104 | ChunAn, 210105 | building heart, 210106 | chicken blood, 210107 |
linan, 210108 | fuyang, 210109 | yuhang
```

图 9-13　查看杭州以下的县市的信息

掌握了以上方法，就可以获得全国省市区的信息了，那么，如何得到某具体城市的天气呢？以杭州市区为例，其县级代号为 210101，访问以下网址：http://www. weather. com. cn/data/list3/city210101. xml。即会返回一个很简单的数据：210101 | 101210101210101 | 101210101，后面就是杭州市区所对应的天气代号，之后通过得到的代号即访问以下网址 http://www. weather. com. cn/data/cityinfo/101210101. html，如图 9-14 所示。

{"weatherinfo":{"city":"杭州","cityid":"101210101","temp1":"5℃","temp2":"20℃","weather":"晴转多云","img1":"n0.gif","img2":"d1.gif","ptime":"18:00"}}{" weatherinfo ": {" city" : "hangzhou", "cityid" : "101210101", "temp1" : "5 ℃", "temp2" : "20 ℃", "weather" : "clear to overcast," "img1" : "n0. GIF", "img2" : "d1. GIF", "ptime" : "18:00"}}

图 9-14　获取杭州市区天气信息

注意，这个网址的后缀是 html，不是 xml，编写代码的时候不要写错了，完成操作后，服务器即会把杭州市区的天气信息以 JSON 格式返回给我们，如下所示。

{"weatherinfo":{"city":"杭州","cityid":"101210101","temp1":"5℃","temp2":"20℃","weather":"晴转多云","img1":"n0.gif","img2":"d1.gif","ptime":"18:00"}}{" weatherinfo ": {" city" : "hangzhou", "cityid" : "101210101", "temp1" : "5 ℃", " temp2" : "20 ℃", "weather" : "clear to overcast," "img1" : "n0. GIF", "img2" : "d1. GIF", "ptime" : "18:00"}}

下面是 Android 使用 OkHttp 获取天气的过程。在 Android 2. 3 中创建应用项目：OkHttp_Wheather。

（1）在内层的 build. gradle 文件，也就是 app/build. gradle 文件的 dependencies 中增加如下内容。

```
dependencies {
  compile fileTree(dir:'libs', include: ['*.jar'])
  androidTestCompile('com.android.support.test.espresso:espresso-core:2.2.2', {
      exclude group: 'com.android.support', module:'support-annotations'
  })
compile 'com.android.support:appcompat-v7:26.+'
compile 'com.squareup.okhttp3:okhttp:3.8.1'
compile 'com.google.code.gson:gson:2.8.0'
}
```

（2）在项目配置文件 app/src/AndroidManifest. xml 中添加网络权限，代码如下。

```
<uses-permission android:name="android.permission.INTERNET"/>
```

（3）res/layout 目录下主布局文件 activity_main. xml 的布局如图 9-15 所示。

图 9-15　主布局文件

（4）主 Activity 文件 MainActivity. java 的代码如下。

```
public class MainActivity extends AppCompatActivity {
private final String url = "http://www.weather.com.cn/data/cityinfo/101210101.html";
private Gson gson = new Gson();
TextView city,temp1,temp2,wheather,ptime;
ImageView gif1;
Button btn;
private OkHttpClient client;
JSONObject jsonObject1,jsonObject2;
@Override
protected void onCreate(Bundle savedInstanceState) {
  super.onCreate(savedInstanceState);
  setContentView(R.layout.activity_main);
  city = (TextView) findViewById(R.id.tvCity);
  temp1 = (TextView) findViewById(R.id.temp1);
  temp2 = (TextView) findViewById(R.id.temp2);
  wheather = (TextView) findViewById(R.id.wheather);
  ptime = (TextView) findViewById(R.id.ptime);
  gif1 = (ImageView) findViewById(R.id.imageView1);
  btn = (Button) findViewById(R.id.btn);
  client = new OkHttpClient();
  btn.setOnClickListener(new OnClickListener(){
    @Override
    public void onClick(View view) {
```

```
        Request request = new Request.Builder().url(url).cacheControl(CacheCon-
trol.FORCE_NETWORK).build();
        client.newCall(request).enqueue(new Callback() {
            @Override
            public void onFailure(Call call, IOException e) {
              Toast.makeText(MainActivity.this, e.getMessage(), Toast.LENGTH_SHORT).
show();
            }
        @Override
    public void onResponse(Call call, Response response) throws IOException
    {
            //获取服务器返回的json字符串
            final String responseString = response.body().string();
        //在主线程中修改UI
        runOnUiThread(new Runnable() {
          @Override
          public void run() {
          try{
            jsonObject1 = new JSONObject(responseString);
          jsonObject2 =new JSONObject(jsonObject1.getString(
            "weatherinfo"));
                        city.setText("城市:" +jsonObject2.getString("city"));
                        temp1.setText("低温:" +jsonObject2.getString("temp1"));
                        temp2.setText("高温:" +jsonObject2.getString("temp2"));
                        wheather.setText("天气:" +jsonObject2.getString("weather"));
                        ptime.setText("时间:" +jsonObject2.getString("ptime"));
                        String path = Environment.getExternalStorageDirectory() +
                                    File.separator +jsonObject2.getString("imag1");
                        Bitmap bm = BitmapFactory.decodeFile(path);
                        gif1.setImageBitmap(bm);
                        }
                      catch(Exception e) {
                    }
                  }
                });
              }
            });
          }
        });
    }
}
```

（5）项目运行结果如图 9-16 所示。

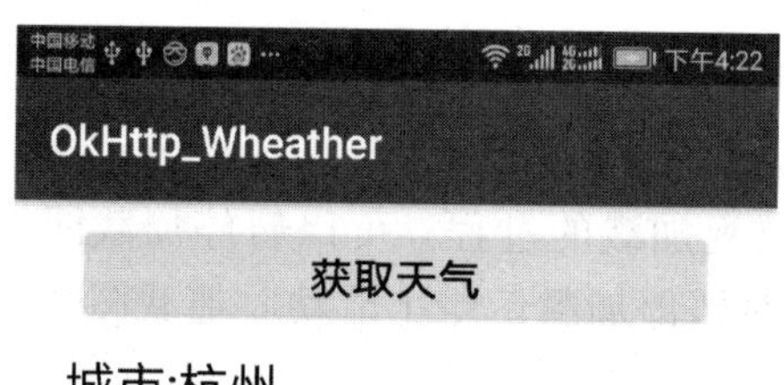

图 9-16　项目运行结果

9.4　使用 Universal-Image-Loader 加载图片

Universal-Image-Loader 可以说是目前使用最广泛的图片开源库之一。在主流的应用中基本都能看到它的身影，它就像个图片加载守护者，默默地守护着图片加载。

Universal-Image-Loader 是一个开源的 UI 组件程序，该项目的目的是为异步图像加载、缓存和显示提供一个可重复使用的仪器。所以，如果程序里需要这个功能的话，不妨尝试用一用。其中已经封装好了一些类和方法，我们可以直接拿来使用，而不用重复去写。其实，写一个这方面的程序是比较麻烦的，要考虑多线程缓存、内存溢出等很多方面的问题。Universal-Image-Loader 包含三大组件：DisplayImageOptions、ImageLoader 和 ImageLoaderConfiguration。

Universal-Image-Loader 具有如下功能特性。

（1）多线程异步加载和显示图片（图片来源于网络、SD 卡、assets 文件夹、drawable 文件夹或新增加载视频缩略图）

例如：

```
"http://site.com/image.png" // from Web
"file:///mnt/sdcard/image.png" // from SD card
"file:///mnt/sdcard/video.mp4" // from SD card(video thumbnail)
"content://media/external/images/media/13" // from content provider
"content://media/external/video/media/13" // from content provider(video thumb-
```

```
nail)
    "assets://image.png" // from assets
    "drawable://" + R.drawable.img // from drawables (non-9patch images)
```

（2）支持通过 listener 监视加载的过程，可以暂停加载图片，在经常使用的 ListView、GridView 中，可以设置滑动时暂停加载，停止滑动时加载图片（便于节约流量，在一些优化中可以使用）。

（3）缓存图片至内存时，可以更加高效地工作。

（4）高度可定制化（可以根据自己的需求进行各种配置，如线程池、图片下载器、内存缓存策略等）。

（5）支持图片的内存缓存、SD 卡（文件）缓存。

（6）在网络速度较慢时，可以对图片进行加载并设置下载监听。

下面是实现过程，完整内容参考本章 Android Studio 项目：Universal_Image_Loader_Example。

要使用 Universal-Image-Loader，需要从网上下载 Universal-Image-Loader 项目，然后编译成库，Universal-Image-Loader 项目的下载地址：https://github.com/nostra13/Android-Universal-Image-Loader，将编译好的 universal-image-loader-1.9.5.jar 文件复制到项目的 libs 目录下，然后增加为库，如图 9-17 所示，这样就可以使用 Universal-Image-Loader 了。

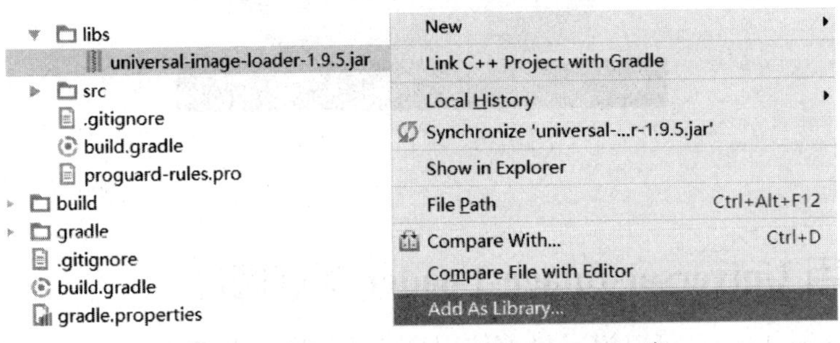

图 9-17　使用 Universal-Image-Loader.jar 库

Universal-Image-Loader 的使用步骤如下。

（1）创建 ImageLoader 配置参数，代码如下。

```
ImageLoaderConfiguration configuration = ImageLoaderConfiguration.createDefault(this);
```

当然,也可以自己定制配置,代码如下。

```
ImageLoaderConfiguration config = new ImageLoaderConfiguration.Builder(context)
        .memoryCacheExtraOptions(480, 800) // default = device screen dimensions
        .diskCacheExtraOptions(480, 800, CompressFormat.JPEG, 75, null)
        .taskExecutor(...)
        .taskExecutorForCachedImages(...)
        .threadPoolSize(4)
        .threadPriority(Thread.NORM_PRIORITY - 1)
        .tasksProcessingOrder(QueueProcessingType.FIFO) // default
        .denyCacheImageMultipleSizesInMemory()
        .memoryCache(new LruMemoryCache(2 * 1024 * 1024))
```

```
        .memoryCacheSize(2 * 1024 * 1024)
        .memoryCacheSizePercentage(13) // default
        .diskCache(new UnlimitedDiscCache(cacheDir)) // default
        .diskCacheSize(50 * 1024 * 1024)
        .diskCacheFileCount(100)
        .diskCacheFileNameGenerator(new HashCodeFileNameGenerator()) // default
        .imageDownloader(new BaseImageDownloader(context)) // default
        .imageDecoder(new BaseImageDecoder()) // default
        .defaultDisplayImageOptions(DisplayImageOptions.createSimple()) // default
        .writeDebugLogs()
        .build();
```

（2）使用配置参数初始化 ImageLoader，代码如下。

```
ImageLoader.getInstance().init(configuration);
```

（3）加载图片。主要使用 ImageLoader 的 loadImage( ) 和 displayImage( ) 方法，这两个方法都是重载的方法，可以根据需要进行选择。

典型的 loadImage 方法如下。

```
loadImage(String uri, ImageSize targetImageSize, DisplayImageOptions options,
ImageLoadingListener listener)
```

其中，uri 为图片的 URL 地址，targetImageSize 为显示图像的大小，options 为显示图像的配置，listener 用于图片下载情况的监听，在调用时实现。

实例代码如下。

```
final ImageView mImageView = (ImageView) findViewById(R.id.image);
String imageUrl =
http://imgsrc.baidu.com/imgad/pic/
item/622762d0f703918fb11c1bef5b3d269758eec4c1.jpg
ImageSize mImageSize = new ImageSize(400, 400);
//显示图片的配置
DisplayImageOptions options = new DisplayImageOptions.Builder()
    .cacheInMemory(true)
    .cacheOnDisk(true)
    .bitmapConfig(Bitmap.Config.RGB_565)
    .build();
ImageLoader.getInstance().loadImage(imageUrl, mImageSize, options, new Sim-
pleImageLoadingListener() {
@Override
public void onLoadingComplete(String imageUri, View view,
Bitmap loadedImage) {
    super.onLoadingComplete(imageUri, view, loadedImage);
    mImageView.setImageBitmap(loadedImage); //将图像设置到界面控件
  }
});
```

监听器 SimpleImageLoadingListener 需要实现回调方法 onLoadingComplete( )，在此方法中

将 loadedImage 设置到 ImageView 上。加载结果如图 9-18 所示。

图 9-18　loadImage( )方法加载网络图像

典型的 displayImage 方法如下。

```
displayImage(String uri, ImageView imageView, DisplayImageOptions options,)
```

其中，uri 为图片的 URL 地址，imageView 为显示图像的控件，options 为显示图像的配置。

实例代码如下。

```
final ImageView mImageView = (ImageView) findViewById(R.id.image);
String imageUrl =
"http://imgsrc.baidu.com/imgad/pic/item/9345d688d43f8794317d4c0dd81b0ef41bd53a16.jpg";
//显示图片的配置
DisplayImageOptions options = new DisplayImageOptions.Builder()
    .showImageOnLoading(R.drawable.ic_stub)
    .showImageOnFail (R.drawable.ic_error)
    .cacheInMemory (true)
    .cacheOnDisk (true)
    .bitmapConfig (Bitmap.Config.RGB_565)
    .build();
ImageLoader.getInstance().displayImage (imageUrl, mImageView, options);
```

加载结果如图 9-19 所示。

图 9-19　displayImage( )方法加载网络图像

在项目配置文件 app/src/AndroidManifest. xml 中添加网络权限和写存储器权限，代码如下。

```
<uses-permission android:name = "android.permission.INTERNET" />
    <! -- Include next permission if you want to allow UIL to cache images on SD card -- >
<uses-permission android:name = "android.permission.WRITE_EXTERNAL_STORAGE" />
```

另外，Universal-Image-Loader 也可以使用 GirdView 和 ListView 加载图片，这样可显示大量的图片，如果希望停止图片的加载且在 GridView 和 ListView 停止滑动的时候加载当前界面的图片，也可以使用这个框架来实现，使用方法也很简单，它提供了 PauseOnScrollListener 这个类来控制 ListView 和 GridView 滑动过程中停止加载图片。

9.5　使用 Volley 加载网络图片

为了更方便地编写网络操作的程序，一些 Android 网络通信框架应运而生，例如 AsyncHttpClient，它把 HTTP 所有的通信细节全部封装在了内部，只需要简单几行代码就可以完成网络操作。另一个常用的框架就是前面介绍的 Universal-Image-Loader，它使界面上显示网络图片的操作极度简单，开发者不用关心如何从网络上获取图片，也不用关心开启线程、回收图片资源等细节，Universal-Image-Loader 已经把一切都做好了。

Android 开发团队也意识到了有必要将 HTTP 的通信操作进行简单化，并在 2013 年

Google I/O 大会上推出了新的网络通信框架 Volley。Volley 可以说是把 AsyncHttpClient 和 Universal-Image-Loader 的优点集于一身，既可以像 AsyncHttpClient 一样非常简单地进行 HTTP 通信，也可以像 Universal-Image-Loader 一样轻松加载网络上的图片。除了简单易用之外，Volley 在性能方面也进行了大幅度的调整，它的设计目标就是进行数据量不大，但通信频繁的网络操作。

若要使用 Volley，需要从网络下载 Volley 项目，然后编译成库，Volley 项目的下载地址：https://github.com/mcxiaoke/android-volley，将编译好的 volley.jar 文件复制到项目的 libs 目录下，然后增加为库，如图 9-20 所示，这样即可使用 Volley 了。

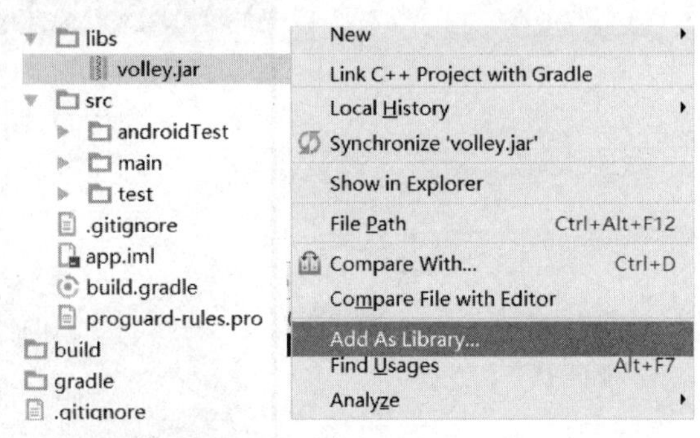

图 9-20　使用 volley.jar 库

下面是实现过程，完整内容参考本章 Android Studio 项目：Volley_Example。

9.5.1　使用 ImageRequest 对象加载图片

具体的实现步骤如下。

(1) 创建一个 RequestQueue 对象，代码如下。

```
RequestQueue mQueue = Volley.newRequestQueue(context);
```

不必为每一次 HTTP 请求都创建一个 RequestQueue 对象，这是非常浪费资源的，一般在每一个需要和网络交互的 Activity 中创建一个 RequestQueue 对象就足够了。

(2) 创建一个 ImageRequest 对象，代码如下。

```
ImageRequest imageRequest = new ImageRequest(
"http://imgsrc.baidu.com/imgad/pic/item/838ba61ea8d3fd1f7515e2193a4e251f95ca5ff0.jpg",
        new Response.Listener<Bitmap>() {
            @Override
            public void onResponse(Bitmap response) {
                imageView.setImageBitmap(response);
            }
          }, 0, 0, Bitmap.Config.RGB_565, new Response.ErrorListener() {
        @Override
      public void onErrorResponse (VolleyError error) {
        imageView.setImageResource (R.mipmap.ic_launcher);
  }
});
```

ImageRequest 的构造函数接收六个参数，第一个参数是图片的 URL 地址。第二个参数是图片请求成功的回调，这里我们把返回的 Bitmap 参数设置到 ImageView 中。第三和第四个参数分别用于指定允许图片最大的宽度和高度，如果指定的网络图片的宽度或高度大于这里的最大值，则会对图片进行压缩，指定成 0，表示不管图片有多大，都不会进行压缩。第五个参数用于指定图片的颜色属性，Bitmap.Config 下的几个常量都可以在这里使用，其中

ARGB_8888 可以展示最好的颜色属性，每个图片像素占据 4 个字节的大小，而 RGB_565 则表示每个图片像素占据 2 个字节大小。第六个参数是图片请求失败的回调，这里在请求失败时 ImageView 中显示一张默认图片。

（3）将 ImageRequest 对象添加到 RequestQueue 中，代码如下。

```
mQueue.add(imageRequest);
```

（4）加载结果如图 9-21 所示。

图 9-21　使用 ImageRequest 对象加载图片

9.5.2　使用 ImageLoader 对象加载图片

可能您认为 ImageRequest 已经非常好用了，实际上，Volley 在请求网络图片方面可以做到的还远远不止这些，ImageLoader 就是一个很好的例子。ImageLoader 也可以用于加载网络上的图片，并且它的内部也是使用 ImageRequest 实现的，不过，ImageLoader 明显要比 ImageRequest 更加高效，因为它不仅可以帮我们对图片进行缓存，还可以过滤掉重复的链接，避免重复发送请求。

由于 ImageLoader 不是继承自 Request，所以它的用法和我们之前学到的内容有所不同，总结起来大致可以分为以下四步。

（1）创建一个 RequestQueue 对象，和前面方面相同，代码如下。

```
RequestQueue mQueue = Volley.newRequestQueue(context);
```

（2）创建一个 ImageLoader 对象，代码如下。

```
ImageLoader imageLoader = new ImageLoader(mQueue, new ImageCache() {
    @Override
    public void putBitmap(String url, Bitmap bitmap) {
    }
```

```
        @Override
        public Bitmap getBitmap(String url) {
            return null;
        }
    });
```

ImageLoader 的构造函数接收两个参数，第一个参数是 RequestQueue 对象，第二个参数是一个 ImageCache 对象，这里我们先创建出一个空的 ImageCache 的实现即可。

（3）获取一个 ImageListener 对象，代码如下。

```
ImageListener listener = ImageLoader.getImageListener (imageView,
R.drawable.default_image, R.drawable.failed_image);
```

ImageLoader 的 getImageListener()方法能够获取一个 ImageListener 对象，getImageListener()方法接收三个参数，第一个参数指定用于显示图片的 ImageView 控件，第二个参数指定加载图片过程中显示的图片，第三个参数指定加载图片失败的情况下显示的图片。

（4）调用 ImageLoader 的 get()方法加载网络上的图片，代码如下。

```
imageLoader.get("http://imgsrc.baidu.com/imgad/pic/item/5fdf8db1cb
1349547f37fd405c4e9258d1094aec.jpg", listener);
```

get()方法接收两个参数，第一个参数是图片的 URL 地址，第二个参数是刚刚获取到的 ImageListener 对象。

（5）加载结果如图 9-22 所示。

图 9-22　使用 ImageLoader 对象加载图片

另外，Volley 也可以从网络获取字符串，例如天气预报的信息，这时使用的对象是 StringRequest 对象，创建代码如下。

```
StringRequest stringRequest = new StringRequest("http://www.baidu.com",
```

```
    new Response.Listener<String>() {
        @Override
    public void onResponse(String response) {
        Log.d("TAG", response);
    }
    }, new Response.ErrorListener() {
    @Override
    public void onErrorResponse(VolleyError error) {
      Log.e("TAG", error.getMessage(), error);
    }
});
```

StringRequest 的构造函数需要接收三个参数，第一个参数是目标服务器的 URL 地址，第二个参数是服务器响应成功的回调，第三个参数是服务器响应失败的回调。本例中，目标服务器地址填写的是百度的首页，在响应成功的回调里打印出服务器返回的内容，在响应失败的回调里打印出失败的详细信息。

其他步骤与加载图片过程相同。

9.6 使用 xUtils 实现网络文件下载

xUtils 是一个目前功能比较完善的 Android 开源框架，分为 4 个功能模块：DbUtils、HttpUtils、ViewUtils、BitmapUtils，最近发布的 xUtil3.0，在增加新功能的同时，提高了框架的性能，xUtil3.0 下载地址：https://github.com/wyouflf/xUtils3，xUtils3 的特点如下。

- xUtils 包含了很多实用的 Android 工具。
- xUtils 支持超大文件（超过 2G）上传，具有更全面的 Http 请求协议支持，拥有更加灵活的 ORM，具有更多的事件注解支持且不受混淆影响。
- xUtils 最低兼容 Android 4.0（API level 14）。
- xUtils3 变化较多，旧版（https://github.com/wyouflf/xUtils）已不再继续维护。

xUtils3 一共有 4 大功能：注解模块、网络模块、图片加载模块、数据库模块。使用 xUtils 需要在项目 libs 文件夹中加入一个 jar 包，如果对服务器返回的数据进行封装，还需要导入一个 Gson 的 jar 包。

下面是使用 xUtils3 的网络模块下载网络文件的实例，使用 xUtils3 的网络模块下载网络文件程序非常简单，因为 xUtils3 的库已经将操作封装好了。

在 Android 2.3 中创建应用项目：XUtils_Demo。

图 9-23　主布局文件

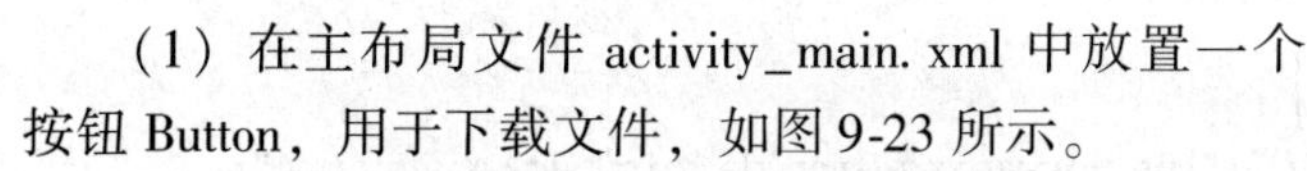

(1) 在主布局文件 activity_main.xml 中放置一个按钮 Button，用于下载文件，如图 9-23 所示。

(2) 在源代码目录下新建应用程序文件 BaseApplication.java，代码如下。

```
public class BaseApplication extends Application {
  @Override
  public void onCreate() {
```

```
        super. onCreate();
        x. Ext. init(this);
    }
}
```

(3) 在项目配置文件 app/src/AndroidManifest. xml 中添加网络权限，声明应用程序，代码如下。

```
<uses-permission android:name = "android. permission. INTERNET" / >
<uses-permission android:name = "android. permission. WRITE_EXTERNAL_STORAGE" / >
<application
  android: name = " . BaseApplication"
…
</application>
```

(4) 主 Activity 文件 MainActivity. java 的代码如下。

```
public class MainActivity extends AppCompatActivity {
    private static final String BASE_URL =
        " http: //clips. vorwaerts-gmbh. de/big_buck_bunny. mp4";
    private static final String BASE_PATH =
        Environment. getExternalStorageDirectory(). getPath() +File. separator;
    private Button buttonDownloadFile;
    private ProgressDialog progressDialog;
    @Override
    protected void onCreate (Bundle savedInstanceState) {
      super. onCreate (savedInstanceState);
      setContentView (R. layout. activity_main);
       buttonDownloadFile = (Button) findViewById (R. id. bt_downloadFile);
      buttonDownloadFile. setOnClickListener (new View. OnClickListener() {
      @Override
        public void onClick (View v) {
        if (ContextCompat. checkSelfPermission (MainActivity. this,
          Manifest. permission. WRITE_EXTERNAL_STORAGE) ! =
            PackageManager. PERMISSION_GRANTED)
          {
            ActivityCompat. requestPermissions (MainActivity. this,
new String [] {Manifest. permission. WRITE_EXTERNAL_STORAGE}, 1);
          }
          else {
            String url = " http: //clips. vorwaerts-gmbh. de/big_buck_bunny. mp4";
            String path = BASE_PATH + " code. mp4";
            downloadFile (url, path);
            }
        }
```

```
        });
    }
private void downloadFile (final String url, String path) {
    progressDialog = new ProgressDialog (this);
    RequestParams requestParams = new RequestParams (url);
    requestParams.setSaveFilePath (path);
    x.http().get (requestParams, new Callback.ProgressCallback<File>() {
      @Override
      public void onWaiting() {        }
      @Override
      public void onStarted() {        }
      @Override
      public void onLoading (long total, long current, boolean isDownloading) {
         progressDialog.setProgressStyle (ProgressDialog.STYLE_HORIZONTAL);
          progressDialog.setMessage (" 正在下载中... ");
          progressDialog.show();
          progressDialog.setMax ( (int) total);
        progressDialog.setProgress ( (int) current);
      }
      @Override
      public void onSuccess (File result) {
        Toast.makeText (MainActivity.this, " 下载成功",
              Toast.LENGTH_SHORT).show();
          progressDialog.dismiss();
      }
      @Override
      public void onError (Throwable ex, boolean isOnCallback) {
          ex.printStackTrace();
          Toast.makeText (MainActivity.this, " 下载失败，请检查网络和 SD 卡",
           Toast.LENGTH_SHORT).show();
           progressDialog.dismiss();
      }
      @Override
      public void onCancelled (CancelledException cex) {        }
      @Override
      public void onFinished() {          }
    });
  }
}
```

（5）项目运行结果如图 9-24 所示。

图 9-24　项目运行结果

9.7 本章小结

网络请求是 Android 客户端很重要的内容，本章介绍了 Android 权限机制，特别是 Android 6.0 运行时的权限；讲解了 JSON 格式数据的构造和解析方法；介绍了三个开源库：Universal-Image-Loader、Volley、xUtils，并分别给出了应用实例。

第十章

更方便的通信——Android无线通信

Android 无线控制是 Android 应用的重要组成部分，目前广泛运用的移动终端无线通信有 WiFi、NFC、蓝牙等。本章介绍 Android 的 Wifi 应用、蓝牙传输数据和 NFC 通信的实现。

10.1 Android Wifi 应用——获取 Wifi 列表

在 Android 中，操作 Wifi 是很简单的，主要使用以下几个类对象或变量。

➢ private WifiManager wifiManager：声明管理对象 OpenWifi。

➢ private WifiInfo wifiInfo：Wifi 信息。

➢ private List <ScanResult> scanResultList：扫描网络连接列表。

➢ private List <WifiConfiguration> wifiConfigList：网络配置列表。

➢ private WifiLock wifiLock：Wifi 锁。

其中最重要是 WifiManager 类。

获取 Wifi 列表的主要步骤如下。

(1) 要想操作 Wifi 设备，需要先获取 Context. getSystemService (Context. WIFI_SERVICE)，以获取 WifiManager 对象，并通过这个对象来管理 Wifi 设备。

例如：wifiManager =(WifiManager) getSystemService (Context. WIFI_SERVICE)。

(2) 使用 WifiManager 的 isWifiEnabled( ) 判断 Wifi 设备是否打开。例如：wifiManager. isWifiEnabled( )。

(3) 使用 WifiManager 的 setWifiEnabled (true) 打开 Wifi，例如：wifiManager. setWifiEnabled (true)。

(4) 使用 WifiManager 的 getScanResults( )返回获取扫描测试的结果，返回的结果是 ScanResult 的列表。例如：List <ScanResult> list;

list = wifiManager. getScanResults( );

ScanResult 的重要属性有以下几个。

➢ BSSID：接入点的地址。

➢ SSID：网络的名字，区别 Wifi 网络的唯一名字

➢ Capabilities：网络接入的性能。

➢ Frequency：当前 Wifi 设备附近热点的频率 (MHz)。

➢ Level 所发现的 Wifi 网络信号强度。

（5）连接 Wifi 热点。

通过 WifiManager. getConfiguredNetworks( )方法返回 WifiConfiguration 对象的列表，然后调用 WifiManager. enableNetwork( )方法，就可以连接上指定的热点。

（6）查看已经连接上的 Wifi 信息。

WifiInfo 是专门用来表示连接的对象，这个对象可以通过 WifiManager. getConnectionInfo( )来获取。WifiInfo 中包含了当前连接的相关信息。

WifiInfo 的主要成员如下。

➢ getBSSID( )：获取 BSSID 属性。

➢ getDetailedStateOf( )：获取客户端的连通性。

➢ getHiddenSSID( )：获取 SSID 是否被隐藏。

➢ getIpAddress( )：获取 IP 地址。

➢ getLinkSpeed( )：获取连接的速度。

➢ getMacAddress( )：获取 Mac 地址。

➢ getRssi( )：获取 802. 11n 网络的信号。

➢ getSSID( )：获取 SSID。

➢ getSupplicanState( )：获取具体客户端状态的信息。

在 Android 2. 3 中创建应用项目：Wifi_List。

（1）在主布局文件 activity_main. xml 中放置一个 ListView 控件，用于显示 Wifi 列表，如图 10-1 所示。

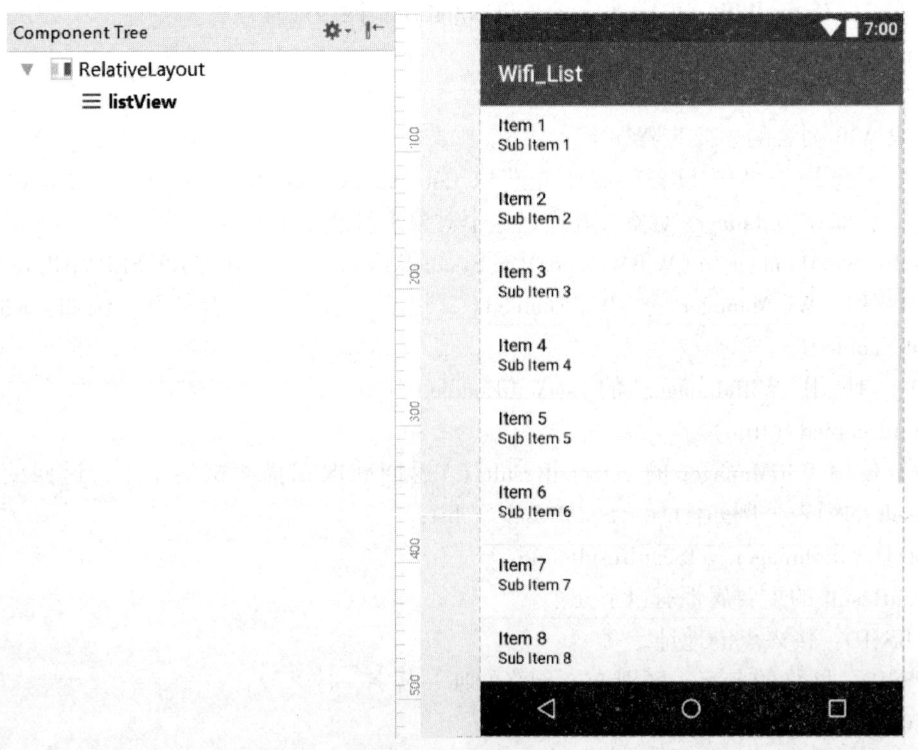

图 10-1　主布局文件

(2) 在项目配置文件 app/src/AndroidManifest. xml 中添加相应权限，代码如下。

```
<!-- GPS 定位权限 -->
<uses-permission android:name="android.permission.ACCESS_COARSE_LOCATION" />
<uses-permission android: name=" android.permission.ACCESS_FINE_LOCATION" />
<!-- Wifi 权限 -->
<uses-permission android: name=" android.permission.ACCESS_NETWORK_STATE" />
<uses-permission android: name=" android.permission.CHANGE_WIFI_STATE" />
<uses-permission android: name=" android.permission.ACCESS_WIFI_STATE" />
<uses-permission android: name=" android.permission.CHANGE_WIFI_MULTICAST_STATE" />
<uses-permission android: name=" android.permission.INTERNET" />
```

(3) 主 Activity 文件 MainActivity. java 的代码如下。

```
public class MainActivity extends AppCompatActivity {
  private WifiManager wifiManager;
  List<ScanResult> list;
  ListView listView;
  private SimpleAdapter adapter;
  String[] listk;
  @Override
  protected void onCreate(Bundle savedInstanceState) {
      super.onCreate(savedInstanceState);
      setContentView(R.layout.activity_main);
          if (ContextCompat.checkSelfPermission (this, Manifest.permission_
group.LOCATION) != PackageManager.PERMISSION_GRANTED)
        {
         ActivityCompat.requestPermissions (this, new String [] {
                 Manifest.permission.ACCESS_FINE_LOCATION,
                 Manifest.permission.ACCESS_COARSE_LOCATION,
                 Manifest.permission.ACCESS_WIFI_STATE,
         }, 1);
     }
   listView = (ListView) findViewById (R.id.listView);
   wifiManager = (WifiManager)
   getApplicationContext(). getSystemService (Context.WIFI_SERVICE);
   if (! wifiManager.isWifiEnabled()) {
      wifiManager.setWifiEnabled (true);
   }
   wifiManager.startScan();
   list = wifiManager.getScanResults();
   listk=new String [list.size()];
   for (int i=0; i<list.size(); i++)
    {
      ScanResult scanResult = list.get (i);
      listk [i] =String.valueOf (scanResult.SSID);
```

```
    }
    ArrayAdapter<String> adapter = new ArrayAdapter<String> (MainActivity.this,
    android.R.layout.simple_list_item_1, listk);
    listView.setAdapter (adapter);
    }
}
```

（4）项目运行结果如图 10-2 所示。

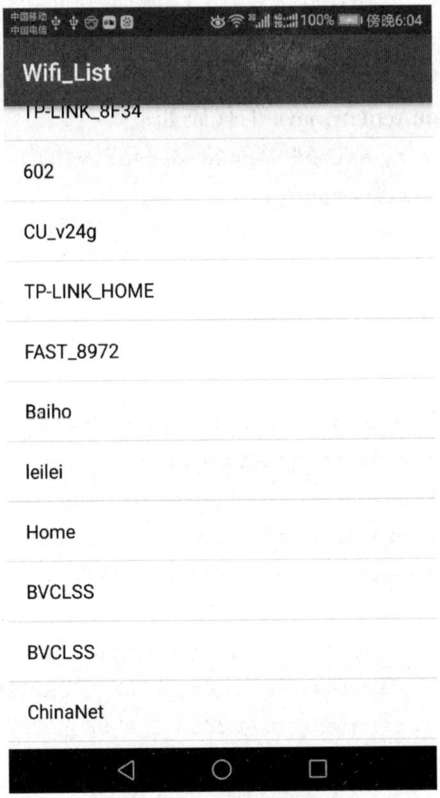

图 10-2 项目运行结果

10.2 Android 蓝牙——查找蓝牙设备

蓝牙（Bluetooth）是一种无线技术标准，可实现固定设备、移动设备和楼宇个人域网之间的短距离数据交换（使用 2.4-2.485GHz 的 ISM 波段的 UHF 无线电波）。蓝牙技术最初由电信巨头爱立信公司于 1994 年创制，当时是作为 RS232 数据线的替代方案。蓝牙可连接多个设备，克服了数据同步的难题。

蓝牙能在包括移动电话、PDA、无线耳机、笔记本电脑、相关外设等众多设备之间进行无线信息交换。蓝牙采用分散式网络结构以及快跳频和短包技术，支持点对点及点对多点通信，其数据速率为 1Mbps。

2010 年 7 月，以低功耗为特点的蓝牙 4.0 标准推出，蓝牙大中华区技术市场经理吕荣良将其看作蓝牙第二波发展高潮的标志，他表示："蓝牙可以跨领域应用，主要有 4 个生态系统，分别是智能手机与笔记本电脑等终端市场、消费电子市场、汽车前装市场和健身运动器材市场。”

NFC 和 UWB 曾经是十分受关注的短距离无线接入技术，但其发展已经日渐式微。业内专家认为，无线频谱的规划和利用在短距离通信中日益重要。短距离通信技术目前主要采用 2.4GHz 的开放频谱，但随着物联网的发展和大量短距离通信技术的应用，频谱需求会快速增长，视频、图像等大数据量的通信正在寻求更高频段的解决方案。

蓝牙 4.0 是蓝牙 3.0 + HS 规范的补充，专门面向对成本和功耗都有较高要求的无线方案，可广泛用于卫生保健、体育健身、家庭娱乐、安全保障等诸多领域。它支持两种部署方式：双模式和单模式。双模式中，低功耗蓝牙功能集成在现有的经典蓝牙控制器中，或在现有经典蓝牙技术（2.1 + EDR/3.0 + HS）芯片上增加低功耗堆栈，整体架构基本不变，因此成本增加有限。单模式面向高度集成、紧凑的设备，使用一个轻量级连接层（Link Layer）提供超低功耗的待机模式操作、简单设备恢复和可靠的点对多点数据传输，还能让联网传感器在蓝牙传输中安排好低功耗蓝牙流量的次序，同时还有高级节能和安全加密连接。

蓝牙 4.0 将传统蓝牙技术、高速技术和低耗能技术三种规格集一体，与 3.0 版本相比，最大的不同就是低功耗。“4.0 版本的功耗较老版本降低了 90%，更省电，”蓝牙技术联盟大中华区技术市务经理吕荣良表示，“随着蓝牙技术由手机、游戏、耳机、便携电脑和汽车等传统应用领域向物联网、医疗等新领域的扩展，对低功耗的要求会越来越高。4.0 版本强化了蓝牙在数据传输上的低功耗性能。”

低功耗版本使蓝牙技术得以延伸到采用纽扣电池供电的一些新兴市场。蓝牙低耗能技术是基于蓝牙低耗能无线技术核心规格的升级版，为开拓钟表、远程控制、医疗保健及运动感应器等广大新兴市场的应用奠定基础。

这项技术将应用于每年出售的数亿台蓝牙手机、个人电脑及掌上电脑。以最低耗能提供持久的无线连接，有效扩大相关应用产品的覆盖距离，开辟全新的网络服务。低耗能无线技术的特点在于超低的峰期、平均值及待机耗能；使装置配件和人机界面装置（HIDs）具备超低成本和轻巧的特性；更能使手机及个人电脑相关配件的成本降至最低、体积缩至更小。

蓝牙 4.0 在个人健身和健康市场的影响很大，Fitbit 无线师、耐克公司的新 Fuelband，摩托罗拉 MOTACTV 和时尚的基带，都是可见的例子。而且，健身手表也承诺使用蓝牙跟踪体力活动和心率。

另外，蓝牙 4.0 依旧向下兼容，包含经典蓝牙技术规范和最高速度 24Mbps 的蓝牙高速技术规范。三种技术规范可单独使用，也可同时运行。

Android 和蓝牙相关的接口类有 BluetoothSocket、BluetoothServerSocket、BluetoothAdapter、BluetoothClass. Service、BluetoothClass. Device，其中最重要的类是 BluetoothAdapter。各接口类的含义如下。

（1）BluetoothAdapter：代表本地的蓝牙设备。

（2）BluetoothDevice：代表远程的蓝牙设备。

（3）BluetoothSocket：一种类似于 TCP Socket 的接口，让当前程序与其他程序通过蓝牙设备实现数据交换的切入点。

（4）BluetoothServerSocket：类似于 ServerSocket，用来监听接入请求，两个 Android 程序

要想链接在一起，必须通过这个类打开一个 ServerSocket，当远程的蓝牙设备请求这个蓝牙设备的时候，如果请求被接受了，BluetoothServerSocket 将返回一个已经连接的 BluetoothSocket。

10.2.1 Android 蓝牙开发步骤

BluetoothAdapter 类简单来说代表了本设备（手机、电脑等）的蓝牙适配器对象，通过它可以操作蓝牙设备，主要有如下功能：(1) 开关蓝牙设备；(2) 扫描蓝牙设备；(3) 设置/获取蓝牙状态信息，例如蓝牙状态值、蓝牙 Name、蓝牙 Mac 地址等。

蓝牙操作的步骤如下。

(1) 获得蓝牙适配器实例，代码如下。

```
public static synchronized BluetoothAdapter getDefaultAdapter()
```

如果设备具备蓝牙功能，则返回 BluetoothAdapter 实例；否则，返回 Null 对象。

例如：BluetoothAdapter mBluetoothAdapter = BluetoothAdapter.getDefaultAdapter()

(2) 打开蓝牙。

- 直接调用 BluetoothAdapter 类成员函数 enable()打开蓝牙设备。
- 系统 API 打开蓝牙设备，该方式会弹出一个对话框样式的 Activity，供用户选择是否打开蓝牙设备。需要注意的是如果蓝牙已经开启，不会弹出该 Activity 界面。

例如：

```
//第一种打开方法：调用 enable
  boolean result = mBluetoothAdapter.enable();
//第二种打开方法:调用系统 API 打开蓝牙
if(! mBluetoothAdapter.isEnabled()) //未打开蓝牙,才需要打开蓝牙
{
  Intent intent = new Intent(BluetoothAdapter.ACTION_REQUEST_ENABLE);
  startActivityForResult (intent, 1);
  //会以 Dialog 样式显示一个 Activity,可以在 onActivityResult()方法中处理返回值
}
```

(3) 关闭蓝牙。

直接调用 BluetoothAdapter 类函数即 disable()即可。该函数若返回 True，表示关闭操作成功；返回 False，表示蓝牙操作失败。

(4) 扫描蓝牙设备。

直接调用 BluetoothAdapter 类函数即 startDiscovery()即可，返回值为 Boolean，需要注意的是，如果蓝牙没有开启，该方法会返回 False，即不会开始扫描过程。

要获得此搜索的结果需要先注册，以获取一个 BroadcastReceiver。先注册再获取信息，然后进行处理，代码如下。

```
//注册,当一个设备被发现时调用 onReceive
IntentFilter filter = new IntentFilter(BluetoothDevice.ACTION_FOUND);
this.registerReceiver (mReceiver, filter);
//当搜索结束后调用 onReceive
filter = new IntentFilter (BluetoothAdapter.ACTION_DISCOVERY_FINISHED);
```

```
this.registerReceiver (mReceiver, filter);
//.......
private BroadcastReceiver mReceiver = new BroadcastReceiver() {
@Override
public void onReceive (Context context, Intent intent) {
    String action = intent.getAction();
    if (BluetoothDevice.ACTION_FOUND.equals (action)) {
    BluetoothDevice device = intent.getParcelableExtra (BluetoothDevice. EXTRA
_DEVICE);
  // 已经配对的则跳过
    if (device.getBondState() ! = BluetoothDevice.BOND_BONDED) {
      mNewDevicesArrayAdapter.add (device.getName() +" \n" +
        device.getAddress());    //保存设备地址与名字
                }
  } else if (BluetoothAdapter.ACTION_DISCOVERY_FINISHED.equals (action))
{  //搜索结束
        if (mNewDevicesArrayAdapter.getCount() = = 0) {
                mNewDevicesArrayAdapter.add (" 没有搜索到设备");
        }
    }
  }
};
```

（5）获取蓝牙相关信息。

public String getName()用于获取蓝牙设备名称。

public String getAddress()用于获取蓝牙设备的硬件地址（MAC 地址），例如：00:11:22:AA:BB:CC。

public String getScanMode()用于获取蓝牙设备的扫描模式。

public static boolean checkBluetoothAddress（String address）用于验证蓝牙设备 MAC 地址是否有效。所有设备地址的英文字母必须大写，且为 48 位，例如 00:43:A8:23:10:F1。

返回值为 True，表示设备地址有效；返回值为 False，表示设备地址无效。

例如：btDesc. setText（" Name ：" + mBluetoothAdapter. getName（）+" Address ：" + mBluetoothAdapter. getAddress() +" Scan Mode --" + mBluetoothAdapter. getScanMode())。

（6）获取与本机绑定的蓝牙信息，代码如下。

```
public Set<BluetoothDevice> getBondedDevices()
```

获取与本机蓝牙所有绑定的远程蓝牙信息，以 BluetoothDevice 类实例返回。如果蓝牙为开启状态，该函数会返回一个空集合。

BluetoothDevice 对象代表一个远程的蓝牙设备，通过这个类可以查询远程设备的物理地址、名称、连接状态等信息。

例如，通过 BluetoothAdapter 类对象的 getBondedDevices()获取连接的蓝牙设备集合，然后加入到 List 数组，代码如下。

```
Set<BluetoothDevice> bts = mBluetoothAdapter.getBondedDevices();
```

```
for(BluetoothDevice device : bts) {
    // Add the name and address to an array adapter to show in a ListView
    list.add(device.getName() + "\n" + device.getAddress());
}
```

(7) 获取给定蓝牙地址的设备，代码如下。

```
public BluetoothDevice getRemoteDevice(String address)
```

该段代码以给定的 MAC 地址创建一个 BluetoothDevice 类实例（代表远程蓝牙实例）。返回 BluetoothDevice 类实例。需要注意的是，如果该蓝牙设备 MAC 地址不能被识别，其蓝牙 Name 为 null。

(8) 在项目配置文件 app/src/AndroidManifest.xml 中添加相应权限。

需要开启定位权限才能搜索到附近的蓝牙设备，代码如下。

```
<uses-permission android:name="android.permission.BLUETOOTH"/>
<uses-permission android:name="android.permission.BLUETOOTH_ADMIN" />
<uses-permission android: name=" android.permission.ACCESS_FINE_LOCATION" />
<uses-permission android: name=" android.permission.ACCESS_COARSE_LOCATION" />
```

10.2.2 Android 查找蓝牙设备

在 Android 2.3 创建应用项目：Bluetooth_Find。

(1) 在主布局文件 activity_main.xml 中放置两个 Button 控件，再放置一个 TextView 控件，以显示本机蓝牙信息，以及一个 ListView 控件，用于显示查找的蓝牙信息，如图 10-3 所示。

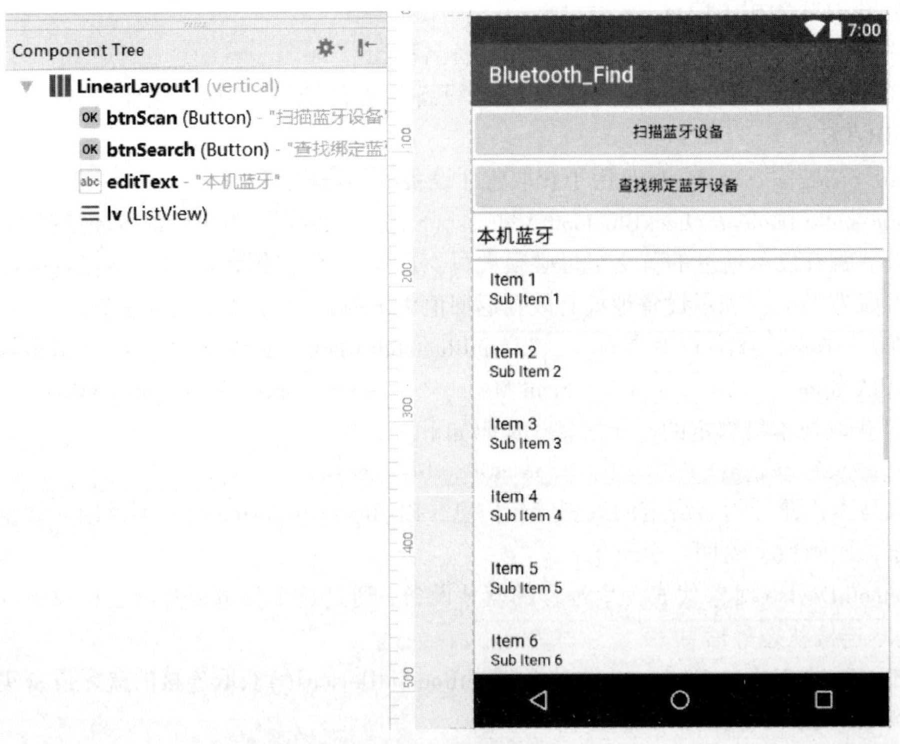

图 10-3　主布局文件

（2）主 Activity 文件 MainActivity. java 的代码如下。

```
public class MainActivity extends AppCompatActivity {
  Button bt1,bt2;
  ListView listView;
  BluetoothAdapter mBluetoothAdapter;
  EditText et;
  List < String > list = new ArrayList < String > ();
  ArrayAdapter < String > adapter;
  @Override
  protected void onCreate(Bundle savedInstanceState) {
      super. onCreate(savedInstanceState);
      setContentView(R. layout. activity_main);
      bt1 = (Button) findViewById (R. id. btnSearch);
      listView = (ListView) findViewById (R. id. lv);
      et = (EditText) findViewById (R. id. editText);
      adapter = new ArrayAdapter < String > (MainActivity. this, android. R. layout. simple_
list_item_1, list);
      listView. setAdapter (adapter);
      IntentFilter intent = new IntentFilter();
      intent. addAction (BluetoothDevice. ACTION_FOUND);
      intent. addAction (BluetoothDevice. ACTION_BOND_STATE_CHANGED);
      registerReceiver (mReceiver, intent);
      mBluetoothAdapter = BluetoothAdapter. getDefaultAdapter();
      if (mBluetoothAdapter. getState() = = BluetoothAdapter. STATE_OFF) //打开蓝牙
          mBluetoothAdapter. enable();
      bt1. setOnClickListener (new View. OnClickListener()
     {
    @Override
    public void onClick (View view) {
      et. setText (" Name : " +mBluetoothAdapter. getName() + " Address : "
         + mBluetoothAdapter. getAddress() + " Scan Mode --" +
mBluetoothAdapter. getScanMode());
        //打印出当前已经绑定成功的蓝牙设备
Set < BluetoothDevice > bts = mBluetoothAdapter. getBondedDevices();
    for (BluetoothDevice device : bts) {
      // Add the name and address to an array adapter to show in a ListView
        list. add (device. getName() + " \n" +device. getAddress());
    }
  }
});
bt2 = (Button) findViewById (R. id. btnScan);
```

```
bt2. setOnClickListener (new View. OnClickListener() {
    @Override
    public void onClick (View view) {
      mBluetoothAdapter. startDiscovery();
    }
});
}
private final  BroadcastReceiver mReceiver = new BroadcastReceiver() {
  @Override
  public void onReceive (Context context, Intent intent) {
      String action = intent. getAction();
      //发现设备
      if (BluetoothDevice. ACTION_FOUND. equals (action)) {
      //从 Intent 中获取蓝牙设备
      BluetoothDevice device =
      intent. getParcelableExtra (BluetoothDevice. EXTRA_DEVICE);
      //添加名字和地址到 list 中
      if (device. getBondState() = = BluetoothDevice. BOND_NONE) {
        String str = " 未配对完成  " + device. getName() + "    "
                                        + device. getAddress();
        if (list. indexOf (str) = = -1) //防止重复添加
            list. add (str);
        adapter. notifyDataSetChanged();
        }
      }
    }
  };
  @Override
  protected void onDestroy()
   {
      super. onDestroy();
      unregisterReceiver (mReceiver);
  }
}
```

(3) 在项目配置文件 app/src/AndroidManifest. xml 中添加相应权限，代码如下。

```
<uses-permission android:name="android. permission. BLUETOOTH"/>
<uses-permission android:name="android. permission. BLUETOOTH_ADMIN" />
<uses-permission android: name=" android.permission.ACCESS_FINE_LOCATION" />
<uses-permission android: name=" android.permission.ACCESS_COARSE_LOCATION" />
```

(4) 项目运行结果如图 10-4 所示。

图 10-4　项目运行结果

10.3 实例：蓝牙控制智能小车

在蓝牙的应用中，经常需要使用手机通过无线控制远程对象，例如通过蓝牙控制小车，本节介绍 Android 手机蓝牙控制智能小车的实现过程。

本节选择的 Arduino 智能小车为 HJduino 可编程蓝牙遥控小车机器人，如图 10-5 所示。

图 10-5　Arduino 智能小车

智能小车采用的主板为 Arduino UNO 类型板，如图 10-6 所示。

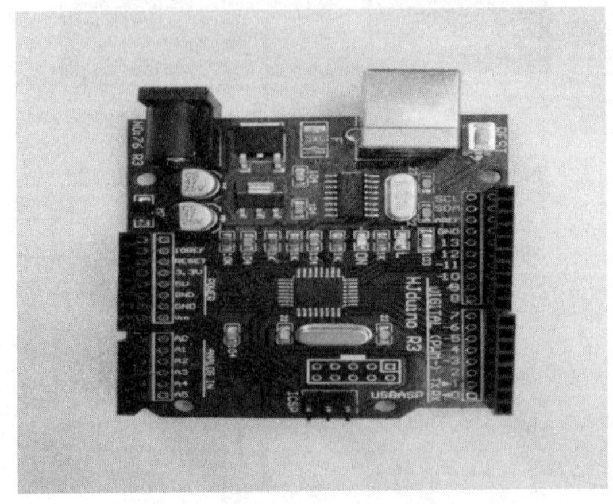

图 10-6　智能小车主板

为了更方便地使用智能小车，对 Arduino UNO 主板的引脚进行扩展，如图 10-7 所示。其中，P4 口接蓝牙，P4 口的引脚为 VCC、GND、TXD、RXD。

图 10-7　智能小车扩展板

智能小车的蓝牙模块为 HC-05 蓝牙模块，支持无线蓝牙串口透传，它的四个引脚为 VCC、GND、TXD、RXD，分别连接到 P4 口的对应引脚，如图 10-8 所示。

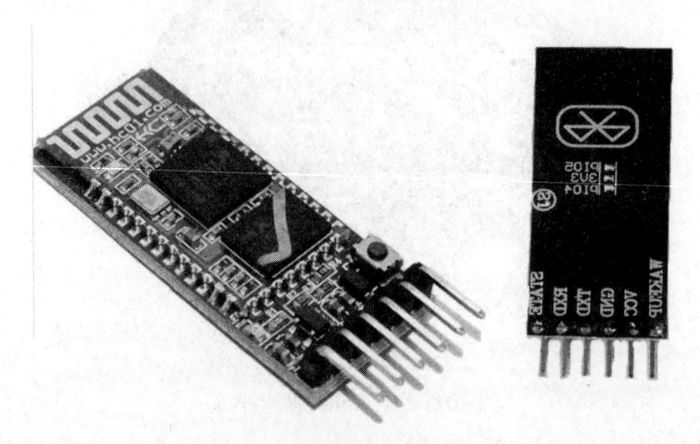

图 10-8　HC-05 蓝牙模块

智能小车的 HC-05 蓝牙模块与扩展板的连接如图 10-9 所示。

图 10-9　智能小车的 HC-05 蓝牙模块与扩展板的连接

手机控制智能小车的结构图如图 10-10 所示

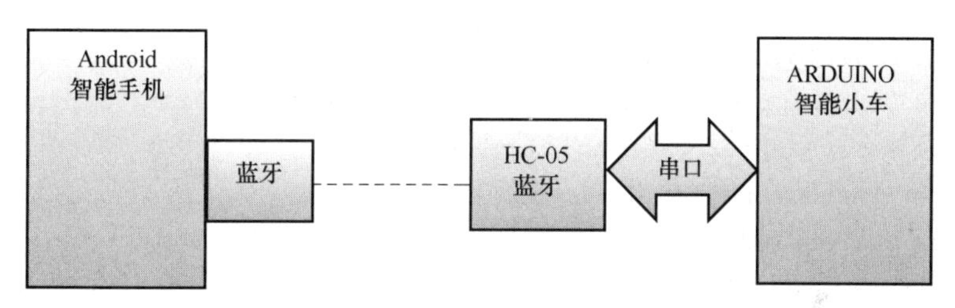

图 10-10　手机控制智能小车的结构图

智能小车控制协议如表 10-1 所示。

表 10-1　智能小车控制协议

前　进	后　退	左　转	右　转	停　止
W	S	A	D	Q

下面是手机端的实现方法，新建项目 Blue _Control，操作步骤如下。

(1) 在 AndridManifest. xml 文件中增加权限，代码如下。

```
<uses-permission android:name = "android. permission. BLUETOOTH_ADMIN" / >
<uses-permission android: name = " android. permission. BLUETOOTH" / >
```

(2) 启动界面布局文件 activity_main. xml，内容如下。

```
<RelativeLayout xmlns:android = "http://schemas. android. com/apk/res/android"
    xmlns:tools = "http://schemas. androd. com/tools"
    android:layout_width = " match_parent"
    android: layout_height = " match_parent"
    tools: context = " . MainActivity" >
    <Button
        android: id = " @ + id/button1"
        android: layout_width = " match_parent"
```

```
        android: layout_height =" wrap_content"
        android: layout_alignParentLeft =" true"
        android: layout_alignParentTop =" true"
        android: text =" 单击搜素蓝牙设备" />
    <ListView
        android: id =" @ +id/list_search"
        android: layout_width =" match_parent"
        android: layout_height =" wrap_content"
        android: layout_alignParentLeft =" true"
        android: layout_below =" @ +id/button1" >
    </ListView >
</RelativeLayout >
```

（3）布局文件 activity_main. xml 对应的处理文件 MainActivity. java 的代码如下。

```
public class MainActivity extends AppCompatActivity  implements OnItemClickListener{
    private static final String TAG = "Main";
    private ListView lvDevices;
    Button b1;
    private BluetoothAdapter bluetoothAdapter;
    private List <String > bluetoothDevices  = new ArrayList <String > ();
    private ArrayAdapter <String > arrayAdapter;
    private BluetoothDevice device;
    private SharedPreferences sp;
    public ArrayList <String > list  =new ArrayList <String > ();
    Set <BluetoothDevice > bondDevices;
      @ Override
    protected void onCreate (Bundle savedInstanceState) {
        super. onCreate (savedInstanceState);
        setContentView (R. layout. activity_main);
        bluetoothAdapter  = BluetoothAdapter. getDefaultAdapter ();
        lvDevices  = (ListView) findViewById  (R. id. list_search);
        sp  = getSharedPreferences  (" config", MODE_PRIVATE);
        Set <BluetoothDevice > pairedDevices  = bluetoothAdapter
                . getBondedDevices ();
         if  (pairedDevices. size ()  > 0)
        {
           for  (BluetoothDevice device : pairedDevices)
            {
                bluetoothDevices. add  (device. getName () +":"
                        + device. getAddress () +"  \n");
            }
        }
    arrayAdapter  = new ArrayAdapter <String >  (this, android. R. layout.
    simple_list_item_1, android. R. id. text1, bluetoothDevices);
```

```
        b1 = (Button) findViewById (R. id. button1);
        b1. setOnClickListener (new OnClickListener ()
         {
          @ Override
          public void onClick (View v) {
            // TODO Auto-generated method stub
            setProgressBarIndeterminateVisibility (true);
            setTitle (" 正在扫描... ");
            if (bluetoothAdapter. isDiscovering ()) {
                  bluetoothAdapter. cancelDiscovery ();
              }
              list. clear ();
              bondDevices = bluetoothAdapter. getBondedDevices ();
              for (BluetoothDevice device : bondDevices) {
                  String str = " 已配对完成" +device. getName () +""
                   + device. getAddress ();
                  list. add (str);
                  arrayAdapter. notifyDataSetChanged ();
              }
              bluetoothAdapter. startDiscovery ();
        }
      });
      lvDevices. setAdapter (arrayAdapter);
      lvDevices. setOnItemClickListener (this);
      IntentFilter filter = new IntentFilter (BluetoothDevice. ACTION_FOUND);
      this. registerReceiver (receiver, filter);
      filter = new IntentFilter (BluetoothAdapter. ACTION_DISCOVERY _FINISHED);
      this. registerReceiver (receiver, filter);
    }
    @ Override
public void onItemClick (AdapterView <? > parent, View view, int position, long id)
     {
        String s = arrayAdapter. getItem (position);
        String address = s. substring (s. indexOf (":") +1). trim ();

        Log. v (TAG, " -------------> " +address);
        Toast. makeText (MainActivity. this, address, 1). show ();
        Editor editor = sp. edit ();
        editor. putString (" address", address);
        editor. commit ();
        Intent intent = new Intent (MainActivity. this, Lanyakongzhi. class);
        startActivity (intent);
        finish ();
```

```
    }
    private final BroadcastReceiver receiver = new BroadcastReceiver()
    {
     @Override
     public void onReceive (Context context, Intent intent)
        {
        String action = intent.getAction();
        if (BluetoothDevice.ACTION_FOUND.equals (action))
          {
          BluetoothDevice device = intent
          .getParcelableExtra (BluetoothDevice.EXTRA_DEVICE);
          if (device.getBondState() != BluetoothDevice.BOND_BONDED)
           {
             bluetoothDevices.add (device.getName() + ":"
                       + device.getAddress() + " \n");
             arrayAdapter.notifyDataSetChanged();
          }
        } else if (BluetoothAdapter.ACTION_DISCOVERY_FINISHED.equals (action))
         {
          setProgressBarIndeterminateVisibility (false);
          setTitle (" 连接蓝牙设备");
        }
      }
    };
  }
```

（4）第二界面布局文件 kongzhi.xml 的内容如下。

```
<?xml version="1.0" encoding="utf-8"?>
<LinearLayout xmlns:android="http://schemas.android.com/apk/res/android"
      android:layout_width=" match_parent"
      android: layout_height=" match_parent"
      android: orientation=" vertical" >
      <LinearLayout
         android: layout_width=" match_parent"
         android: layout_height=" wrap_content"
         android: gravity=" center_horizontal"
         android: orientation=" horizontal" >
         <Button
            android: id=" @+id/btnF"
            android: layout_width=" wrap_content"
            android: layout_height=" wrap_content"

            android: text=" Forward" />
      </LinearLayout>
```

```
        <LinearLayout
            android: layout_width = " match_parent"
            android: gravity = " center_horizontal"
            android: layout_height = " wrap_content" >
            <Button
                android: id = " @ + id/btnL"
                android: layout_width = " wrap_content"
                android: layout_height = " wrap_content"
                android: text = " Left" / >
            <Button
                android: id = " @ + id/btnS"
                android: layout_width = " wrap_content"
                android: layout_height = " wrap_content"
                android: text = " Stop" / >
            <Button
                android: id = " @ + id/btnR"
                android: layout_width = " wrap_content"
                android: layout_height = " wrap_content"
                android: text = " Right" / >
        < /LinearLayout >
        <LinearLayout
            android: layout_width = " match_parent"
            android: layout_height = " wrap_content"
            android: gravity = " center_horizontal"
            android: orientation = " vertical" >
        <Button
            android: id = " @ + id/btnB"
            android: layout_width = " wrap_content"
            android: layout_height = " wrap_content"
            android: gravity = " center_horizontal"
            android: text = " Back" / >
        < /LinearLayout >
< /LinearLayout >
```

（5）第二界面布局文件 kongzhi. xml 对应的处理文件 Lanyakongzhi. java 的内容如下。

```
public class Lanyakongzhi extends Activity {
    private static final String TAG = "BLUEZ_CAR";
    private static final boolean D = true;
    private BluetoothAdapter mBluetoothAdapter = null;
    private BluetoothSocket btSocket = null;
    private OutputStream outStream = null;
    private SharedPreferences sp;
    private String address;
```

```
        Button mButtonF;
        Button mButtonB;
        Button mButtonL;
        Button mButtonR;
        Button mButtonS;
        private static final UUID MY_UUID = UUID.fromString (" 00001101-0000-1000-
8000-00805F9B34FB");
// Intent intent = new Intent();
//String result = intent.getStringExtra (" textViewLabel");
// String address = result;
        // private static String address = " result"; // < = =要连接的蓝牙设备 MAC 地址
        /* * Called when the activity is first created. * /
        @Override
        public void onCreate (Bundle savedInstanceState) {
        super.onCreate (savedInstanceState);
        setContentView (R.layout.kongzhi);
        sp = getSharedPreferences (" config", MODE_PRIVATE);
        address =  sp.getString (" address", null);
        if (TextUtils.isEmpty (address)) {
            Toast.makeText (Lanyakongzhi.this, " 蓝牙地址为空", 1). show();
        }
    ///获取蓝牙数据
    // Intent intent = new Intent();
    // address = intent.getStringExtra (" BTname");
    //Toast.makeText (Lanyakongzhi.this, address, Toast.LENGTH_SHORT). show();
        //前进
        mButtonF = (Button) findViewById (R.id.btnF);
        mButtonF.setOnTouchListener (new Button.OnTouchListener() {
          @Override
          public boolean onTouch (View v, MotionEvent event) {
            // TODO Auto-generated method stub
            String message;
            byte [] msgBuffer;
            int action = event.getAction();
          switch (action)
           {
            case MotionEvent.ACTION_DOWN:
            try {
                outStream = btSocket.getOutputStream();
              } catch (IOException e) {
                Log.e (TAG, " ON RESUME: Output stream creation failed. ", e);
              }
              message = " W";
```

```
        msgBuffer = message.getBytes();
        try {
          outStream.write (msgBuffer);
        } catch (IOException e) {
                Log.e (TAG, " ON RESUME: Exception during write.", e);
        }
      break;
    case MotionEvent.ACTION_UP:
        try {
          outStream = btSocket.getOutputStream();
        } catch (IOException e) {
          Log.e (TAG, " ON RESUME: Output stream creation failed.", e);
        }
        message = " 0";
        msgBuffer = message.getBytes();
        try {
          outStream.write (msgBuffer);
          } catch (IOException e) {
            Log.e (TAG, " ON RESUME: Exception during write.", e);
          }
        break;
      }
    return false;
    }
  });
   //后退
  mButtonB = (Button) findViewById (R.id.btnB);
  mButtonB.setOnTouchListener (new Button.OnTouchListener() {
      @Override
      public boolean onTouch (View v, MotionEvent event) {
      // TODO Auto-generated method stub
      String message;
      byte [] msgBuffer;
      int action = event.getAction();
      switch (action)
     {
      case MotionEvent.ACTION_DOWN:
      try {
          outStream = btSocket.getOutputStream();
        } catch (IOException e) {
        Log.e (TAG, " ON RESUME: Output stream creation failed.", e);
        }
     message = " S";
```

```
    msgBuffer = message.getBytes();
    try {
        outStream.write(msgBuffer);
    } catch (IOException e) {
        Log.e(TAG, " ON RESUME: Exception during write.", e);
        }
  break;
case MotionEvent.ACTION_UP:
  try {
        outStream = btSocket.getOutputStream();
    } catch (IOException e) {
      Log.e(TAG, " ON RESUME: Output stream creation failed.", e);
    }
    message = " 0";
    msgBuffer = message.getBytes();
    try {
        outStream.write(msgBuffer);
    } catch (IOException e) {
        Log.e(TAG, " ON RESUME: Exception during write.", e);
        }
    break;
    }
    return false;
  }
  });
//左转
mButtonL = (Button) findViewById(R.id.btnL);
mButtonL.setOnTouchListener(new Button.OnTouchListener() {
@Override
public boolean onTouch(View v, MotionEvent event) {
// TODO Auto-generated method stub
  String message;
  byte [] msgBuffer;
  int action = event.getAction();
  switch (action)
   {
  case MotionEvent.ACTION_DOWN:
  try {
      outStream = btSocket.getOutputStream();
  } catch (IOException e) {
    Log.e(TAG, " ON RESUME: Output stream creation failed.", e);
    }
  message = " A";
```

```
  msgBuffer = message.getBytes();
  try {
      outStream.write (msgBuffer);
    } catch (IOException e) {
        Log.e (TAG, " ON RESUME: Exception during write.", e);
    }
    break;
  case MotionEvent.ACTION_UP:
    try {
        outStream = btSocket.getOutputStream();
    } catch (IOException e) {
      Log.e (TAG, " ON RESUME: Output stream creation failed.", e);
    }
    message = " 0";
    msgBuffer = message.getBytes();
    try {
        outStream.write (msgBuffer);
    } catch (IOException e) {
      Log.e (TAG, " ON RESUME: Exception during write.", e);
    }
    break;
  }
  return false;
    }
  });
  //右转
mButtonR = (Button) findViewById (R.id.btnR);
mButtonR.setOnTouchListener (new Button.OnTouchListener() {
  @Override
  public boolean onTouch (View v, MotionEvent event) {
  // TODO Auto-generated method stub
  String message;
  byte [] msgBuffer;
  int action = event.getAction();
  switch (action)
   {
    case MotionEvent.ACTION_DOWN:
    try {
      outStream = btSocket.getOutputStream();
    } catch (IOException e) {
    Log.e (TAG, " ON RESUME: Output stream creation failed.", e);
  }
 message = " D";
```

```
msgBuffer = message.getBytes();
try {
    outStream.write (msgBuffer);
  } catch (IOException e) {
      Log.e (TAG, " ON RESUME: Exception during write.", e);
  }
  break;
case MotionEvent.ACTION_UP:
  try {
      outStream = btSocket.getOutputStream();
  } catch (IOException e) {
    Log.e (TAG, " ON RESUME: Output stream creation failed.", e);
  }
message = " 0";
msgBuffer = message.getBytes();
  try {
    outStream.write (msgBuffer);
  } catch (IOException e) {
      Log.e (TAG, " ON RESUME: Exception during write.", e);
  }
  break;
  }
  return false;
}
});
//停止
mButtonS = (Button) findViewById (R.id.btnS);
mButtonS.setOnTouchListener (new Button.OnTouchListener() {
  @Override
  public boolean onTouch (View v, MotionEvent event) {
  // TODO Auto-generated method stub
  if (event.getAction() = =MotionEvent.ACTION_DOWN)
  try {
      outStream = btSocket.getOutputStream();
  } catch (IOException e) {
      Log.e (TAG, " ON RESUME: Output stream creation failed.", e);
  }
  String message = " Q";
  byte [] msgBuffer = message.getBytes();
  try {
    outStream.write (msgBuffer);
  } catch (IOException e) {
      Log.e (TAG, " ON RESUME: Exception during write.", e);
```

```
    }
return false;
}
});
if (D)
Log.e (TAG, " +++ ON CREATE +++");
mBluetoothAdapter = BluetoothAdapter.getDefaultAdapter();
if (mBluetoothAdapter = = null) {
Toast.makeText (this, " Bluetooth is not available.", Toast.LENGTH_LONG).show();
        finish();
        return;
}
if (! mBluetoothAdapter.isEnabled()) {
Toast.makeText (this,   " Please enable your Bluetooth and re-run this pro-
gram.", Toast.LENGTH_LONG).show();
        finish();
        return;
}
if (D)
Log.e (TAG, " +++ DONE IN ON CREATE, GOT LOCAL BT ADAPTER +++");
}
@Override
public void onStart() {
    super.onStart();
    if (D) Log.e (TAG, " ++ ON START ++");
}
@Override
public void onResume() {
    super.onResume();
    if (D) {
    Log.e (TAG, " + ON RESUME +");
    Log.e (TAG, " + ABOUT TO ATTEMPT CLIENT CONNECT +");
  }
BluetoothDevice device = mBluetoothAdapter.getRemoteDevice (address);
  try {
     btSocket = device.createRfcommSocketToServiceRecord (MY_UUID);
    } catch(IOException e) {
     Log.e(TAG, "ON RESUME: Socket creation failed.", e);
    }
    mBluetoothAdapter.cancelDiscovery();
    try {
      btSocket.connect();
      Log.e(TAG, "ON RESUME: BT connection established, data transfer link open.");
```

```
    } catch(IOException e) {
    try {
      btSocket.close();
    } catch(IOException e2) {
      Log.e(TAG,"ON RESUME: Unable to close socket during connection failure", e2);
      }
    }
    // Create a data stream so we can talk to server.
    if(D)
      Log.e(TAG, " + ABOUT TO SAY SOMETHING TO SERVER +");
    try {
        outStream = btSocket.getOutputStream();
      } catch(IOException e) {
        Log.e(TAG, "ON RESUME: Output stream creation failed.", e);
      }
    String message = "1";
    byte[] msgBuffer = message.getBytes();
    try {
        outStream.write(msgBuffer);
      } catch(IOException e) {
        Log.e(TAG, "ON RESUME: Exception during write.", e);
      }
    }
    @Override
  public void onPause() {
    super.onPause();
      if(D)
        Log.e(TAG, "- ON PAUSE -");
        if(outStream != null) {
        try {
            outStream.flush();
        } catch(IOException e) {
          Log.e(TAG, "ON PAUSE: Couldn't flush output stream.", e);
        }
        }
      try {
            btSocket.close();
        } catch(IOException e2) {
            Log.e(TAG, "ON PAUSE: Unable to close socket.", e2);
        }
    }
  @Override
  public void onStop() {
```

```
    super.onStop();
    if(D)Log.e(TAG, "-- ON STOP --");
  }
  @Override
  public void onDestroy() {
    super.onDestroy();
    if(D) Log.e(TAG, "--- ON DESTROY ---");
  }
}
```

（6）执行结果如图 10-11 所示。

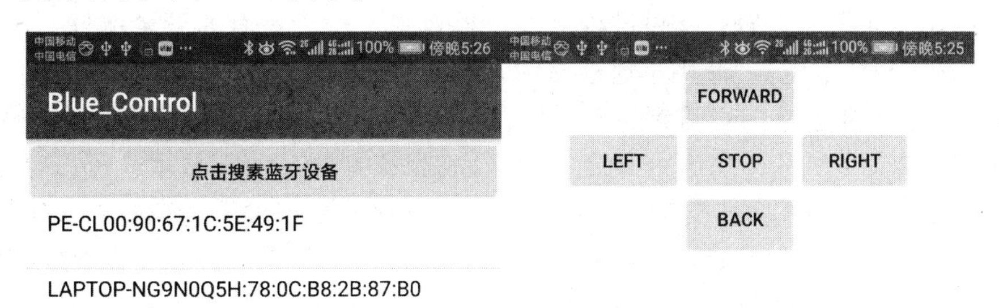

图 10-11　Android 蓝牙控制小车执行结果

基于安全性考虑，设置开启蓝牙可被搜索后，Android 系统会默认给出 120 秒的时间，其他设备可以在这 120 秒内搜索到它。

Android 智能手机控制智能小车的过程，总体来说就是先进行蓝牙连接，然后通过蓝牙传送控制命令。

10.4 AndroidNFC——通过 NFC 读取 MifareClassic 卡信息

NFC，即 Near Field Communication，近距离无线通信技术，是一种短距离的（通常 < = 4cm 或更短）高频（13.56M Hz）无线通信技术，它提供了一种简单、触控式的解决方案，可以让消费者简单直观地交换信息、访问内容与服务。

NFC 的技术优势如下。

➢ 与蓝牙相比：NFC 操作简单，配对迅速。

➢ 与 RFID 相比：NFC 适用范围广泛、可读可写，能直接集成在手机中。

➢ 与红外线相比：数据传输较快、安全性高、能耗低。

➢ 与二维码相比：识别迅速、信息类型多样。

将来与移动支付相结合，势必简化支付的购买流程，重塑消费者的购物模式。

NFC 近场通信全方位的测量精度可以达到厘米。这项技术也促进了其他一些有意思的技术的成长，比如把两个手机碰到一起，可以启动多人游戏，把手机贴近 NFC 读写器可以进行付款。

在 Android 4.4 之前，NFC 支付过程需要借助设备上一个专有的安全部件（Secure Element，可以存在 SIM 卡中），使用本地存储的方式，关联设备本身的某种支付方式，这样的话，其他的 App 很难通过 NFC 进行支付操作，因为这个过程是依靠部分硬件的，也就是 Secure Element。基于主机的卡仿真（HCE）是 Android 4.4 的一项新技术，可以让 App 绕过 Secure Element，然后使用云端支付信息或者其他方式存储的支付信息来模拟 NFC 卡。有了 HCE，任何 App 都可以模拟 NFC 卡，而且任意一台 Android 设备都可以当作 NFC 读写器。

这项技术由免接触式射频识别（RFID）演变而来，由飞利浦公司和索尼公司共同开发，是一种非接触式识别和互联技术，可以在移动设备、消费类电子产品、PC 和智能控件工具间进行近距离无线通信。

近场通信是一种短距高频的无线电技术，在 13.56MHz 频率运行于 20 厘米距离内。其传输速度有 106Kbit/秒、212Kbit/秒和 424Kbit/秒三种。目前近场通信已成为 ISO/IEC IS 18092 国际标准、EMCA-340 标准与 ETSI TS 102 190 标准。

消费者可以使用支持该技术的手机在公交、地铁、超市进行刷卡消费，此项技术早年曾在诺基亚 6131i 等产品上出现，在北京、广州、厦门等城市已有成功使用先例。

NFC 有如下三种工作模式。

（1）卡模式（Card emulation）：这个模式的手机其实就相当于一张采用 RFID 技术的 IC 卡，可以替代现在大量的 IC 卡（包括信用卡）。此种方式有一个极大的优点，就是卡片通过非接触读卡器的 RF 域来供电，即便是寄主设备（如手机）没电也可以工作。

（2）点对点模式（P2P mode）：这个模式和红外线差不多，可用于数据交换，只是传输距离比较短，传输创建速度快很多，传输速度也快一些，功耗低。将两个具备 NFC 功能的设备链接，能实现数据点对点传输，如下载音乐、交换图片或者同步设备地址簿。因此，通过 NFC，多个设备如数字相机、PDA、计算机、手机之间，可以交换资料或者服务。

（3）读卡器模式（Reader/writer mode）：作为非接触读卡器使用，比如从海报或者展览信息电子标签上读取相关信息。

NFC 技术可以说是 RFID 技术的一个延伸，说起 RFID 技术，大家可能摇摇头说没听过。实际上它已经大量地应用在我们的生活当中，城市的公交系统、大学的水卡、饭卡、旅馆的门禁都是 RFID 技术的应用。不过 RFID 只能实现信息的读取以及判定，而 NFC 技术则强调的是信息交互。通俗的说，NFC 就是 RFID 的演进版本，可以近距离交换信息。

在 Android NFC 应用中，Android 手机通常是作为通信中的发起者，也就是作为各种 NFC 卡的读写器。Android 对 NFC 的支持主要在 android.nfc 和 android.nfc.tech 两个包中，如图 10-12

所示。

Android. nfc 包中有如下几种主要类。

（1）NfcManager：可以用来管理 Android 设备中指出的所有 NfcAdapter，但由于大部分 Android 设备只支持一个 NfcAdapter，所以一般直接调用 getDefaultAapater 来获取手机中的 Adapter。

（2）NfcAdapter：相当于一个 NFC 适配器，类似于电脑装了网络适配器才能上网，手机装了 NfcAdapter 才能发起 NFC 通信。

（3）NDEF：NFC Data Exchange Format，即 NFC 数据交换格式。

（4）NdefMessage 和 NdefRecord：NFC forum 定义的数据格式。

（5）Tag：代表一个被动式 Tag 对象，可以代表一个标签、卡片等。当 Android 设备检测到一个 Tag 时，会创建一个 Tag 对象，将其放在 Intent 对象，然后发送到相应的 Activity。

（6）Android. nfc. tech：定义了可以对 Tag 进行读写操作的类，这些类按照其使用的技术类型可以分成不同的类，如 NfcA、NfcB、NfcF，以及 MifareClassic 等，其中 MifareClassic 比较常见。

（7）Ndef：NFC Data Exchange Format，即 NFC 数据交换格式。

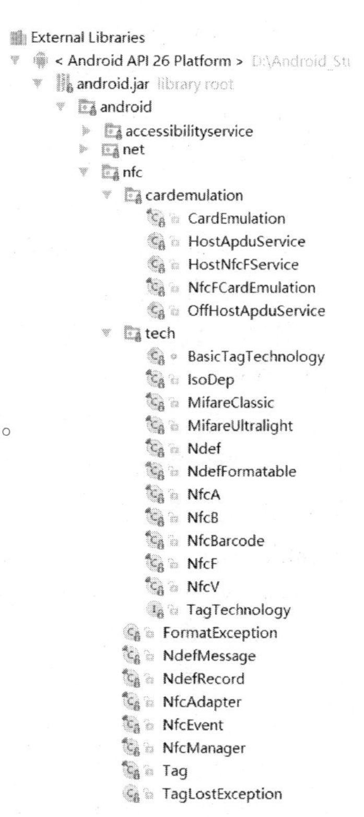

图 10-12　Android 的 NFC 库

Android 支持的 NFC 的数据格式类如表 10-2 所示。

表 10-2　Android 支持的 NFC 的数据格式类

数据格式类	介　绍
TagTechnology	所有的 NFC 标签技术类必须实现的接口
NfcA	提供对 NFC-A（ISO-14443-3A）属性和 I/O 操作的访问
NfcB	提供对 NFC-B（ISO-14443-3B）属性和 I/O 操作的访问
NfcF	提供对 NFC-F（ISO-6319-4）属性和 I/O 操作的访问
NfcV	提供对 NFC-V（ISO-15693）属性和 I/O 操作的访问
IsoDep	提供对 NFC-A（ISO-14443-4）属性和 I/O 操作的访问
Ndef	提供对 NDEF 格式的 NFC 标签上的 NDEF 数据的操作和访问
NdefFormatable	提供对可以被 NDEF 格式的 NFC 标签的格式化操作
MifareClassic	如果 Android 设备支持 MIFARE，那么将提供对经典的 MIFARE 类型标签属性和 I/O 操作的访问
MifareUltralight	如果 Android 设备支持 MIFARE，那么将提供对超薄的 MIFARE 类型标签属性和 I/O 操作的访问

检测到标签后，在 Activity 中的处理流程如下。

（1）在 onCreate( ) 中获取 NfcAdapter 对象，代码如下。

```
NfcAdapter nfcAdapter = NfcAdapter.getDefaultAdapter(this);
```

（2）取出封装在 Intent 中的 Tag，代码如下。

```
Tag tagFromIntent = intent.getParcelableExra(NfcAdapter.EXTRA_TAG);
```

（3）读取 Tag，代码如下。

```
MifareClassic mfc = MifareClassic.get(tagFromIntent);
```

(4) 允许进行标签操作：mfc. connect( )。

(5) 标签的相关操作。

➢ 获取 Tag 的类型：int type = mfc. getType( )。

➢ 获取 Tag 中包含的扇区数：int sectorCount = mfc. getSectorCount( )。

➢ 扇区密码验证：auth = mfc. authenticateSectorWithKeyA (j, MifareClassic. KEY_DEFAULT)。

➢ 读扇区：mfc. readBlock (bIndex)。

➢ 命令读写：mfc. transceive (cmd. getBytes( ))，参数为读写操作的命令。

在本实例中，使用 MifareClassic 卡进行数据读取测试。在 Android 2. 3 中创建应用项目：NEC_TEST。

MifareClassic 卡的数据分为 16 个区 (Sector)，每个区有 4 个块 (Block)，每个块可以存放 16 字节的数据。每个区最后一个块称为 Trailer，主要用来存放读写该区 Block 数据的 Key，可以有 A、B 两个 Key，每个 Key 长度为 6 个字节，缺省的 Key 值一般为全 FF 或是 0，由 MifareClassic. KEY_DEFAULT 定义。因此，读写 Mifare Tag 首先需要有正确的 Key 值 (起到保护的作用)，只有鉴权成功，之后才可以读写该区数据。

具体实现步骤如下。

(1) 总配置文件 AndroidManifest. xml 的代码如下。

```
<?xml version="1.0" encoding="utf-8"?>
  <manifest xmlns:android=http://schemas.android.com/apk/res/android
  package="com.example.hefugui.nec_test" >
  <uses-permission android:name=" android.permission.NFC" />
  <uses-feature android:name=" android.hardware.nfc" android:required=" true" />
  <application
   ……
      <intent-filter>
      <action android:name=" android.nfc.action.TECH_DISCOVERED" />
      </intent-filter>
      <meta-data
          android:name=" android.nfc.action.TECH_DISCOVERED"
          android:resource=" @xml/nfc_tech_filter" />
      </activity>
  </application>
</manifest>
```

(2) 新建文件 res/xml/nfc_tech_filter. xml，代码如下。

```
<?xml version="1.0" encoding="utf-8"?>
<resources xmlns:xliff="urn:oasis:names:tc:xliff:document:1.2">
      <tech-list>
        <tech>android.nfc.tech.MifareClassic</tech>
      </tech-list>
</resources>
```

当手机开启了 NFC，并且检测到一个 Tag 后，Tag 分发系统会自动创建一个封装了 NFC Tag 信息的 Intent。如果多于一个应用程序能够处理这个 Intent，那么手机会弹出一个对话框，让用户选择处理该 Tag 的 Activity。Tag 分发系统定义了 3 种 Intent，按优先级从高到低排列为：NDEF_DISCOVERED、TECH_DISCOVERED、TAG_DISCOVERED

当 Android 设备检测到有 NFC Tag 靠近时，会根据 Action 声明的顺序向对应的 Activity 发送含 NFC 消息的 Intent。

此处，我们使用的 intent-filter 的 Action 类型为 TECH_DISCOVERED，从而可以处理所有类型为 ACTION_TECH_DISCOVERED 并且使用的技术为 nfc_tech_filter. xml 文件中定义的类型的 Tag。

当手机检测到一个 Tag 时，启用 Activity 的匹配过程如图 10-13 所示。

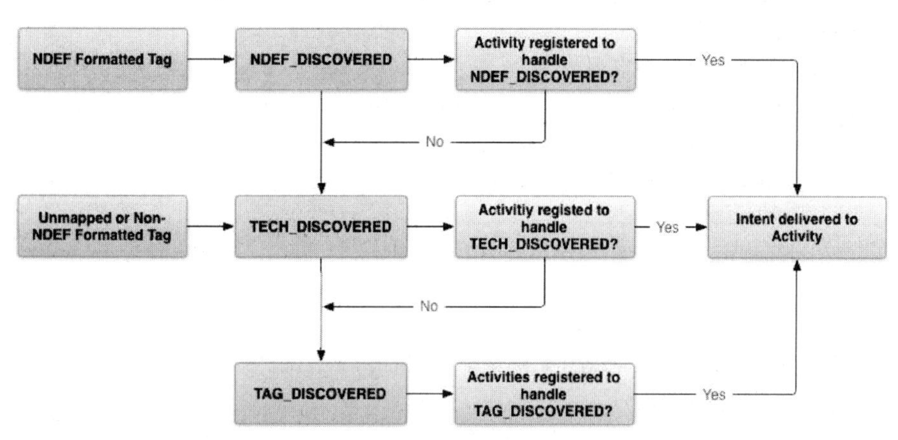

图 10-13　Tag 的 Activity 匹配过程

（3）主布局文件 main. xml 的代码如下。

```
<? xml version ="1.0" encoding ="utf-8"? >
<LinearLayout xmlns:android ="http://schemas.android.com/apk/res/android"
    android:layout_width =" match_parent"
    android: layout_height =" match_parent"
    android: orientation =" vertical" >
    <ScrollView
        android: id =" @ +id/scrollView"
        android: layout_width =" fill_parent"
        android: layout_height =" fill_parent"
        android: background =" @android: drawable/edit_text" >
          <TextView
            android: id =" @ +id/promt"
            android: layout_width =" fill_parent"
            android: layout_height =" wrap_content"
            android: scrollbars =" vertical"
            android: singleLine =" false"
            android: text =" @string/info" />
    </ScrollView>
</LinearLayout>
```

（4）在 strings. xml 文件中增加如下内容。

```
<string name ="app_name" >NFC 测试</string>
    <string name =" info" >扫描中…</string>
```

（5）界面处理文件 MainActivity. java 的代码如下。

```
public class MainActivity extends Activity {
    NfcAdapter nfcAdapter;
    TextView promt;
    @Override
    public void onCreate(Bundle savedInstanceState) {
        super.onCreate(savedInstanceState);
        setContentView(R.layout.main);
        promt = (TextView) findViewById(R.id.promt);
        //获取默认的 NFC 控制器
        nfcAdapter = NfcAdapter.getDefaultAdapter(this);
        if(nfcAdapter == null) {
            promt.setText("设备不支持 NFC!");
            //finish();
            return;
        }
        if(! nfcAdapter.isEnabled()) {
            promt.setText("请在系统设置中先启用 NFC 功能!");
            //finish();
            return;
        }
    }
    @Override
    protected void onResume() {
        super.onResume();
        //得到是否检测到 ACTION_TECH_DISCOVERED 触发
  if(NfcAdapter.ACTION_TECH_DISCOVERED.equals(getIntent().getAction()))
{
            //处理该 intent
            processIntent(getIntent());
        }
    }
    //字符序列转换为 16 进制字符串
    private String bytesToHexString(byte[] src) {
        StringBuilder stringBuilder = new StringBuilder("0x");
        if(src == null || src.length <= 0) {
            return null;
        }
        char[] buffer = new char[2];
        for(int i = 0; i < src.length; i++) {
            buffer[0] = Character.forDigit((src[i] >>> 4) & 0x0F, 16);
            buffer[1] = Character.forDigit(src[i] & 0x0F, 16);
            System.out.println(buffer);
            stringBuilder.append(buffer);
```

```
    }
    return stringBuilder.toString();
}
/* Parses the NDEF Message from the intent and prints to the TextView */
private void processIntent(Intent intent) {
    //取出封装在 intent 中的 TAG
    Tag tagFromIntent = intent.getParcelableExtra(NfcAdapter.EXTRA_TAG);
    for(String tech : tagFromIntent.getTechList()) {
        System.out.println(tech);
    }
    boolean auth = false;
    //读取 TAG
    MifareClassic mfc = MifareClassic.get(tagFromIntent);
    try {
        String metaInfo = "";
        //Enable I/O operations to the tag from this TagTechnology object.
        mfc.connect();
        int type = mfc.getType();//获取 TAG 的类型
        int sectorCount = mfc.getSectorCount();//获取 TAG 中包含的扇区数
        String typeS = "";
        switch(type) {
        case MifareClassic.TYPE_CLASSIC:
            typeS = "TYPE_CLASSIC";
            break;
        case MifareClassic.TYPE_PLUS:
            typeS = "TYPE_PLUS";
            break;
        case MifareClassic.TYPE_PRO:
            typeS = "TYPE_PRO";
            break;
        case MifareClassic.TYPE_UNKNOWN:
            typeS = "TYPE_UNKNOWN";
            break;
        }
  metaInfo += "卡片类型:" + typeS + "\n 共" + sectorCount + "个扇区\n 共"
      + mfc.getBlockCount() + "个块\n 存储空间: " + mfc.getSize() + "B\n";
  for(int j = 0; j < sectorCount; j++) {
    //Authenticate a sector with key A.
  auth = mfc.authenticateSectorWithKeyA(j, MifareClassic.KEY_DEFAULT);
    int bCount;
    int bIndex;
    if(auth) {
      metaInfo += "Sector " + j + ":验证成功\n";
      //读取扇区中的块
```

```
                bCount = mfc.getBlockCountInSector(j);
                bIndex = mfc.sectorToBlock(j);
                for(int i = 0; i < bCount; i ++) {
                    byte[] data = mfc.readBlock(bIndex);
                    metaInfo += "Block " + bIndex + " : "
                                    + bytesToHexString(data) + "\n";
                    bIndex ++;
                    }
                } else {
                    metaInfo += "Sector " + j + ":验证失败\n";
                }
                }
                promt.setText(metaInfo);
            } catch(Exception e) {
                e.printStackTrace();
            }
        }
    }
```

（6）执行结果如图 10-14 所示。

图 10-14　执行结果

10.5 本章小结

本章主要介绍了 Android 目前应用最广泛的无线通信技术：Wifi、蓝牙和 NFC，详细地介绍了它们的使用方法，并给出了具体的应用案例。

第十一章 Android的开源库和开源项目

他山之石，可以攻玉，汲取他人的精华为己所用，对于程序员来讲，最好的学习是多看别人优秀的代码，最高效的开发是利用有效的资源。软件库的存在使得 Android 开发更方便快捷，能更快地实现软件的功能，参考功能完整详细的开源项目，可以更快地成为一名优秀的开发者。

11.1 Android 的开源库

本节介绍几款常用的 Android 开源库。

11.1.1 Android View Animations

Android View Animations 是一个能实现很多很酷的效果的强大动画库，使用这个动画库，可以轻松创建各种动画效果。下载地址：https://github.com/daimajia/AndroidViewAnimations。Android View Animations 开源库的源码目录如图 11-1 所示。

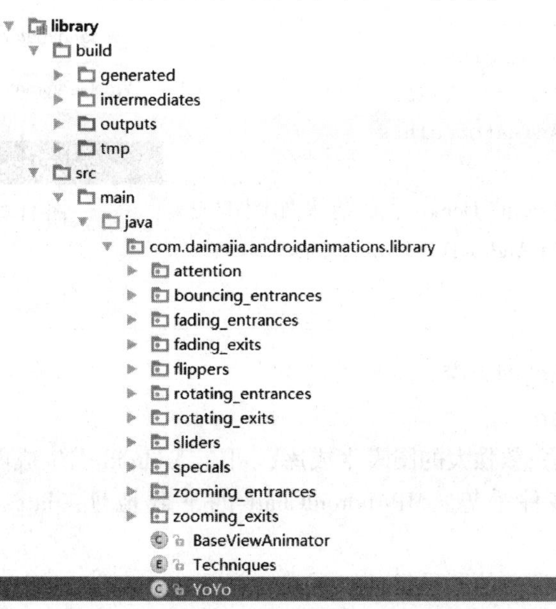

图 11-1　Android View Animations 开源库的源码目录

使用 Android View Animations 库时，首先需要在 app/build. gradle 文件的 dependencies 闭包中添加引用，内容如下。

```
...
dependencies {
    compile fileTree(dir:'libs', include: [' *.jar'])
    compile project(':library')
}
```

Android View Animations 库的调用格式如下。

```
YoYo.YoYoString rope = YoYo.with(Techniques.FadeIn).duration(1000).playOn
(mTarget);
```

其中，playOn（mTarget）为动画的目标，duration（1000）为动画时长，Techniques. FadeIn 为动画种类，定义在库文件 Techniques. java 中，如下所示。

```
public enum Techniques {
DropOut(DropOutAnimator.class),
Landing(LandingAnimator.class),
TakingOff(TakingOffAnimator.class),
Flash(FlashAnimator.class),
Pulse(PulseAnimator.class),
RubberBand(RubberBandAnimator.class),
Shake(ShakeAnimator.class),
Swing(SwingAnimator.class),
Wobble(WobbleAnimator.class),
Bounce(BounceAnimator.class),
Tada(TadaAnimator.class),
StandUp(StandUpAnimator.class),
Wave(WaveAnimator.class),
Hinge(HingeAnimator.class),
RollIn(RollInAnimator.class),
RollOut(RollOutAnimator.class),
...
```

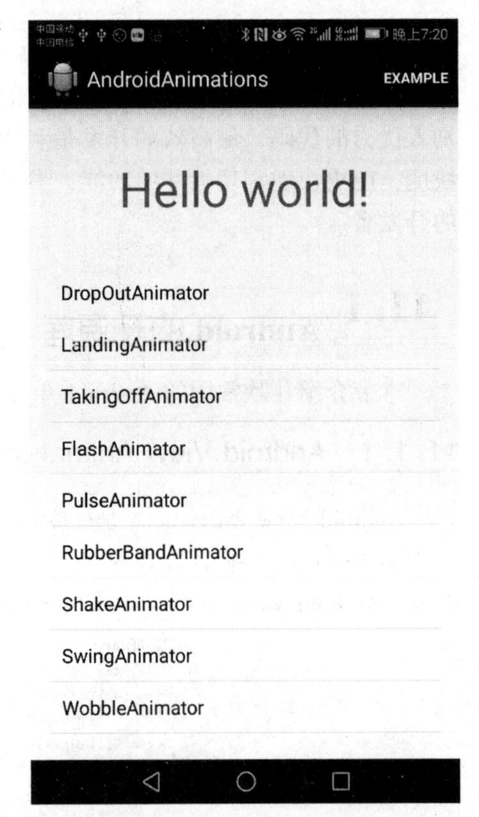

图 11-2　动画运行结果

AndroidViewAnimations 的 Demo 显示结果如图 11-2 所示，详见本书代码项目 AndroidViewAnimations-master。

11.1.2　图表库

本节介绍两款常用的图表库。

1. MPAndroidChart

MPAndroidChart 是一款强大的图表生成库，可在 Android 上生成图表，同时还提供 8 种不同的图表类型和多种手势。MPAndroidChart 的下载地址：https://github.com/PhilJay/MPAndroidChart。

MPAndroidChart 开源库的源码目录如图 11-3 所示。

使用 MPAndroidChart 库时，首先需要在 app/build. gradle 文件的 dependencies 闭包中添加引用，内容如下。

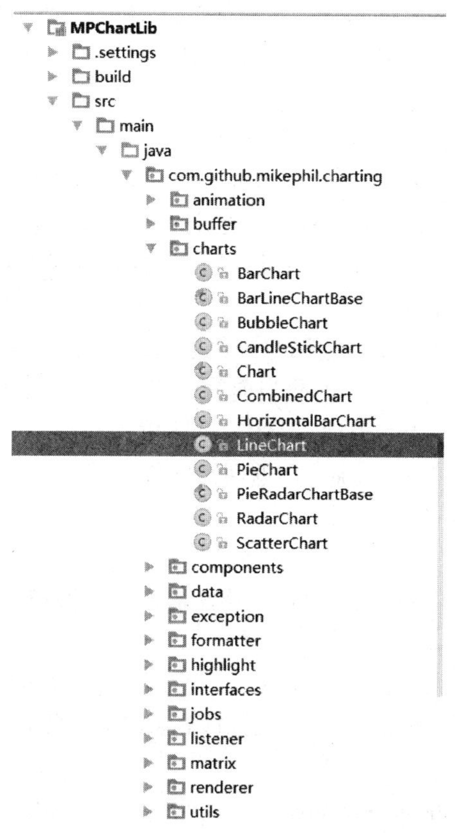

图 11-3　MPAndroidChart 开源库的源码目录

```
dependencies {
    …
    compile 'com.github.PhilJay:MPAndroidChart-Realm:v2.0.2@aar'
    compile project(':MPChartLib')'
}
```

MPAndroidChart 库的调用格式如下。

如果使用 LineChart、BarChart、ScatterChart、CandleStickChart 或 PieChart，可以直接在界面的 xml 中定义。

```
<com.github.mikephil.charting.charts.LineChart
        android:id="@+id/chart"
        android:layout_width=" match_parent"
        android: layout_height=" match_parent" />
```

在界面对应的 Activity 处理代码中实例化。

```
LineChart chart = (LineChart) findViewById(R.id.chart);
```

或者直接在代码中声明和实例化。

```
LineChart chart = new LineChart(Context);
```

主要的 Api 方法如下。

- setDescription（String desc）：设置表格的描述。
- setDescriptionTypeface（Typeface t）：自定义表格中显示的字体。
- setDrawYValues（boolean enabled）：设置是否显示 y 轴的值的数据。
- setValuePaintColor（int color）：设置表格中 y 轴的值的颜色，但是必须设置 setDrawY-Values（true）。
- setValueTypeface（Typeface t）：设置字体。
- setValueFormatter（DecimalFormat format）：设置显示的格式。
- setPaint（Paint p，int which）：自定义笔刷。
- public ChartData getDataCurrent()：返回 ChartData 对象当前显示的图表，包含所有信息的显示值最小和最大值等。
- public float getYChartMin()：返回当前最小值。
- public float getYChartMax()：返回当前最大值。
- public float getAverage()：返回所有值的平均值。
- public float getAverage（int type）：返回平均值。
- public PointF getCenter()：返回中间点。
- public Paint getPaint（int which）：得到笔刷。
- setTouchEnabled（boolean enabled）：设置是否可以触摸，如为 False，则不能进行拖动、缩放等操作。
- setDragScaleEnabled（boolean enabled）：设置是否可以拖拽、缩放。
- setOnChartValueSelectedListener（OnChartValueSelectedListener l）：设置表格上的点被单击时的回调函数。
- setHighlightEnabled（boolean enabled）：设置单击 value 时是否高亮显示。
- public void highlightValues（Highlight [ ] highs）：设置高亮显示。
- saveToGallery（String title）：保存图表到图库中。
- saveToPath（String title，String pathOnSD）：设置保存路径。
- setScaleMinima（float x，float y）：设置最小的缩放。
- centerViewPort（int xIndex，float val）：设置视口。
- fitScreen()：适应屏幕。

所有的图表类型都支持下面三种动画，分别是 x 方向、y 方向、xy 方向。

- animateX（int durationMillis）：x 轴方向。
- animateY（int durationMillis）：y 轴方向。
- animateXY（int xDuration，int yDuration）：xy 轴方向。

例如：

```
mChart.animateX(3000f); // animate horizontal 3000 milliseconds
mChart.animateY(3000f); // animate vertical 3000 milliseconds
mChart.animateXY(3000f, 3000f); // animate horizontal and vertical 3000 milli-
seconds
```

需要注意的是，调用动画方法后，就没有必要调用 invalidate()方法来刷新界面了。

MPAndroidChart 的 Demo 显示结果如图 11-4 所示，详见本书代码项目 MPAndroidChart-master。

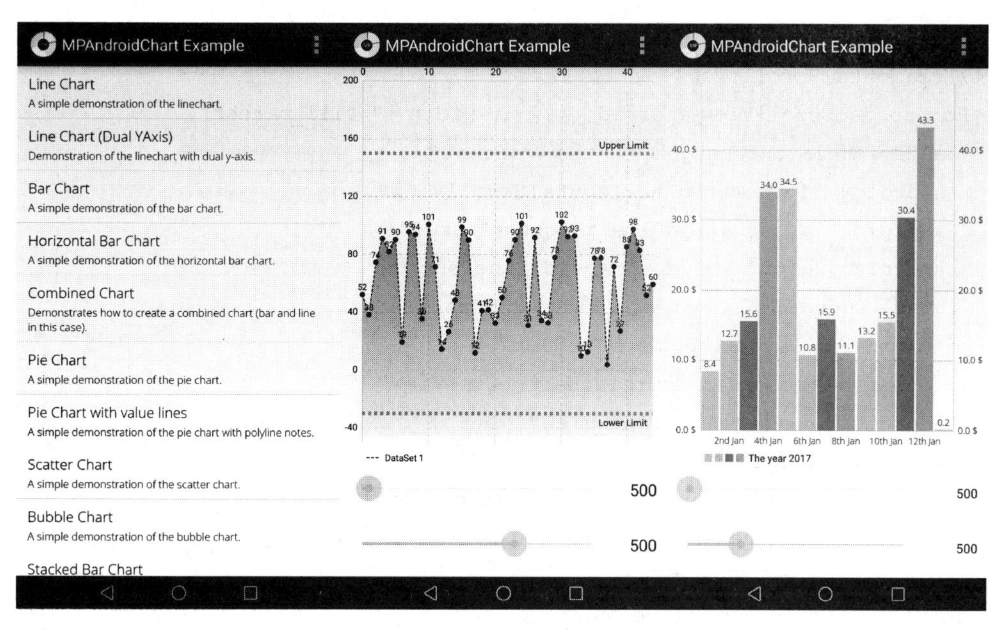

图 11-4　MPAndroidChart 的 Demo 显示结果

2. AndroidCharts

AndroidCharts 是一款简单的图表创建工具，具有自定义的功能，其中包含曲线/折线图、饼图、时钟图、柱状图等创建工具。AndroidCharts 的下载地址：https://github. com/HackPlan/AndroidCharts。

使用 AndroidCharts 库时，首先需要在 app/build. gradle 文件的 dependencies 闭包中添加引用，内容如下。

```
...
dependencies {
     compile project(':AndroidCharts')
}
```

AndroidCharts 库的调用格式如下。

在界面的 xml 中定义如下。

```
<RelativeLayout xmlns:android=http://schemas.android.com/apk/res/android
   android:layout_width=" match_parent"
   android: layout_height=" match_parent"
   android: background=" #ffffff" >
  <HorizontalScrollView
   android: layout_width=" fill_parent"
  android: layout_height=" wrap_content"
   android: id=" @ +id/horizontalScrollViewFloat"
   android: layout_alignParentRight=" true"
   android: layout_above=" @ +id/horizontalScrollView" >
  <view android: layout_width=" wrap_content"
     android: layout_height=" 200dp"
     class=" im.dacer.androidcharts.LineView"
```

```
        android: id=" @ +id/line_view_float" />
    </HorizontalScrollView>
    <HorizontalScrollView android: layout_width=" fill_parent"
        android: layout_height=" wrap_content"
        android: id=" @ +id/horizontalScrollView"
        android: layout_alignParentRight=" true"
        android: layout_above=" @ +id/line_button" >
      <view  android: layout_width=" wrap_content"
          android: layout_height=" 200dp"
          class=" im. dacer. androidcharts. LineView"
        android: id=" @ +id/line_view" />
    </HorizontalScrollView>
    <Button  android: layout_width=" wrap_content"
      android: layout_height=" wrap_content"
      android: text=" Random"
      android: id=" @ +id/line_button"
      android: layout_alignParentBottom=" true"
      android: layout_centerHorizontal=" true" />
    </RelativeLayout>
```

在主程序的使用方法如下。

```
LineView lineView = (LineView)findViewById(R. id. line_view);
lineView. setDrawDotLine (false); //optional
lineView. setShowPopup (LineView. SHOW_POPUPS_MAXMIN_ONLY); //optional
LineView. setBottomTextList (strList);
LineView. setDataList (dataLists);
```

AndroidCharts 的 Demo 显示结果如图 11-5 所示，详见本书代码项目 AndroidCharts-master。

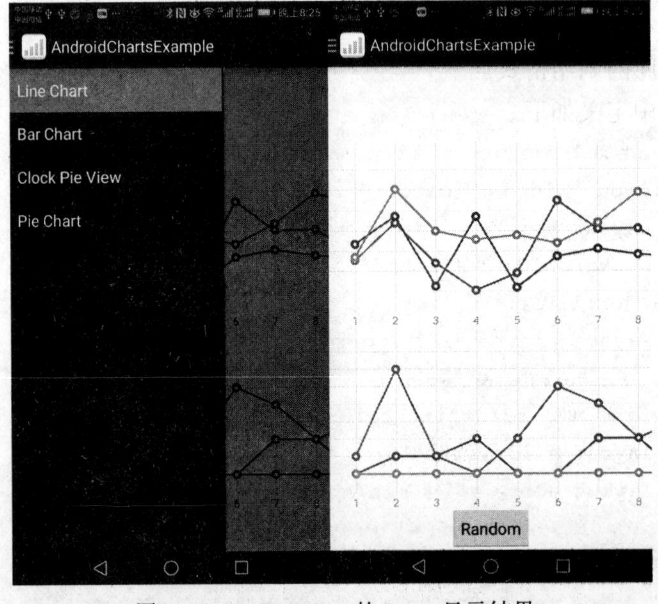

图 11-5　AndroidChart 的 Demo 显示结果

11.1.3　CameraFilter

CameraFilter 为 OpenGL 着色器的实时相机滤镜，下载地址：https://github.com/nekocode/CameraFilter。

11.1.4　Lottie

Lottie 是 Airbnb 开源的支持 Android、iOS 以及 ReactNative 并利用 JSON 文件方式快速实现动画效果的库。通过 Adobe After Effects 做出动画效果，然后通过 Bodymovin（AE 的插件）导出 JSON 数据，通过该库生成原生动画效果。Github 地址：https://github.com/airbnb/lottie-android，Demo 程序的 github 地址：https://github.com/panacena/LottieTest/。

在 Android 2.3 中创建应用项目：LottieAnimation。

（1）在内层的 build.gradle 文件，也就是 app/build.gradle 文件的 dependencies 中增加如下内容。

```
dependencies {
...
    compile 'com.airbnb.android:lottie:1.0.1'
}
```

（2）主布局文件 activity_main.xml 中含有三个 Button 控件，用于显示三个动画界面，这三个动画界面为 activity_first.xml、activity_second.xml、activity_third.xml，如图 11-6 所示。

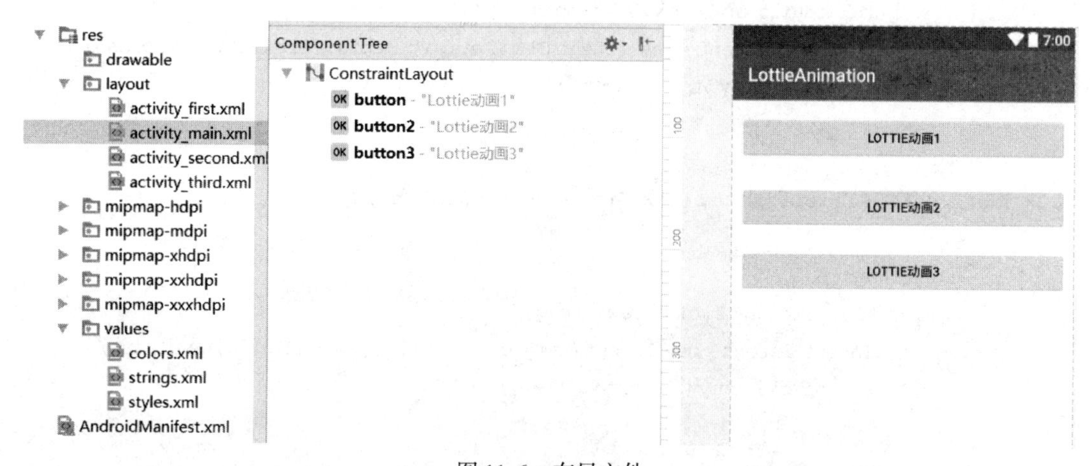

图 11-6　布局文件

布局文件 activity_first.xml 的代码如下。

```
<?xml version="1.0" encoding="utf-8"?>
<LinearLayout xmlns:android="http://schemas.android.com/apk/res/android
    android:layout_width="match_parent"
    android:layout_height="match_parent"
    xmlns:app="http://schemas.android.com/apk/res-auto">
    <com.airbnb.lottie.LottieAnimationView
        android:id="@+id/animation_view"
        android:layout_width="wrap_content"
```

```
        android: layout_height = " wrap_content"
        app: lottie_fileName = " data. json"
        app: lottie_loop = " true"
        app: lottie_autoPlay = " true" />
</LinearLayout>
```

app：lottie_fileName = " data. json"，data. json 是 Adobe After Effects 做出动画效果后通过 Bodymovin（AE 的插件）导出的 JSON 数据。

（3）主 Activity 文件 MainActivity. java 的代码如下。

```
public class MainActivity extends AppCompatActivity {
  Button bt1,bt2,bt3;
  @Override
    protected void onCreate(Bundle savedInstanceState) {
        super. onCreate(savedInstanceState);
        setContentView(R. layout. activity_main);
        bt1 = (Button) findViewById(R. id. button);
        bt2 = (Button) findViewById(R. id. button2);
        bt3 = (Button) findViewById(R. id. button3);
        bt1. setOnClickListener(new View. OnClickListener()
        {
            @Override
            public void onClick(View view) {
                Intent intent =new Intent(MainActivity. this,FirstActivity. class);
                startActivity(intent);
            }
      });
      bt2. setOnClickListener(new View. OnClickListener()
        {
          @Override
          public void onClick(View view) {
              Intent intent =new Intent(MainActivity. this,SecondActivity. class);
              startActivity(intent);
          }
      });
    bt3. setOnClickListener(new View. OnClickListener()
      {
        @Override
        public void onClick(View view) {
            Intent intent =new Intent(MainActivity. this,ThirdActivity. class);
              startActivity(intent);
        }
      });
  }
}
```

（4）第一个界面的 Activity 的处理文件 FirstActivity. java 的代码如下。

```
public class FirstActivity extends AppCompatActivity {
  @Override
    protected void onCreate(Bundle savedInstanceState) {
        super.onCreate(savedInstanceState);
      setContentView(R.layout.activity_first);
}
```

（5）项目运行结果如图 11-7 所示。

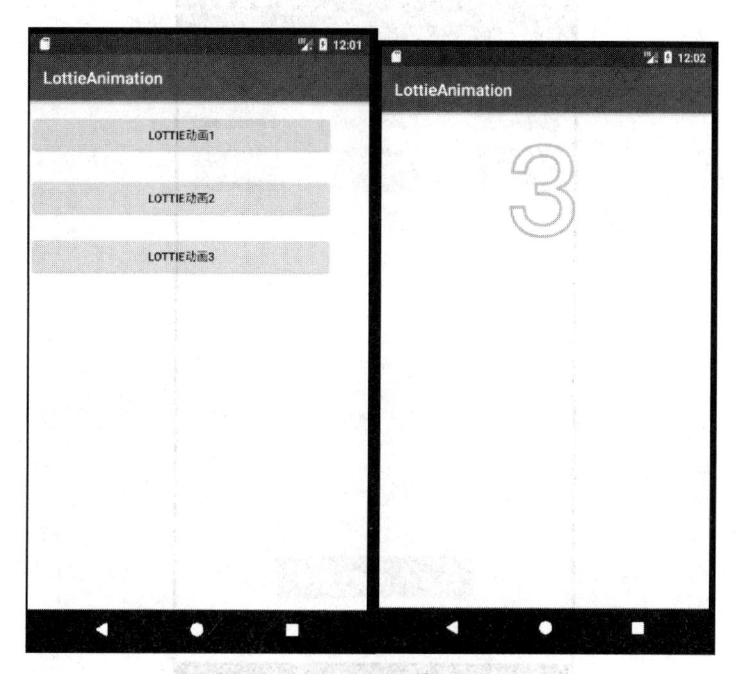

图 11-7　项目运行结果

11.1.5　StyleableToast

StyleableToast 是一个自定义 Toast 库，Github 下载地址：https://github.com/Muddz/StyleableToast

StyleableToast 的使用步骤如下。

（1）在内层的 build. gradle 文件，也就是 app/build. gradle 文件的 dependencies 中增加如下内容。

```
dependencies {
    ...
    compile 'com.muddzdev:styleabletoast:1.0.9'
}
```

（2）在代码中的使用方法如下。

```
private StyleableToast styleableToast;
styleableToast = new StyleableToast
  .Builder(this)
```

```
.icon(R.drawable.ic_overheating)
.text (" Phone is overheating!")
.textBold()
.textColor (Color.parseColor (" #FFDA44"))
.cornerRadius (5)
.build();
```

显示的 Toast 格式如图 11-8 所示。

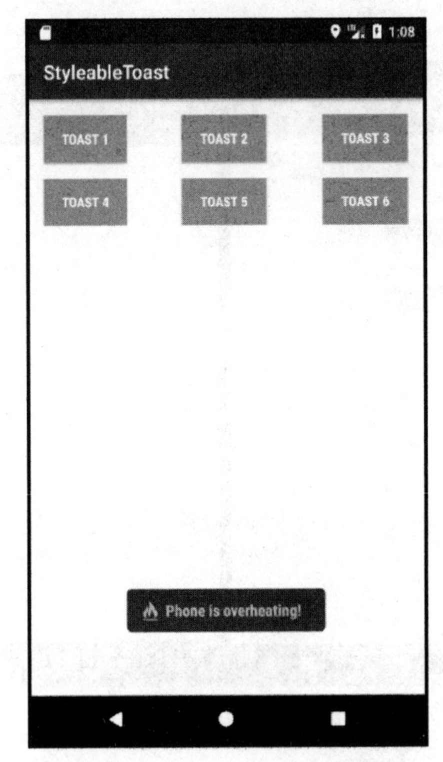

图 11-8　StyleableToast 显示效果

除了 StyleableToast 之外，Toasty 也是一个自定义 Toast 的库，Github 地址：https://github.com/GrenderG/Toasty。

11.1.6　CameraFragment

CameraFragment 可以快速实现打开相机视图，并提供便捷的 API 捕获图片。CameraFragment 的 Github 地址：https://github.com/florent37/CameraFragment。

CameraFragment 的使用步骤如下。

(1) 在内层的 build.gradle 文件，也就是 app/build.gradle 文件的 dependencies 中增加如下内容。

```
dependencies {
…,
    compile 'com.github.florent37:camerafragment:1.0.7'
}
```

（2）设置配置文件，代码如下。

```
final Configuration.Builder builder = new Configuration.Builder();
builder.setCamera(Configuration.CAMERA_FACE_FRONT)
    .setFlashMode (Configuration.FLASH_MODE_ON)
    .setMediaAction (Configuration.MEDIA_ACTION_VIDEO);
```

（3）实例化 CameraFragment 对象。

```
final CameraFragment cameraFragment = CameraFragment.newInstance (builder.build());
```

（4）将 Fragement 增加到指定的容器。

```
getSupportFragmentManager().beginTransaction().replace(R.id.content, cameraFragment, FRAGMENT_TAG).commitAllowingStateLoss();
```

（5）拍照方法的代码如下。

```
cameraFragment.takePhotoOrCaptureVideo(new CameraFragmentResultListener() {
    @Override
    public void onVideoRecorded(String filePath) {
    }
    @Override
    public void onPhotoTaken(byte[] bytes, String filePath) {

    }
  },
"/storage/self/primary",
"photo0");
}
```

（6）关闭/打开闪光灯的代码如下。

```
cameraFragment.toggleFlashMode();
```

（7）摄像头前置/后置切换的代码如下。

```
cameraFragment.switchCameraTypeFrontBack();
```

（8）设置图像/视频的大小（分辨率）的代码如下。

```
cameraFragment.openSettingDialog();
```

（9）设置 Camera 行为（拍照还是录制视频）的代码如下。

```
cameraFragment.switchActionPhotoVideo();
```

（10）在 CameraFragmentResultListener 中得到录制（或者拍照）的结果，代码如下。

```
cameraFragment.setResultListener(new CameraFragmentResultListener() {
    @Override
    public void onVideoRecorded(byte[] bytes, String filePath) {
        //called when the video record is finished and saved
                startActivityForResult (PreviewActivity.newIntentVideo(MainActivity.this, filePath));
    }
    @Override
    public void onPhotoTaken(byte[] bytes, String filePath) {
```

```
            //called when the photo is taken and saved

                startActivity (PreviewActivity.newIntentPhoto (MainActivity.this,
filePath));
        }
    });
```

（11）CameraFragment 实例截图如图 11-9 所示。

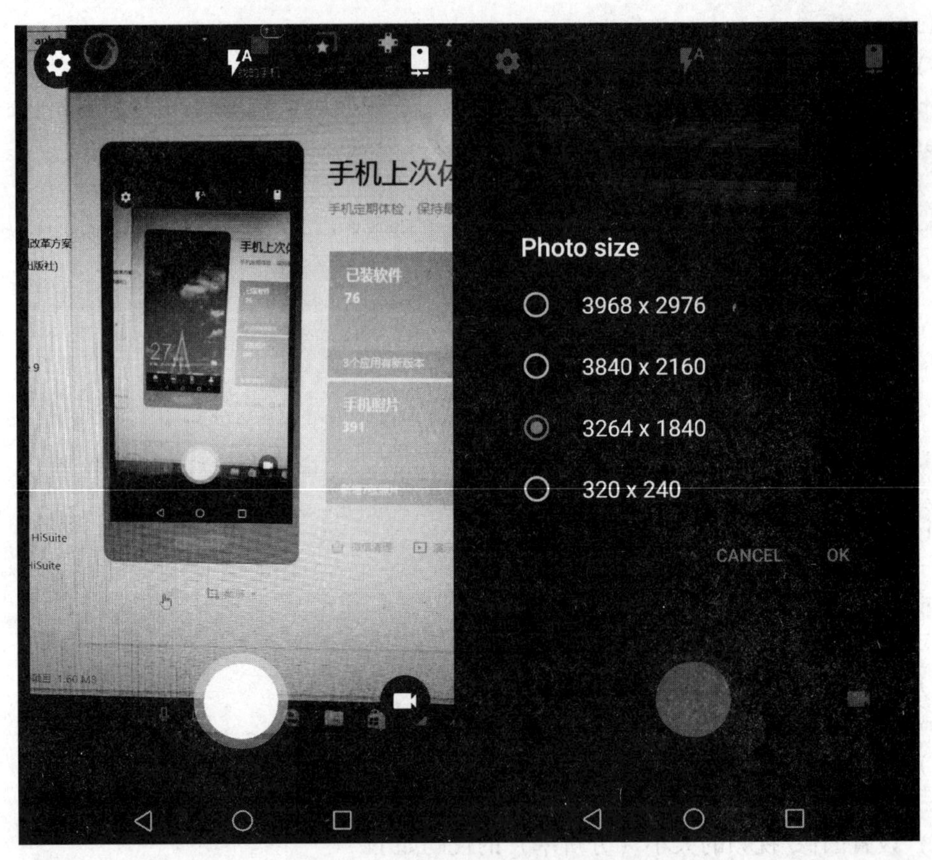

图 11-9 CameraFragment 实例截图

11.2 Android 开源项目

对于程序员来说，最有效的学习多看别人优秀的代码，加以总结、学习和应用。如果您想成为一名优秀的开发者，就必须阅读大量的代码，充分利用好功能完整详细的开源项目资源。

11.2.1 Easy Sound Recorder

Easy Sound Recorder 是一款简单的录音 App。如果您想学习关于录音方面的知识，这个开源项目可以帮助您，其代码非常好理解，并且采用的是 MD 设计，Github 下载地址：https://github.com/dkim0419/SoundRecorder，项目运行结果如图 11-10 所示。

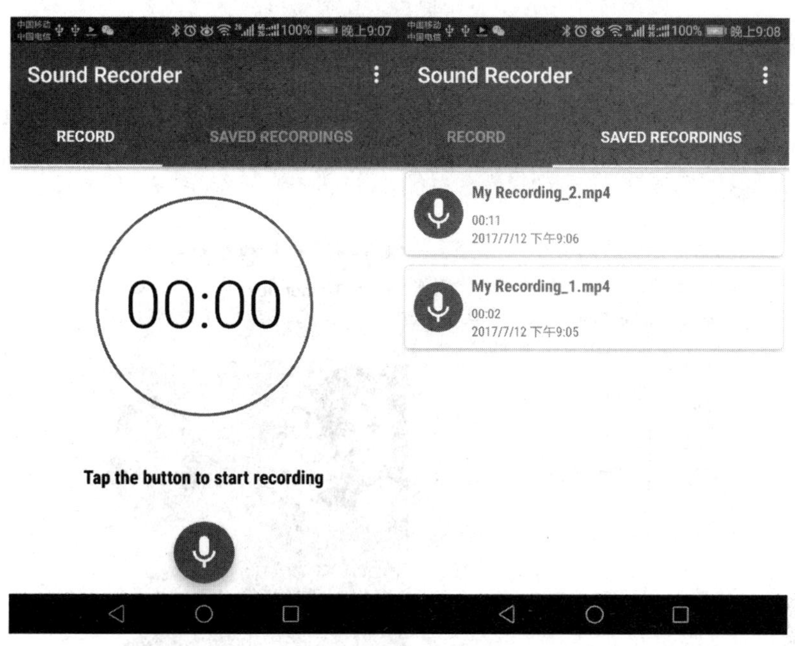

图 11-10　Easy Sound Recorder 项目运行结果

11.2.2　MLManager

MLManager 可帮助管理手机里面的 App。从这个项目中，可以学到如何获取软件的详细信息、导出 APK、卸载软件等。这个项目的编码风格很好，可以借鉴，它的简洁代码设计和 MD 设计都值得参考。Github 下载地址：https://github.com/javiersantos/MLManager。项目运行结果如图 11-11 所示。

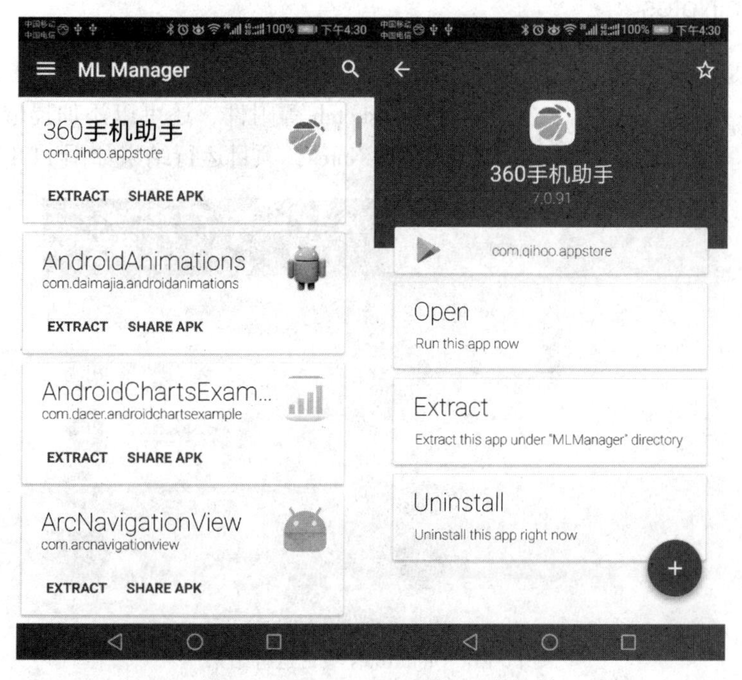

图 11-11　MLManager 项目运行结果

11.2.3 Timber

Timber 是一款设计非常美观的音乐播放器，如果您正在开发一款属于自己的播放器，那么正好可以参考学习。Github 下载地址：https://github.com/naman14/Timber，项目运行结果如图 11-12 所示。

图 11-12　Timber 项目运行结果

11.2.4 OmniNotes

OmniNotes 是一款类似于 Evernote 的笔记类 App。该项目含有大量的功能，比如分享和收缩 note，在 note 中添加图片、视频、音频、sketch 等附件，还可以添加提醒人。Github 下载地址：https://github.com/federicoiosue/Omni-Notes，项目运行结果如图 11-13 所示。

图 11-13　OmniNotes 项目运行结果

11.2.5　Super Clean Master

Super Clean Master 是一款模仿“清理大师”的应用，包括内存加速、缓存清理、自启管理、软件管理等功能，建议仔细研究，Github 下载地址：https://github.com/joyoyao/superCleanMaster，项目运行结果如图 11-14 所示。

图 11-14　Super Clean Master 项目运行结果

11.2.6　Pedometer

Pedometer 是一款传感器计步类的 App。Github 下载地址：https://github.com/j4velin/Pedometer，项目运行结果如图 11-15 所示。

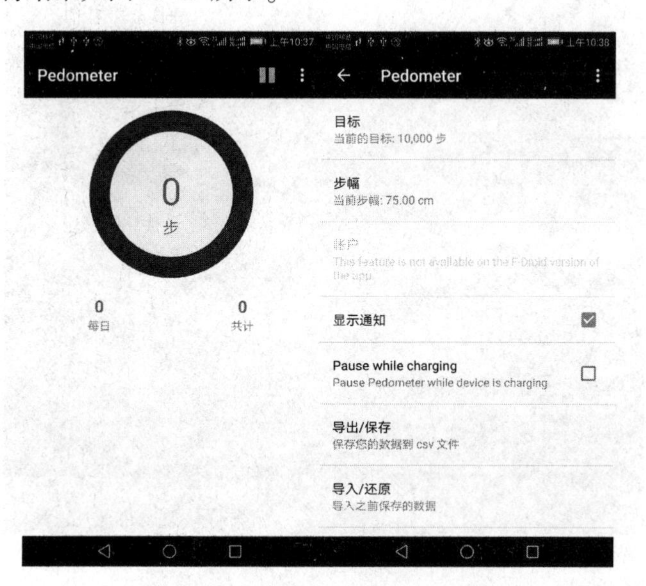

图 11-15　Pedometer 项目运行结果

11.2.7 Traval Mate

如果您正在开发一款重度依赖位置和地图的旅行类 App，那么可以参考 Traval Mate 项目。Github 下载地址：https://github.com/Swati4star/Travel-Mate，项目运行结果如图 11-16 所示。

图 11-16　Traval Mate 项目运行结果

11.2.8 Music-Player

Music-Player 用代码实现音乐列表，具有精美的播放界面 UI 效果，非常值得参考，Github 下载地址：https://github.com/andremion/Music-Player，项目运行结果如图 11-17 所示。

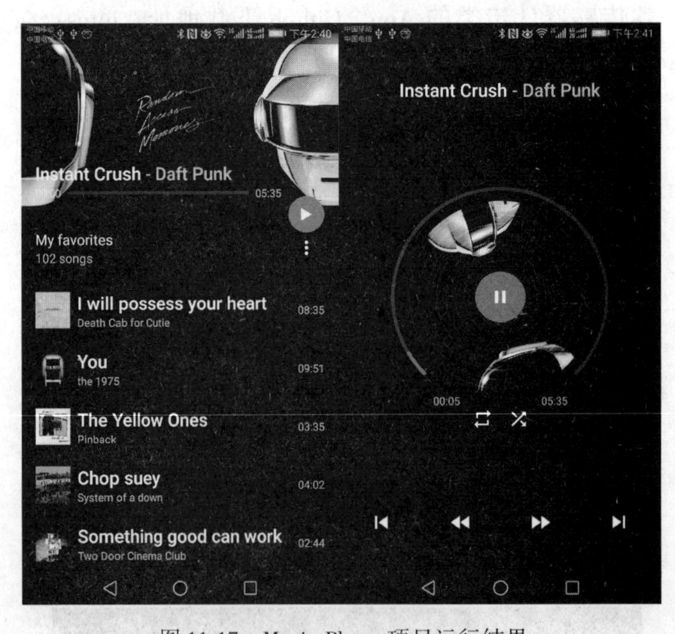

图 11-17　Music-Player 项目运行结果

11.2.9　PLDroidPlayer

PLDroidPlayer 是一个适用于 Android 平台的音视频播放器 SDK，可高度定制化和二次开发，为 Android 开发者提供了简单、快捷的接口，帮助开发者在 Android 平台上快速开发播放器应用。Github 下载地址：https://github.com/pili-engineering/PLDroidPlayer。

PLDroidPlayer 功能和接口说明如表 11-1 所示。

表 11-1　PLDroidPlayer 功能和接口说明

功能/接口	描　　述	版　　本
支持软硬解自动切换	自动解码模式下，优先硬解，硬解失败自动切换到软解	v1.4.1（+）
支持 HTTPS 协议、speex 解码、mp4v 解码	无	v1.4.1（+）
提供接口获取 metadata 信息	用户可以调用接口获取播放的 metadata 信息	v1.4.1（+）
提供接口获取当前的播放状态	用户可以主动调用接口获取当前播放状态	v1.4.0（+）
支持设置封面	在播放开始前显示相关图片信息	v1.4.0（+）
支持带 IP 地址的播放 URL	URL 格式："protocol：//ip/path？domain = xxxx.com"	v1.3.0（+）
支持 DNS 解析优化	支持 DNS 提前解析和缓存管理	v1.3.0（+）
支持直播累积延时优化	优化直播过程中的累积延时	v1.2.3（+）
支持后台播放	支持退到后台，只播放音频	v1.2.3（+）
支持音量设置	设置播放器音量，可实现静音功能	v1.2.2（+）
支持画面镜像翻转	由 PLVideoTextureView 提供，支持播放画面镜像翻转	v1.2.2（+）
支持首屏秒开	在网络条件较好的情况下，可以实现秒开	v1.2.0（+）
PLMediaPlayer	类似于 Android MediaPlayer，提供了播放器的核心功能	v1.2.0（+）
PLVideoView	类似于 Android VideoView，基于 SurfaceView 的播放控件	v1.2.0（+）
PLVideoTextureView	类似于 Android VideoView，基于 TextureView 的播放控件	v1.2.0（+）
支持画面旋转	由 PLVideoTextureView 提供，支持播放画面以 0 度、90 度、180 度，270 度旋转	v1.2.0（+）
支持设置画面预览模式	由 PLVideoView 和 PLVideoTextureView 提供，支持多种画面预览模式，包括原始尺寸、适应屏幕、全屏铺满、16:9、4:3 等	v1.2.0（+）
支持 ARM、X86 芯片体系架构	无	v1.1.3（+）
支持 ARM64v8a 芯片体系架构	无	v1.1.1（+）
AVOptions	用于配置播放器参数，包括超时时间、软硬件编解码	v1.1.1（+）
AudioPlayer	用于纯音频播放，支持后台运行，从 v1.2.0 开始被标记为 Deprecated，并由 PLMediaPlayer 代替	v1.1.0（+）
支持 ARMv7a 芯片体系架构	无	v1.0.0（+）
VideoView	基于 SurfaceView 的播放控件，从 v1.2.0 开始被标记为 Deprecated，并由 PLVideoView 代替	v1.0.0（+）

PLDroidPlayer 的特性如下。

- 支持 RTMP 和 HLS 协议的直播流媒体播放。
- 支持常见的音视频文件播放（MP4、M4A、flv 等）。
- 支持 MediaCodec 硬件解码。
- 提供播放器核心类 PLMediaPlayer。

- 提供 PLVideoView 控件。
- 提供 PLVideoTextureView 控件。
- 支持多种画面预览模式。
- 支持画面旋转（0 度、90 度、180 度、270 度）。
- 支持画面镜像变换。
- 支持播放器音量设置，可实现静音功能。
- 支持纯音频播放。
- 支持后台播放。
- 支持首屏秒开。
- 支持直播累积延时优化。
- 支持带 IP 地址的播放 URL。
- 支持设置封面。
- 支持软硬解自动切换。
- 支持 HTTPS 协议、Speex 解码、MP4V 解码。
- 具有可高度定制化的 MediaController。
- 支持 ARM、ARMv7a、ARM64v8a、X86 主流芯片体系架构。

PLDroidPlayer 的使用方法如下。

（1）在项目的 build. gradle 中加入如下语句。

```
dependencies {
...
  compile 'com.qiniu.pili:pili-android-qos:0.8.+'
}
```

（2）添加网络状态监测的权限，代码如下。

```
<uses-permission android:name="android.permission.ACCESS_NETWORK_STATE" />
```

项目运行结果如图 11-18 所示。

图 11-18　PLDroidPlayer 项目运行结果

11.3 Android 开源网站

GitHub 是全球最大的代码托管网站，任何开源软件都可以免费地将代码提交到 GitHub 中，以零成本的代价进行代码托管，GitHub 的一个最重要的作用就是发现全世界最优秀的开源项目，那么，如何发现优秀的开源项目呢？下面进行详细介绍。

GitHub 的官方网站：https://github.com/，首页如图 11-19 所示。

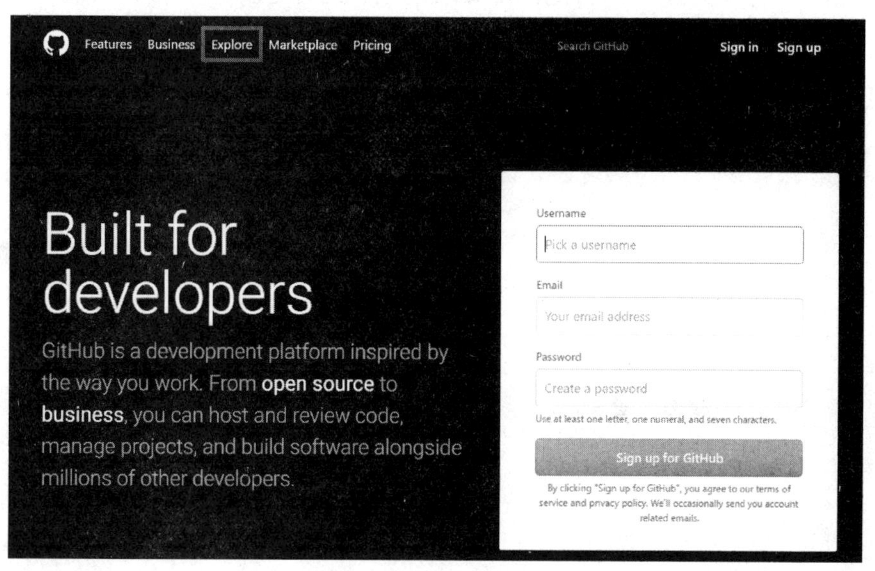

图 11-19　GitHUb 首页

在 GitHub 首页中选择 Explore 选项，显示如图 11-20 的网页，其中显示了一些项目的情况。

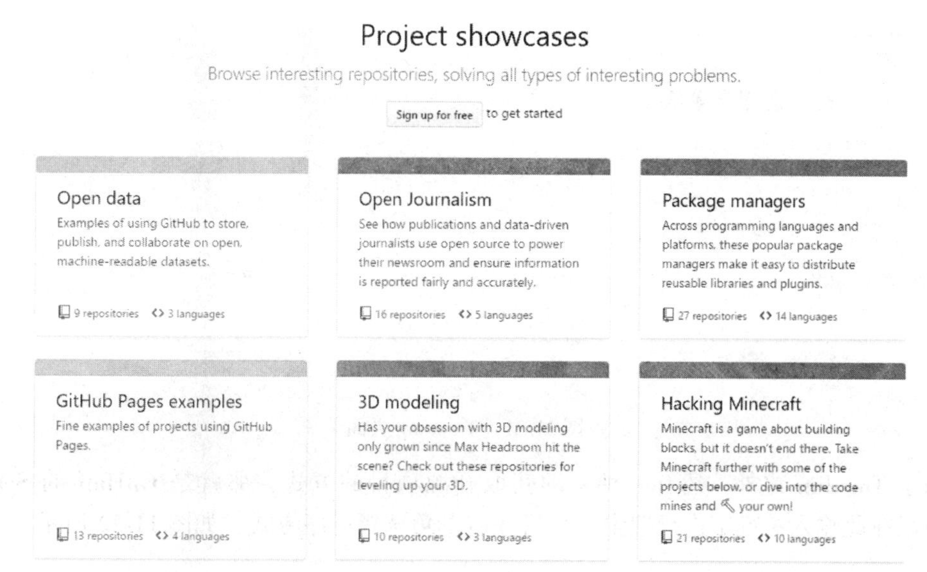

图 11-20　项目情况

在这个网页中，页面右上方有一个 Trending 选项，这个 Trending 选项页面页面有什么作用呢？在这个页面中可以看到最近一些热门的开源项目，这个页面是很多人主动获取开源项目最好的途径，可以选择当天、一周之内或一月之内的热门项目进行查看，还可以分语言类别进行查看，比如查看最近热门的 Android 项目，如图 11-21 所示。

图 11-21　Trending 页面

除了 Trending 之外，还有一种主动获取开源项目的方式，那就是 GitHub 的 Search 功能，可以在此输入关键字进行搜索，然后在右上角选择排序方式，如图 11-22 所示。

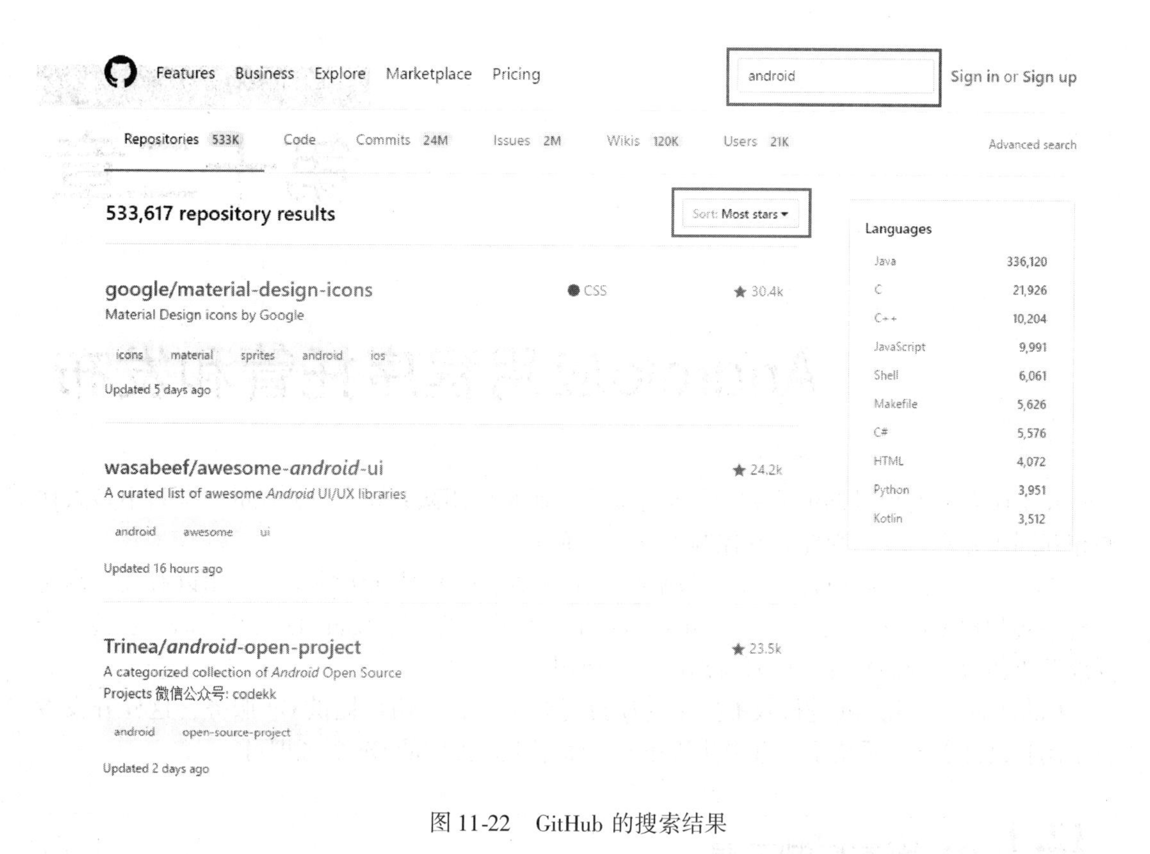

图 11-22　GitHub 的搜索结果

11.4 本章小结

本章主要介绍了一些典型 Android 开源库的获取和使用方法，以及一些 Android 开源项目的的功能，最后介绍了在 GitHub 网站中获取 Android 开源资源的方法。

第十二章

Android应用程序托管和发布

几乎所有大的项目都不是由一个人完成的，而是由团队共同合作完成的，项目中多人代码相互同步异常重要，因此版本控制工具应运而生。

Git 是一个开源的分布式版本控制系统，跟 SVN、CVS 是同级的概念，用以有效、高速地管理各种规模的项目版本，它是 Linux Torvalds 为了帮助管理 Linux 内核开发而开发的一个开放源码的版本控制软件，后来得到广泛的使用。

Github 是一个用 Git 进行版本控制的项目托管平台，为用户提供 Git 服务，这样开发者就不用自己部署 Git 系统了，直接注册账号，使用平台提供的 Git 服务即可。

12.1 Git 版本控制工具

Git 是一款免费、开源的分布式版本控制系统，用于高效处理任何大小的项目。

Torvalds 着手开发 Git 是为了作为一种过渡方案来替代 BitKeeper，后者之前一直是 Linux 内核开发人员在全球使用的主要源代码工具。开放源码社区中的有些人觉得 BitKeeper 的许可证并不适合开放源码社区的工作，因此 Torvalds 决定着手研究许可证更为灵活的版本控制系统。尽管最初 Git 的开发是为了辅助 Linux 内核开发的过程，但是在很多其他自由软件项目中也使用了 Git，例如，很多 Freedesktop 的项目迁移到了 Git 上。

分布式相比于集中式的最大区别在于开发者可以提交到本地，每个开发者通过克隆（git clone），可以在本地机器上复制一个完整的 Git 仓库。

从一般开发者的角度来看，Git 有以下功能。

（1）可以从服务器上克隆完整的 Git 仓库（包括代码和版本信息）到单机上。

（2）可以在自己的机器上根据不同的开发目的，创建分支、修改代码。

（3）可以在单机上自己创建的分支上提交代码。

（4）可以在单机上合并分支。

（5）可以获取服务器上最新版的代码，然后跟自己的主分支合并。

（6）可以生成补丁（patch），并把补丁发送给主开发者。

（7）可以根据主开发者的反馈，进行相应处理。如果主开发者发现两个一般开发者之间有冲突（他们之间可以合作解决的冲突），就会要求先解决冲突，然后再由其中一人提交。如果主开发者可以自己解决，或者没有冲突，则通过。

（8）开发者之间可以使用 pull 命令解决冲突，解决完冲突之后再向主开发者提交补丁。

从主开发者的角度（假设主开发者不用开发代码）看，Git 有以下功能。

（1）可以查看邮件或者通过其他方式查看一般开发者的提交状态。

（2）可以打上补丁，解决冲突，可以自己解决，也可以要求开发者之间解决，以后再重新提交，如果是开源项目，还要决定哪些补丁有用，哪些不用。

（3）可以向公共服务器提交结果，然后通知所有开发人员。

12.1.1　安装 Git

Git 的下载网址：https://git-scm.com/download/，如图 12-1 所示，可以看到有 Windows、Linux、Mac OS、Solaris 四种版本，这里选择常用的 Windows 版本。

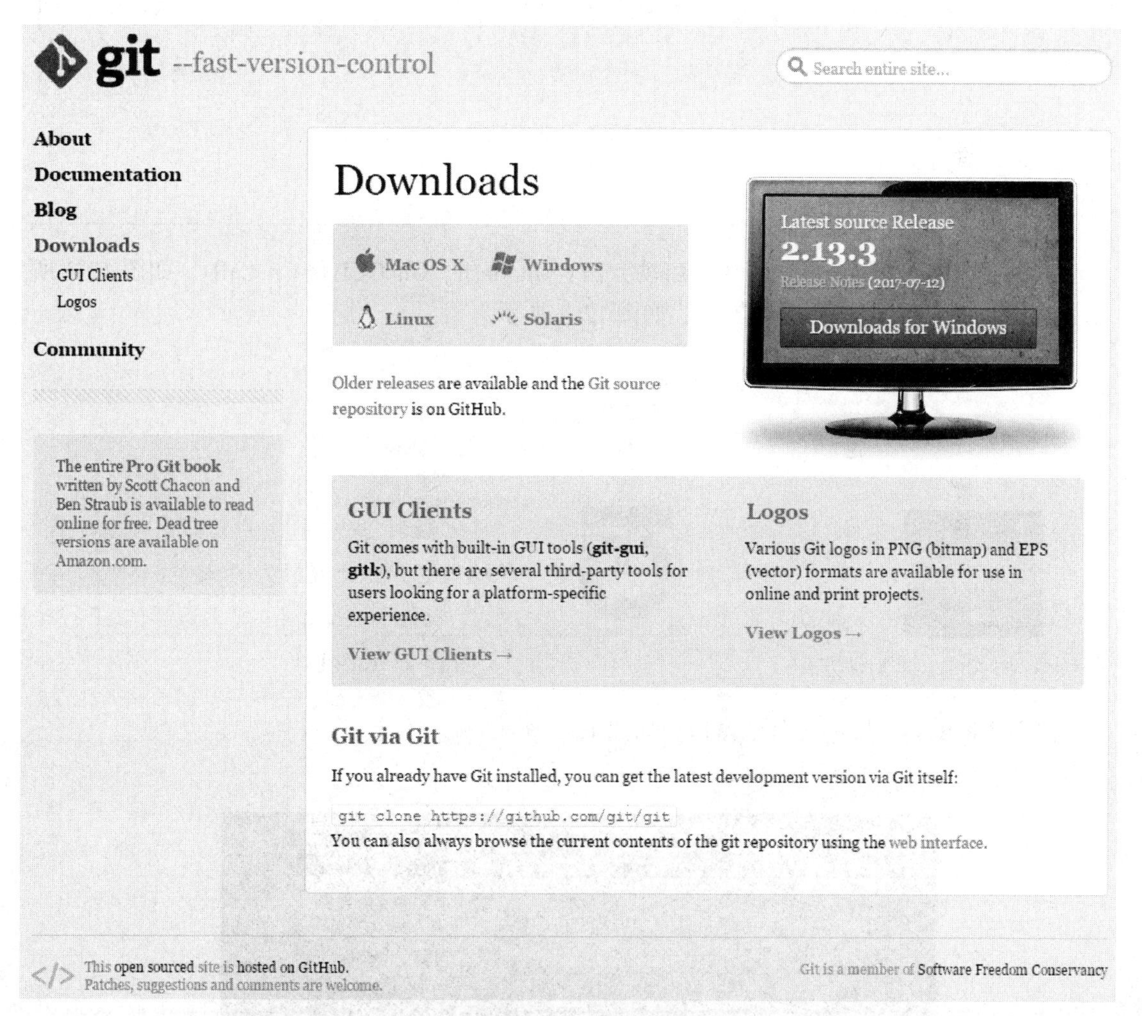

图 12-1　Git 下载页面

目前最新的版本是 2.13.3，对应的文件名为 Git-2.13.3-32-bit.exe 和 Git-2.13.3-64-bit.exe，这里选择 64 位版本。

单击文件名 Git-2. 13. 3-64-bit. exe 开始安装，一直单击 Next 按钮，完成安装，如图 12-2 所示。

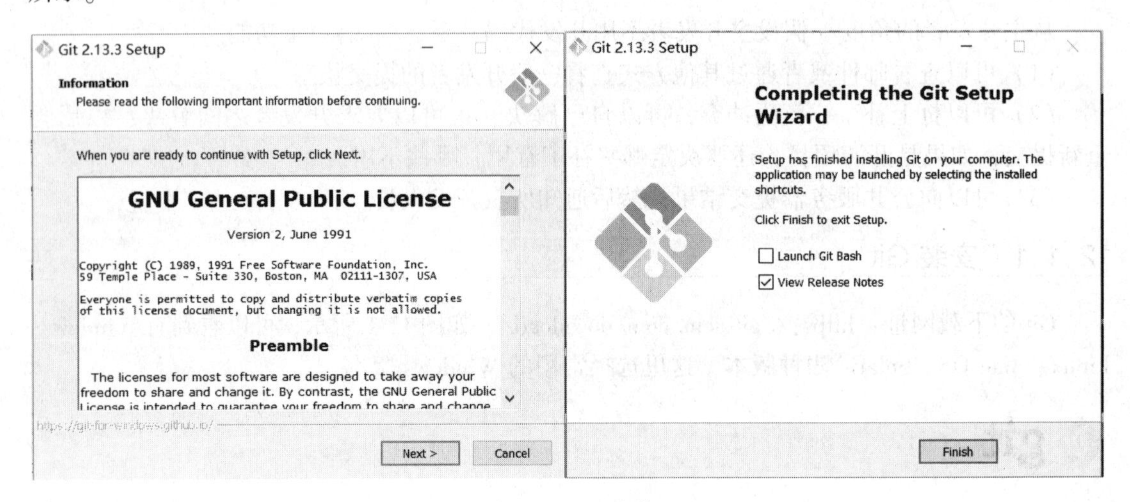

图 12-2　Git 安装

12. 1. 2　创建代码仓库

安装完成 Git 后，可以采用 3 种启动方法：Git Bash、Git GUI 和 Git CMD，如图 12-3 所示。选择 Git GUI，将进入图形界面，如图 12-4 所示。

图 12-3　Git 启动方式

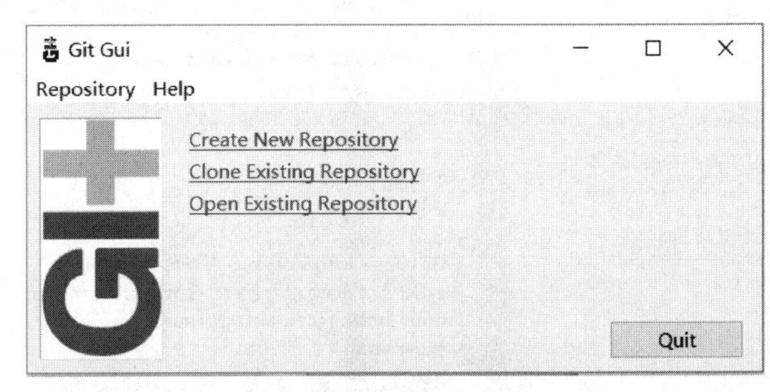

图 12-4　Git GUI

选择 Git Bash，将进入命令窗口，如图 12-5 所示。

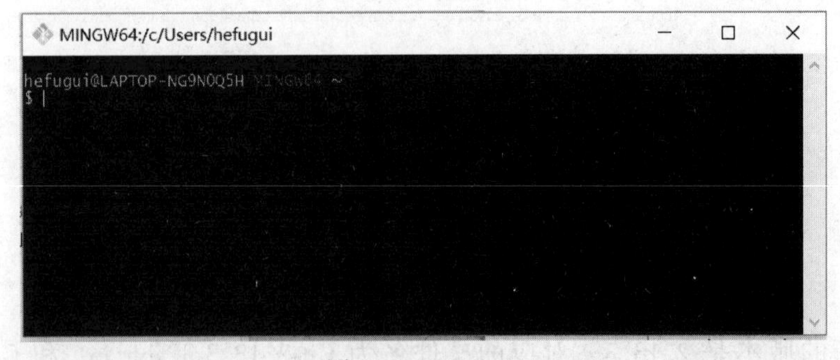

图 12-5　Git Bash

以命令形式启动有助于掌握其主要内容，这里以 Git Bash 为例说明操作过程。

（1）首先配置身份，这样在提交代码时就知道是谁提交的代码，命令如下。

```
Git config--global  user.name  mike
Git config  --global  user.email  mike@163.com
```

（2）配置完成后，验证是否配置成功，只需要将最后的名字和邮箱地址去掉即可，如图 12-6 所示。

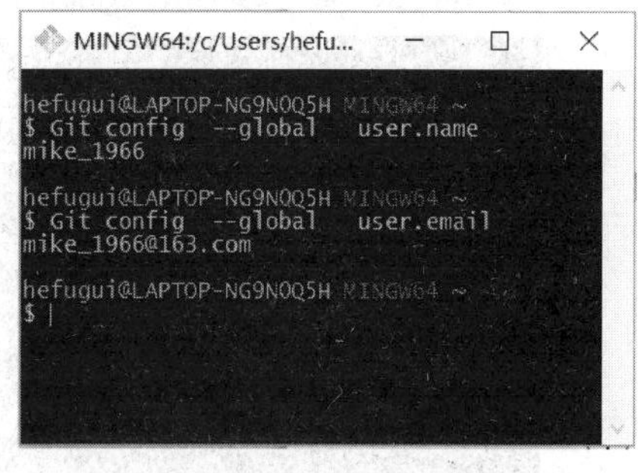

图 12-6 验证配置

（3）然后就可以创建代码仓库了，仓库（Repository）用于保存版本管理信息，所有本地提交的代码会被提交到代码仓库中，如果需要，可以推送到远程仓库中。

这里我们尝试为 BluetoothChat 项目建立一个代码仓库。先建立一个 BluetoothChat 项目的目录，如图 12-7 所示。

图 12-7 建立 BluetoothChat 目录

（4）切换到 BluetoothChat 目录中，如图 12-8 所示。

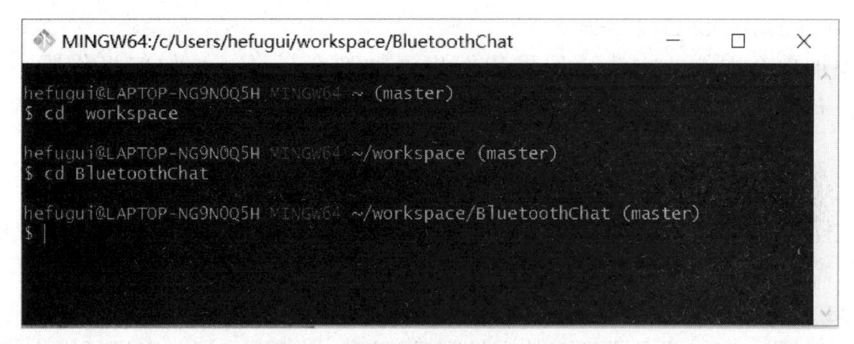

图 12-8 切换到 BluetoothChat 目录中

（5）然后在这个目录中输入如下命令：git init，这就完成了创建代码仓库的操作，如图 12-9 所示。

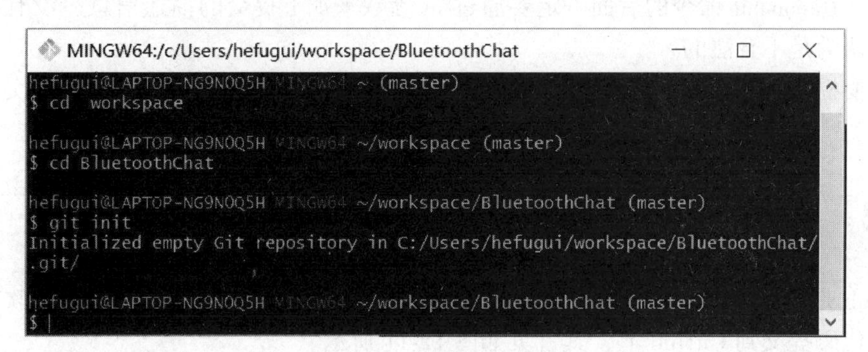

图 12-9 创建代码仓库

创建完成后，在 BluetoothChat 项目的目录下生成一个隐藏的 . git 文件夹，这个文件夹就是用来记录本地所有的 Git 操作的，可以通过 ls -al 命令进行查看，如图 12-10 所示。

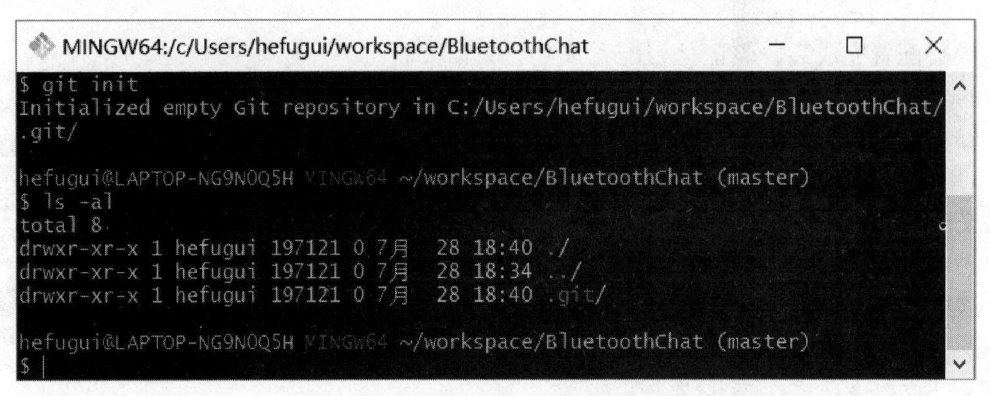

图 12-10　查看 . git 文件夹

如果想删除本地仓库，只需要删除这个文件夹即可。

12.1.3　提交本地代码

代码仓库建立完之后，就可以提交代码了，提交代码的方法非常简单，只需要使用 add 和 commit 命令即可。add 用于把想要提交的代码添加进来，而 commit 则是真正地执行提交操作。比如，我们想添加 AndroidManifest. xml 文件，输入如下命令即可。

```
git add AndroidManifest.xml
```

这是添加单个文件的方法，如果我们想添加某个目录，该如何操作呢？只需要在 add 后面加上目录名即可。比如，将整个 src 目录下的所有文件添加进来，输入如下命令即可。

```
git add src
```

可是这样一个个地添加还是比较麻烦，有没有一次性把所有文件都添加完成的办法呢？当然有，只需要在 add 的后面加上一个点，就表示添加所有的文件，命令如下。

```
git add.
```

现在，BluetoothChat 项目下所有的文件都已经添加完成，我们可以进行提交了，输入如下命令。

```
git commit -m "First commit."
```

注意，在 commit 命令的后面一定要通过-m 参数来加上提交的描述信息，没有描述信息的提交被认为是不合法的。

如果要提交到 Github，需要在 Github 创建一个版本库，然后使用 Git 命令提交，请参见下节内容。

12.2　GitHub

GitHub 是全球最大的代码托管网站，主要借助 Git 进行版本控制。任何开源软件都可以免费地将代码提交到 GitHub 上，其首页如图 12-11 所示。

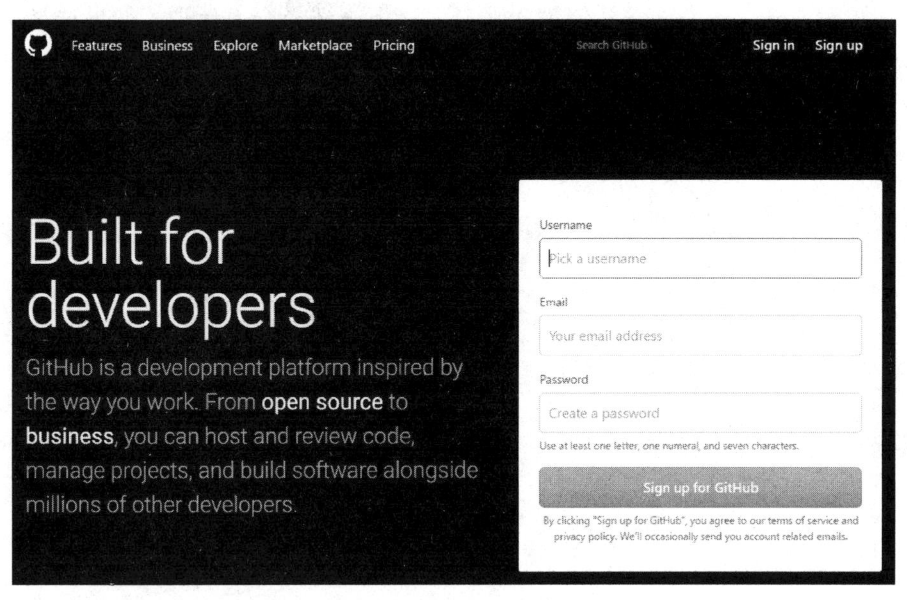

图 12-11　GitHub 的首页

12.2.1　在 GitHub 中注册创建版本库

本节介绍 GitHub 中注册创建版本库的方法。

1. 注册 GitHub 账号

（1）使用 GitHub 代码托管时，首先要有一个 GitHub 账号，单击 Sign up for GitHub 按钮进行注册，输入用户名、邮箱和密码，如图 12-12 所示。

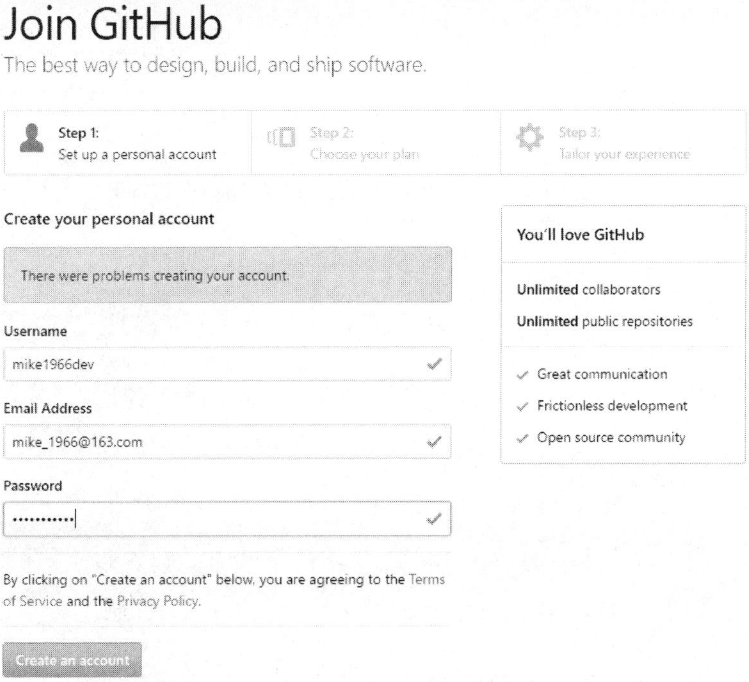

Join GitHub
The best way to design, build, and ship software.
Step 1: Set up a personal account
Step 2: Choose your plan
Step 3: Tailor your experience
Create your personal account
There were problems creating your account.
Username
mike1966dev
Email Address
mike_1966@163.com
Password
By clicking on "Create an account" below, you are agreeing to the Terms of Service and the Privacy Policy.
Create an account
You'll love GitHub
Unlimited collaborators
Unlimited public repositories
Great communication
Frictionless development
Open source community

图 12-12　注册账号

（2）单击 Create an account 按钮创建账号，接下来选择个人计划，如果选择了收费计划，则有创建个人版本库的权限，如图 12-13 所示。

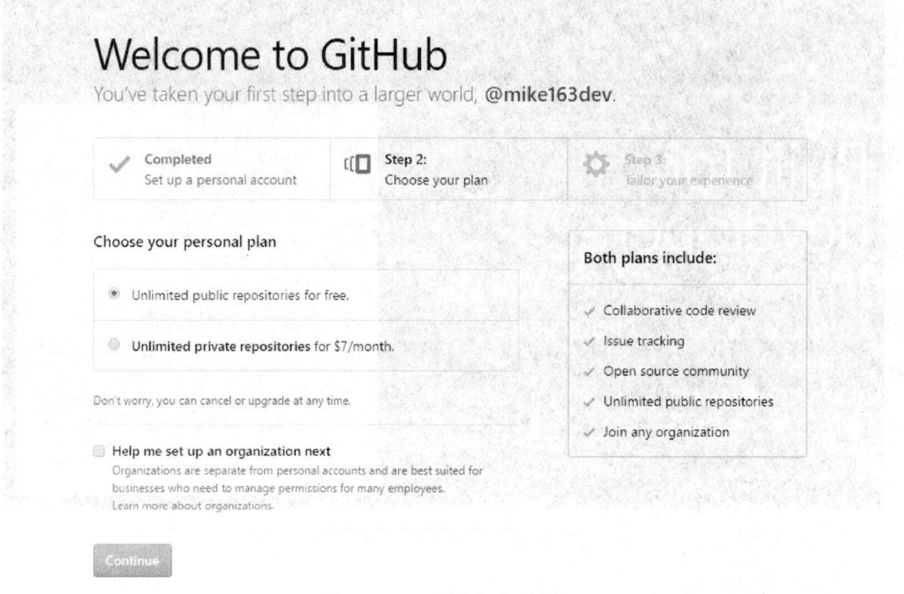

图 12-13　选择免费计划

（3）单击 Continue 进入一个问卷调查页面，如图 12-14 所示，如果不想填写，则单击下方的 skip this step 跳过此步即可。

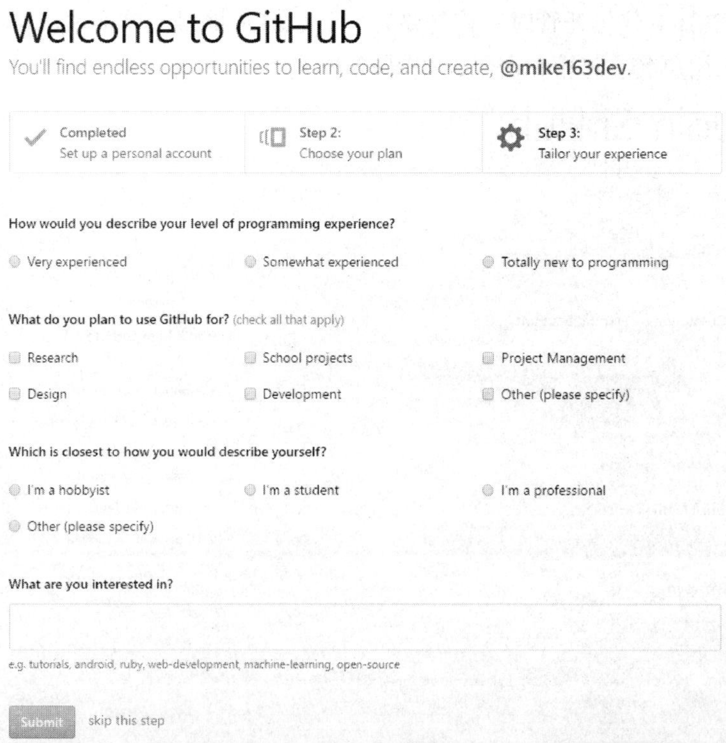

Welcome to GitHub

You'll find endless opportunities to learn, code, and create, @mike163dev.

Completed
Set up a personal account

Step 2:
Choose your plan

Step 3:
Tailor your experience

How would you describe your level of programming experience?

Very experienced　Somewhat experienced　Totally new to programming

What do you plan to use GitHub for? (check all that apply)

Research　School projects　Project Management

Design　Development　Other (please specify)

Which is closest to how you would describe yourself?

I'm a hobbyist　I'm a student　I'm a professional

Other (please specify)

What are you interested in?

e.g. tutorials, android, ruby, web-development, machine-learning, open-source

Submit　skip this step

图 12-14　问卷调查

（4）这样就把账号注册好了，此时会自动跳转到 GitHub 的主页，如图 12-15 所示。

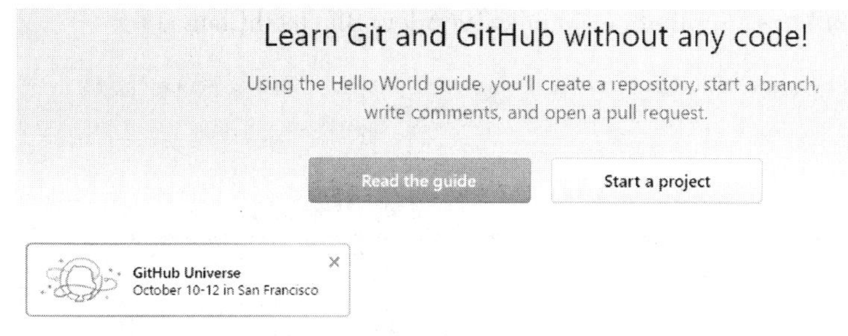

图 12-15　GitHub 个人主页

2. 创建版本库

（1）在图 12-15 中单击 Start a Project 按钮，由于刚注册完，需要进行邮箱验证，如图 12-16 所示。

Please verify your email address

Before you can contribute on GitHub, we need you to verify your email address.
An email containing verification instructions was sent to **mike@163.com**.

Didn't get the email? Resend verification email or change your email settings.

图 12-16　验证邮箱

（2）验证以后即开始创建。这里将版本库命名为 BluetoothChat，然后添加一个 Android 项目类型的 .gitignore 文件，并使用 Apache License 2.0 作为项目的开源协议，如图 12-17 所示。

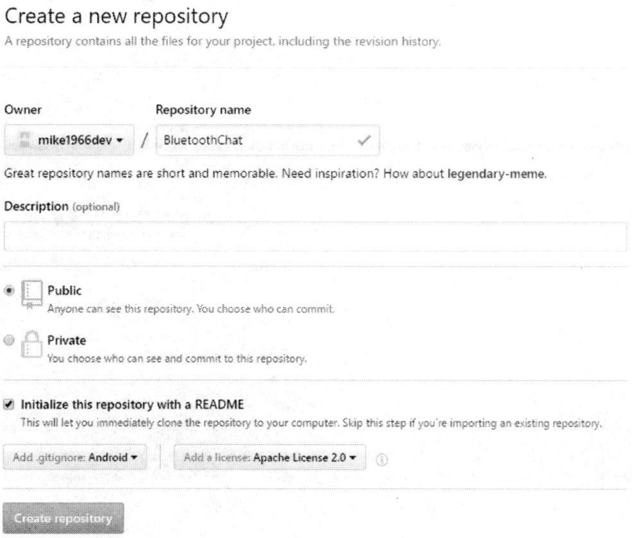

Create a new repository
A repository contains all the files for your project, including the revision history.

Owner　　Repository name
mike1966dev ▾ / BluetoothChat ✓

Great repository names are short and memorable. Need inspiration? How about **legendary-meme**.

Description (optional)

Public
Anyone can see this repository. You choose who can commit.

Private
You choose who can see and commit to this repository.

Initialize this repository with a README
This will let you immediately clone the repository to your computer. Skip this step if you're importing an existing repository.

Add .gitignore: Android ▾　　Add a license: Apache License 2.0 ▾

Create repository

图 12-17　创建版本库

（3）单击 Creat repository 按钮，就创建完成了 BluetoothChat 版本库，如图 12-18 所示，版本库主页地址为 https://github.com/mike1966dev/BluetoothChat。

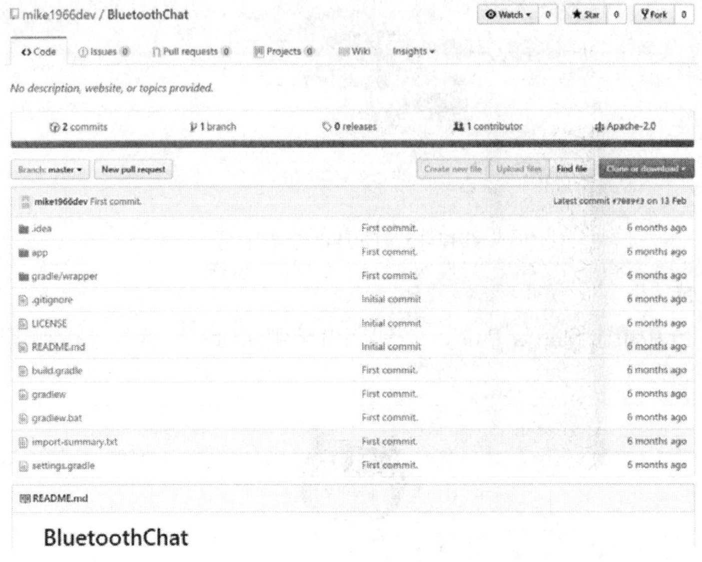

图 12-18　版本库主页

可以看到，GitHub 已经创建了 .gitignore、LICENSE、README.md 这 3 个文件，其中，编辑 README.md 文件的内容，可以修改版本库主页的描述。

12.2.2　将代码托管到 GitHub

打开刚才创建的版本库主页 https://github.com/mike1966dev/BluetoothChat，单击 Clone or download 按钮，下面的选项卡内就是远程版本库的 Git 地址，单击右边的复制按钮将其复制到剪贴板，本项目的 Git 地址为 https://github.com/mike1966dev/BluetoothChat.git，如图 12-19 所示。

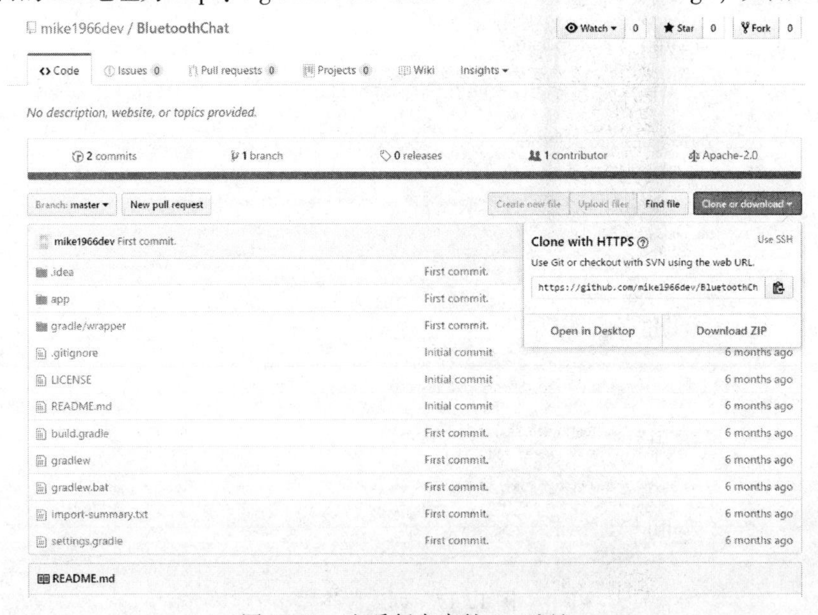

图 12-19　查看版本库的 Git 地址

然后打开 Git Bash，并切换到 BluetoothChat 项目目录下，输入下列命令。

```
git  clone  https://github.com/mike1966dev/BluetoothChat.git
```

把远程版本复制到本地，如图 12-20 所示，复制结束后，可以看到在项目目录下增加了一个 BluetoothChat 目录。

```
MINGW64:/c/Users/hefugui/workspace/BluetoothChat
drwxr-xr-x 1 hefugui 197121 0 7月  28 18:40 ./
drwxr-xr-x 1 hefugui 197121 0 7月  28 18:34 ../
drwxr-xr-x 1 hefugui 197121 0 7月  28 18:40 .git/

hefugui@LAPTOP-NG9N0Q5H MINGW64 ~/workspace/BluetoothChat (master)
$ git  clone  https://github.com/mike1966dev/BluetoothChat.git
Cloning into 'BluetoothChat'...
remote: Counting objects: 268, done.
remote: Total 268 (delta 0), reused 0 (delta 0), pack-reused 268
Receiving objects: 100% (268/268), 651.46 KiB | 123.00 KiB/s, done.
Resolving deltas: 100% (10/10), done.

hefugui@LAPTOP-NG9N0Q5H MINGW64 ~/workspace/BluetoothChat (master)
$ ls
BluetoothChat/

hefugui@LAPTOP-NG9N0Q5H MINGW64 ~/workspace/BluetoothChat (master)
$ |
```

图 12-20　将远程版本复制到本地

进入新建的 BluetoothChat 目录，应用查看命令 ls-al，BluetoothChat 项目的目录结构如图 12-21 所示。

```
MINGW64:/c/Users/hefugui/workspace/BluetoothChat
hefugui@LAPTOP-NG9N0Q5H MINGW64 ~/workspace/BluetoothChat/BluetoothChat (master)
$ cd ..

hefugui@LAPTOP-NG9N0Q5H MINGW64 ~/workspace/BluetoothChat (master)
$ ls -al
total 53
drwxr-xr-x 1 hefugui 197121     0 2月  13 09:28 ./
drwxr-xr-x 1 hefugui 197121     0 2月  12 18:10 ../
drwxr-xr-x 1 hefugui 197121     0 2月  13 09:27 .git/
-rw-r--r-- 1 hefugui 197121   505 2月  13 08:33 .gitignore
drwxr-xr-x 1 hefugui 197121     0 2月  12 18:10 .gradle/
drwxr-xr-x 1 hefugui 197121     0 2月  12 18:10 .idea/
drwxr-xr-x 1 hefugui 197121     0 2月  12 18:10 app/
-rw-r--r-- 1 hefugui 197121   867 2月  12 18:04 BluetoothChat.iml
drwxr-xr-x 1 hefugui 197121     0 2月  12 18:10 build/
-rw-r--r-- 1 hefugui 197121   316 2月  12 18:02 build.gradle
drwxr-xr-x 1 hefugui 197121     0 2月  12 18:10 gradle/
-rwxr-xr-x 1 hefugui 197121  4971 2月  12 18:02 gradlew*
-rw-r--r-- 1 hefugui 197121  2404 2月  12 18:02 gradlew.bat
-rw-r--r-- 1 hefugui 197121  1459 2月  12 18:02 import-summary.txt
-rw-r--r-- 1 hefugui 197121 11558 2月  13 08:33 LICENSE
-rw-r--r-- 1 hefugui 197121   262 2月  12 18:02 local.properties
-rw-r--r-- 1 hefugui 197121    15 2月  13 08:33 README.md
-rw-r--r-- 1 hefugui 197121    16 2月  12 18:02 settings.gradle

hefugui@LAPTOP-NG9N0Q5H MINGW64 ~/workspace/BluetoothChat (master)
$ |
```

图 12-21　BluetoothChat 项目的目录结构

接下来，把 BluetoothChat 项目中的文件提交到 GitHub 中，先将所有文件添加到版本控制，如下所示。

```
git  add  .
```

先在本地执行提交操作，如下所示。

```
git commit -m "First commit."
```

最后将提交的内容同步到远程版本库，也就是 GitHub 中，如下所示。

```
git  push  origin  master
```

在提交的过程中，需要校验 GitHub 注册的用户名和密码，如图 12-22 所示。

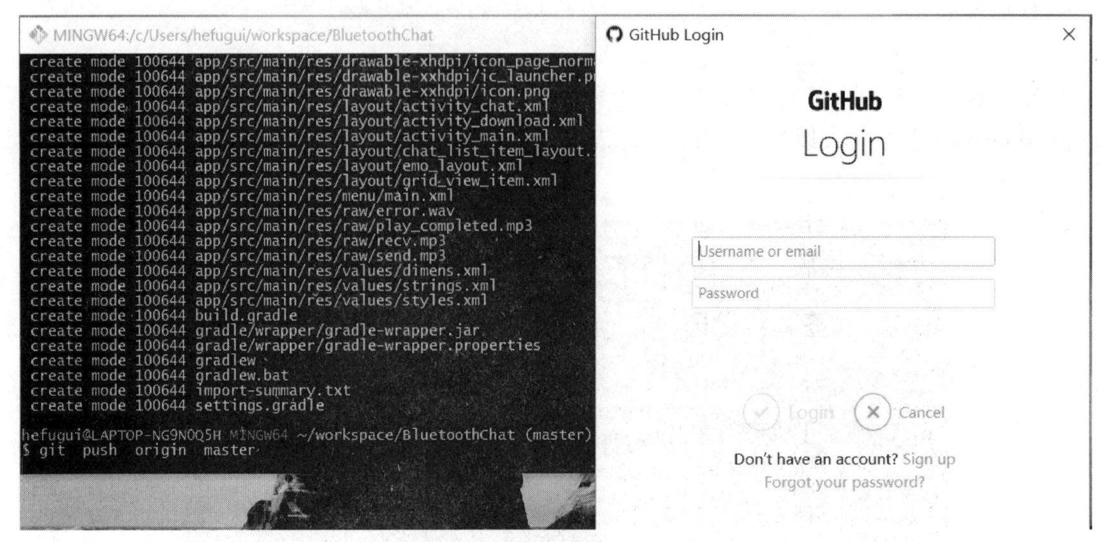

图 12-22　输入用户名和密码

执行 git　push　origin　master 命令，如图 12-23 所示。

```
MINGW64:/c/Users/hefugui/workspace/BluetoothChat
hefugui@LAPTOP-NG9NOQ5H MINGW64 ~/workspace/BluetoothChat (master)
$ git push origin master
Counting objects: 263, done.
Delta compression using up to 4 threads.
Compressing objects: 100% (251/251), done.
Writing objects: 100% (263/263), 647.16 KiB | 0 bytes/s, done.
Total 263 (delta 9), reused 0 (delta 0)
remote: Resolving deltas: 100% (9/9), done.
To https://github.com/mike1966dev/BluetoothChat.git
   7fb4c8b..f7089f3  master -> master

hefugui@LAPTOP-NG9NOQ5H MINGW64 ~/workspace/BluetoothChat (master)
$
```

图 12-23　执行命令

这样，项目的代码上传即已完成，现在打开版本库的主页 https://github.com/mike1966dev/BluetoothChat，可以看到刚才提交的文件，如图 12-24 所示。

mike1966dev First commit.		Latest commit f7089f3 5 minutes ago
.idea	First commit.	5 minutes ago
app	First commit.	5 minutes ago
gradle/wrapper	First commit.	5 minutes ago
.gitignore	Initial commit	10 hours ago
LICENSE	Initial commit	10 hours ago
README.md	Initial commit	10 hours ago
build.gradle	First commit.	5 minutes ago
gradlew	First commit.	5 minutes ago
gradlew.bat	First commit.	5 minutes ago
import-summary.txt	First commit.	5 minutes ago
settings.gradle	First commit.	5 minutes ago

图 12-24　在 GitHub 上查看提交的内容

12.3 将应用程序发布到 360 应用商店

应用程序开发完成以后，需要将应用程序发布到某个商店中，用户可以通过商店找到和下载应用程序。

除了 Google 官方推出的 Google Play 之外，我国还有像 360、豌豆荚、百度、应用宝等知名的应用商店，这些商店提供的功能比较相似。

12.3.1 生成正式签名的 APK 文件

Android Studio 将程序安装到手机中的时候，会将程序代码打包成一个 APK 文件，然后将这个文件传输到手机上，最后执行安装操作。

不是所有的 APK 文件都能安装到手机上，Android 系统要求只有签名后的 APK 文件才可以安装，前面介绍的程序能安装到手机上，是因为 Android Studio 使用了一个默认的 keystore 文件帮助我们自动进行了签名。keystore 文件位置：C:\Users\<用户名>\.android\debug.keystore。

不过这仅仅适合于开发阶段，若要发布应用，要使用一个正式的 keystore 文件签名。

下面介绍如何使用 Android Studio 来生成正式签名的 APK 文件。

（1）在 Android Studio 项目完成以后，单击菜单栏中 Build > Generate Signed APK 命令，如图 12-25 所示。

（2）弹出如图 12-26 的对话框。

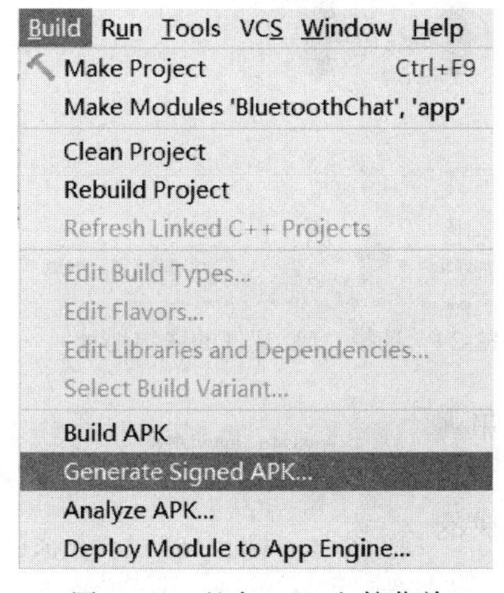

图 12-25　签名 APK 文件菜单

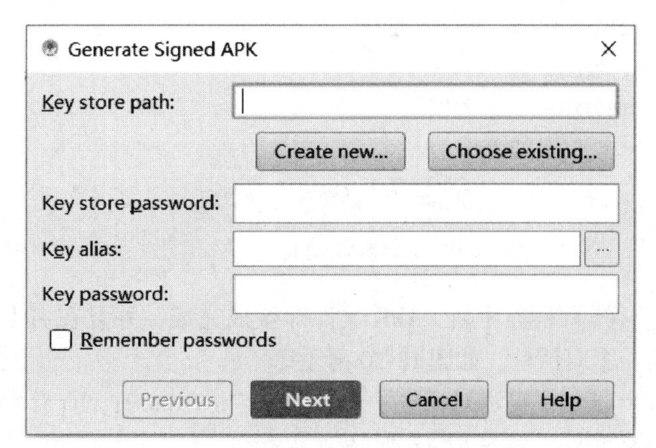

图 12-26　创建签名 APK 对话框

（3）目前还没有正式的 keystore 文件，单击 Create new 按钮，弹出如图 12-27 的对话框，填写创建 keystore 文件必要的信息，根据实际情况填写。Key store path 是生成 keystore 文件的保存路径，填写完成后，单击 OK 按钮。

（4）弹出如图 12-28 所示的对话框，勾选 Remember passwords 复选框，之后就不用再输密码了，然后单击 Next 按钮，选择 APK 文件的输出地址，如图 12-29 所示，单击 Finish 按钮。

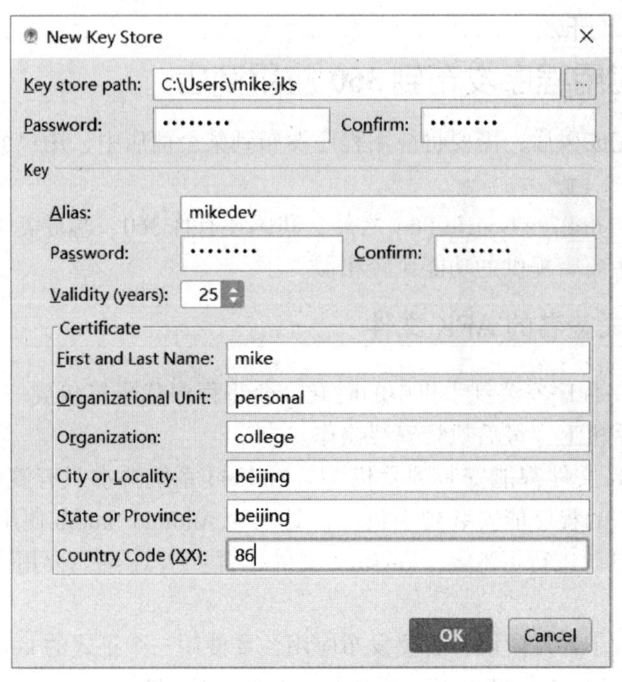

图 12-27　填写 keystore 文件信息

图 12-28　生成签名 APK

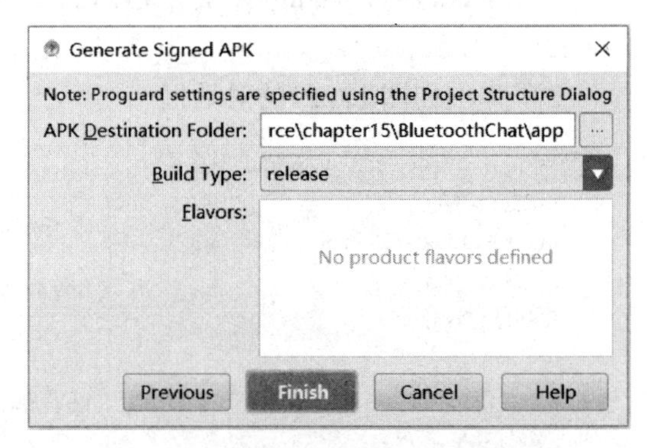

图 12-29　选择 APK 文件的输出地址

（5）稍等片刻，APK 文件即生成完成，并且在右上角弹出一个对话框，如图 12-30 所示。

（6）在提示对话框中单击 Show in Explorer，查看生成的 APK 文件，如图 12-31 所示。

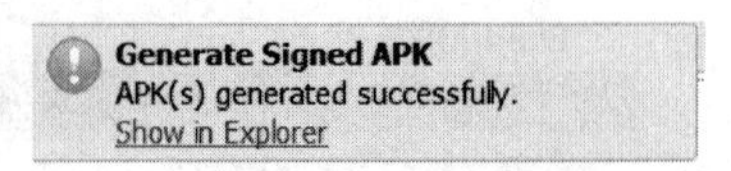

图 12-30　提示 APK 文件生成成功

build	2017/2/13 12:51	文件夹
libs	2017/2/13 8:16	文件夹
src	2017/2/13 8:16	文件夹
.gitignore	2017/2/13 8:16	文本文档
app.iml	2017/2/13 19:28	IML 文件
app-release.apk	2017/2/13 19:26	APK 文件
build.gradle	2017/2/13 8:16	GRADLE 文件
proguard-rules.pro	2017/2/13 8:16	PRO 文件

图 12-31　查看生成的 APK 文件

12.3.2　申请 360 开发账号

生成签名的 APK 以后，需要把安装包发布到应用商店中，如果要发布到 360 应用商店，还需要申请一个 360 开发者账号，申请地址：http://dev.360.cn，如图 12-32 所示。

图 12-32　360 移动开放平台

（1）在页面顶部有“登录”和“注册”按钮，如果尚未注册，则需要先进行注册，如果已注册，则可直接登录。单击“注册”按钮，填入相关信息，如图 12-33 所示。

图 12-33　360 移动开放平台注册

（2）单击“马上注册”按钮，跳转到申请开发者类型界面，如图 12-34 所示。

图 12-34　选择开发者类型

（3）选择“个人开发者”，弹出如图 12-35 所示界面，填写基本信息和联系方式。

注册个人开发者账号

基本信息

注册账号：mike_1966
用此账号进行登录。

开发者姓名：
请与身份证信息保持一致。

出品人：
填写网站或团队名称，会出现在应用介绍信息。

上传证件照：上传
请上传手持身份证照片，jpg、png、gif格式的图片（不超过1M）。查看示例

个人身份证件：护照
国内开发者请填写15位或18位大陆身份证号码，国外开发者请填写护照号码

图 12-35　填写基本信息和联系方式

（4）全部填写完成后，单击屏幕最下方的“同意并注册开发者”按钮，完成注册，如图 12-36 所示。

图 12-36　完成开发者注册

这样就成为一名 360 开发者了。

12.3.3　发布应用程序

接下来要发布 BluetoothChat 这个应用，在浏览器中访问地址：http://dev.360.cn，登录账号，出现如图 12-37 所示界面。单击“软件发布”，即会显示如图 12-38 所示的界面，这里单击“软件”按钮。

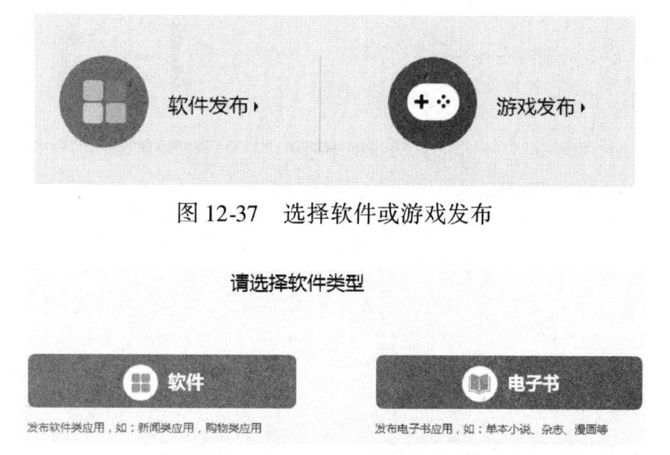

图 12-37　选择软件或游戏发布

图 12-38　选择软件类型

（1）弹出的界面如图 12-39 所示，填写发布软件的相关信息和上传发布的 APK，选择软件分类和应用简介。

图 12-39　选择应用分类

（2）接着滚动屏幕，填写版本信息，如图 12-40 所示。

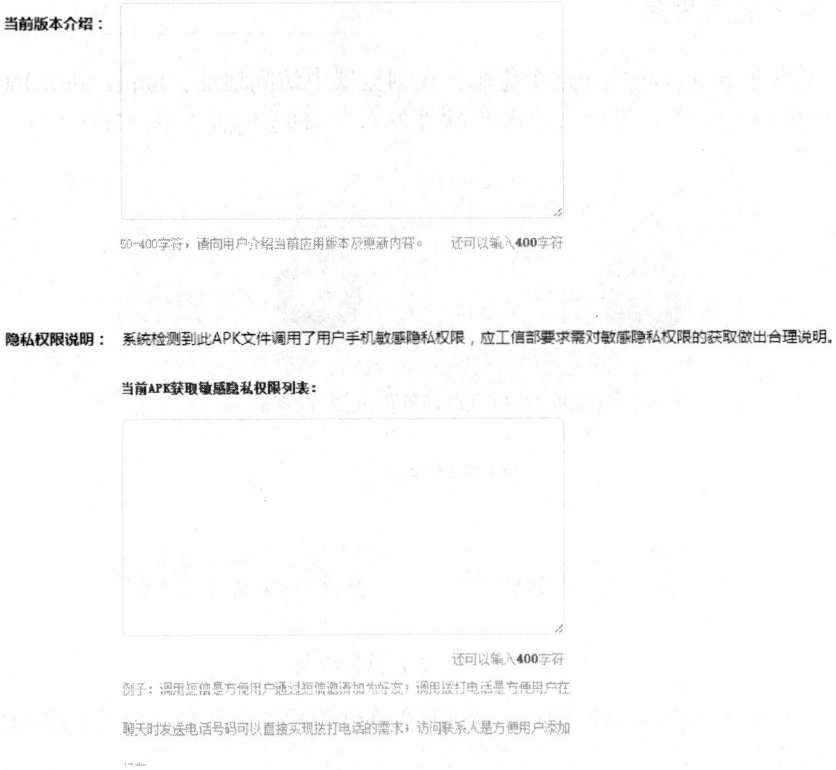

当前版本介绍：

50~400字符，请向用户介绍当前应用版本及更新内容。　还可以输入**400**字符

隐私权限说明：　系统检测到此APK文件调用了用户手机敏感隐私权限，应工信部要求需对敏感隐私权限的获取做出合理说明。

当前APK获取敏感隐私权限列表：

还可以输入**400**字符

例子：调用短信是方便用户通过短信邀请加为好友；调用拨打电话是方便用户在聊天时发送电话号码可以直接实现拨打电话的需求；访问联系人是方便用户添加

图 12-40　填写版本信息

（3）接着向上滚动，提交 5 张应用程序的截图，如图 12-41 所示。

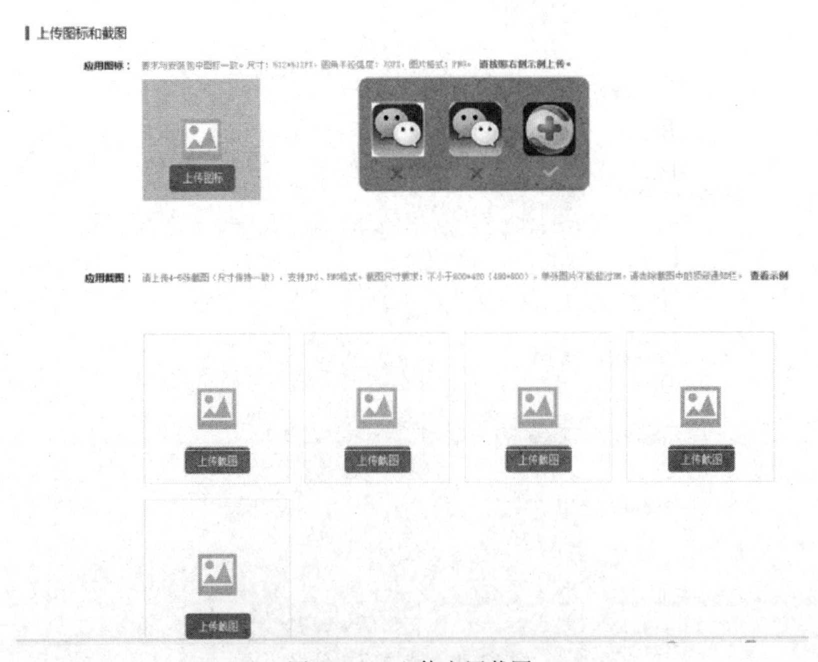

图 12-41　上传应用截图

（4）向下滚动，选择一些选项，完成后，单击屏幕最下方的“提交审核”按钮，等待审核，如图 12-42 所示。

是否进行云测试：◉ 是 ○ 否

由 Testin云测 提供专业的应用机型测试，选择后将自动为您的应用进行云测试，并发送测试报告。

进行网络友好度测试：◉ 是 ○ 否

由 智測雲 smarterapps.cn 提供 专业的网络有好度测试，选择后将自动为您的应用进行测试，并发送测试报告。

发布时间：◉ 审核后立即发布 ○ 定时发布

提交审核

图 12-42　提交审核

12.3.4　嵌入广告

国内的 Android 应用基本都是免费的，那么开发者如何获得收入呢？在应用中插入广告是一种比较常用的盈利手段。本节主要讲解如何在 Android 应用中插入广告。

Android 应用程序应用的流程可分为三部分：应用开发、嵌入广告平台的 SDK、发布到安卓市场，第一部分是核心，但是后面两部分对收入来说却是最重要的。

在 Android 应用程序开发完成以后，就可以在 Android 应用程序中嵌入广告了。能实现嵌入广告的国内广告平台有很多，用户数量比较多的有万普世纪、有米传媒、多盟、腾讯广告联盟、百度联盟及亿动智道等。

12.3.4.1　Android 应用程序嵌入广告

本节以万普世纪和有米广告为例，介绍在应用程序中嵌入广告的方法。

1. 万普世纪

万普世纪传媒隶属于北京万普世纪科技有限公司，简称“万普世纪”（WAPS），是国内领先的移动营销服务提供商，致力于为全球广告客户提供基于移动互联网的效果广告及整合营销服务，打造中国最大的智能移动广告联播网络。

万普世纪始建于 2005 年，是中国第一家专业提供移动互联网营销平台的公司，并首创积分墙广告与多位一体的效果营销模式，10 余年来长期专注于效果计费型移动广告及营销服务，目前已发展成为国内最大的移动广告平台及移动虚拟货币交易平台，业务重点是为移动开发者及广告主提供基于智能移动应用（Android 及 iOS 应用）的效果计费型广告服务。

万普世纪的官网主页如图 12-43 所示。

万普世纪实现嵌入广告的步骤如下。

（1）注册平台账户，在万普世纪官网单击“注册”按钮，完成注册，如图 12-44 所示。

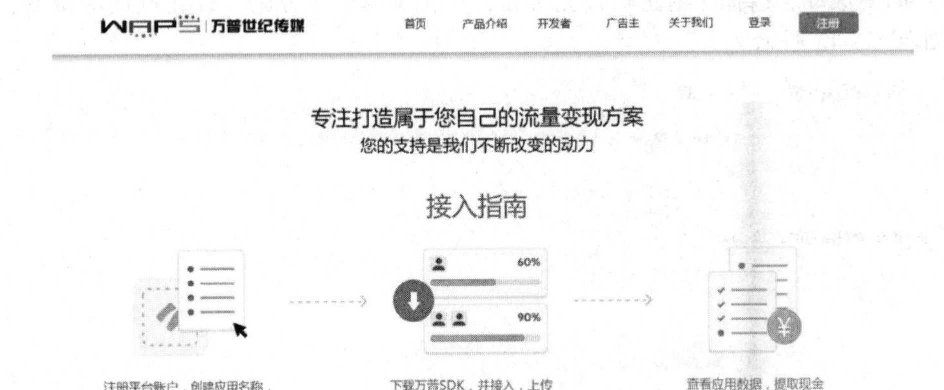

图 12-43　万普平台

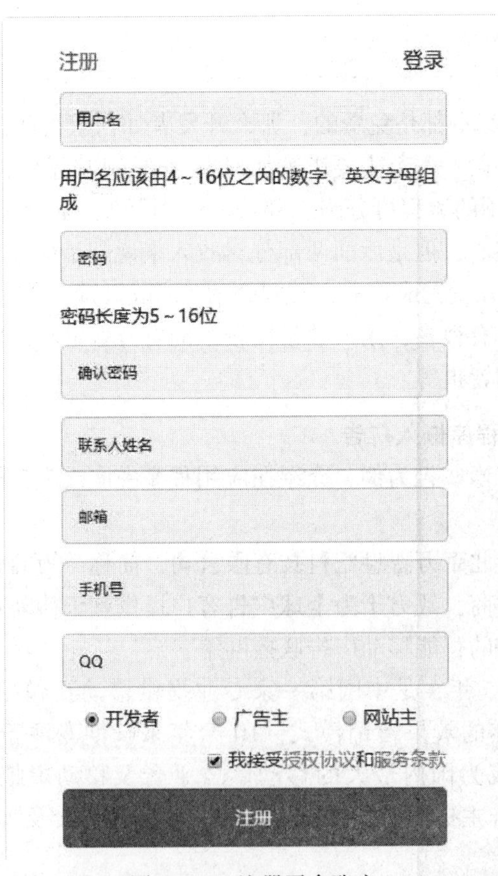

图 12-44　注册平台账户

（2）注册以后登录账户，如图 12-45 所示，单击“添加应用”按钮，输入应用名称，选择应用平台为 Android，单击“下一步”按钮。

WAPS 万普世纪传媒

应用管理　＋ 添加应用

应用中心
下载SDK
数据报表
开发工具
财务中心

添加应用

创建应用　获取APP_ID　完善应用详情

应用名称：

应用平台：　Android　iOS　Web

下一步

图 12-45　用户登录

（3）进入如图 12-46 所示的窗口，获取相应的 App_ID。

图 12-46　获取相应 App_ID

（4）在当前界面中单击“下载 SDK”按钮，将会弹出“新建下载任务”对话框，如

图 12-47 所示。

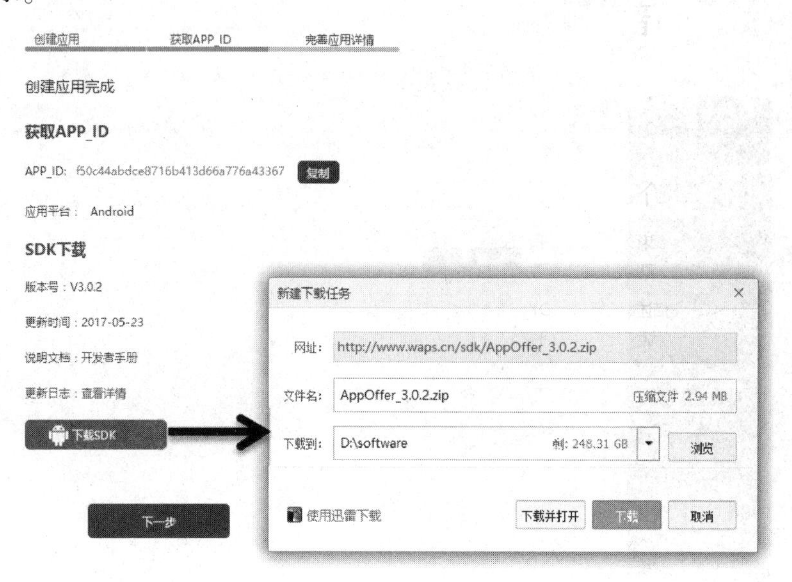

图 12-47　下载 SDK

(5) 下载的 SDK 文件名为“AppOffer_3. 0. 2. zip”，这是一个压缩文件，文件中包含了嵌入广告使用的库、Demo 项目和开发者手册，如图 12-48 所示。

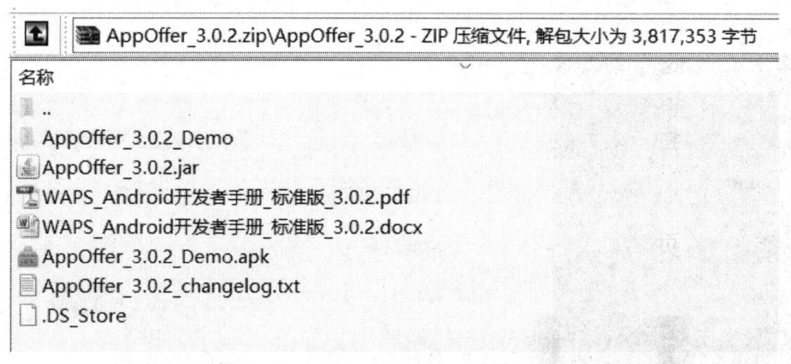

图 12-48　SDK 文件中包含的内容

(6) 打开开发者手册文件“WAPS_Android 开发者手册_标准版_3. 0. 2. docx”，如图 12-49 所示，里面详细介绍了使用过程，可参照开发者手册实现广告嵌入，也可参考其中的 Demo 程序。

2. 有米广告

有米广告是有米科技股份有限公司旗下的国内第一批综合性移动广告平台，成立于 2010 年 4 月，总部位于广州，在北京、上海、香港设有分支机构及客户服务团队。有米广告致力于运用精准投放技术，通过海量媒体的人群细分覆盖，为广告主提供优质的品牌营销与产品推广服务，同时助力开发者获得丰富稳定的广告收益。

基于突破性的双驱动 DSP + Ad Network 体系，有米广告深入对接移动端全景流量，专注技术创新与数据积累，推出以程序化购买为基础、精准人群定向为核心的专业营销解决方案。依托成长式的智能机器学习算法，有米广告将大数据挖掘与人工专家的优化调控相结合，从而保障广告投放的精准度，直击目标受众。借助深度聚合的优质社媒资源，有米广告打造出立体触达的跨维营销新模式，帮助广告主实现移动整合营销的品效兼具。

万普平台 Android 版 SDK 开发者手册 (标准版 Ver3.0.2)

·平台简介

万普世纪移动营销服务平台(以下称为“万普平台”)的 Android 版 SDK 提供了一套现成的开发包及 Demo 源代码，便于开发者在 Android 应用中方便的集成万普平台的各项功能。

本文档描述了标准版 SDK 的用途与用法，并提供了示例代码。您仅需要在现有的应用中加入少量新代码，就可以集成万普平台的各项功能，轻松获得用户量和收入的倍增。

目录

图 12-49　开发者手册

有米广告的官网主页如图 12-50 所示。

图 12-50　有米广告

有米广告的部分开发者成功案例如图 12-51 所示。

图 12-51　有米广告的部分开发者成功案例

有米广告实现嵌入广告的步骤如下。

（1）注册平台账户，在有米广告官网单击“注册”按钮，完成注册，如图 12-52 所示。

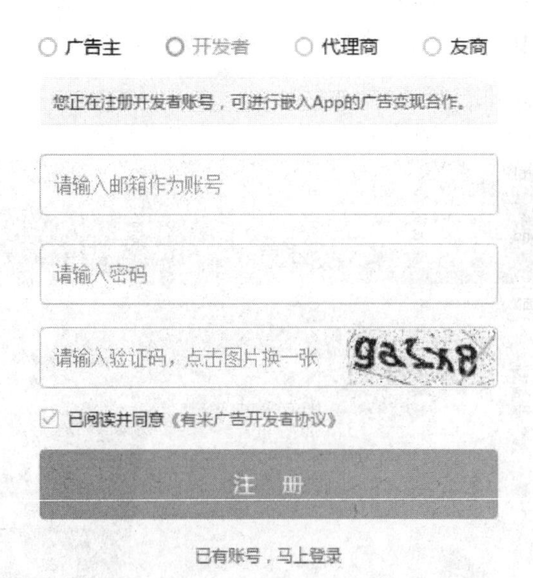

图 12-52　注册平台账户

（2）注册平台账户后登录账户，登录后需完善个人信息，完善信息后单击“添加应用”标签，如图 12-53 所示。

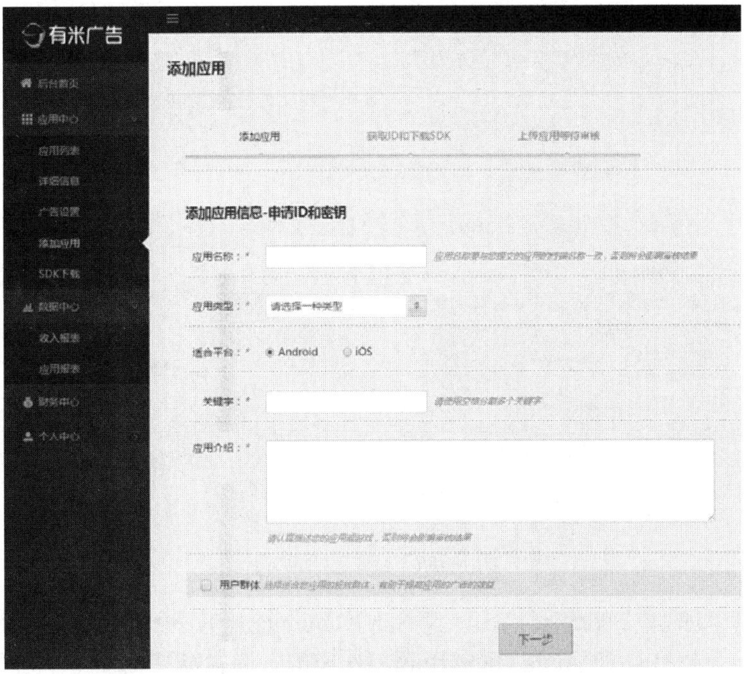

有米广告

后台首页
应用中心
应用列表
详细信息
广告设置
添加应用
SDK下载
数据中心
收入报表
应用报表
财务中心
个人中心

添加应用

添加应用 获取ID和下载SDK 上传应用等待审核

添加应用信息-申请ID和密钥

应用名称：* 应用名称要与您提交的应用的终端名称一致，否则将会影响审核结果

应用类型：* 请选择一种类型

适合平台：* Android iOS

关键字：* 请使用空格分割多个关键字

应用介绍：*

请认真描述您的应用或游戏，否则将会影响审核结果

用户群体 选择适合您应用的投放群体，有助于提高应用的广告的效益

下一步

图 12-53 添加应用

（3）填写应用信息后单击“下一步”按钮，如图 12-54 所示。

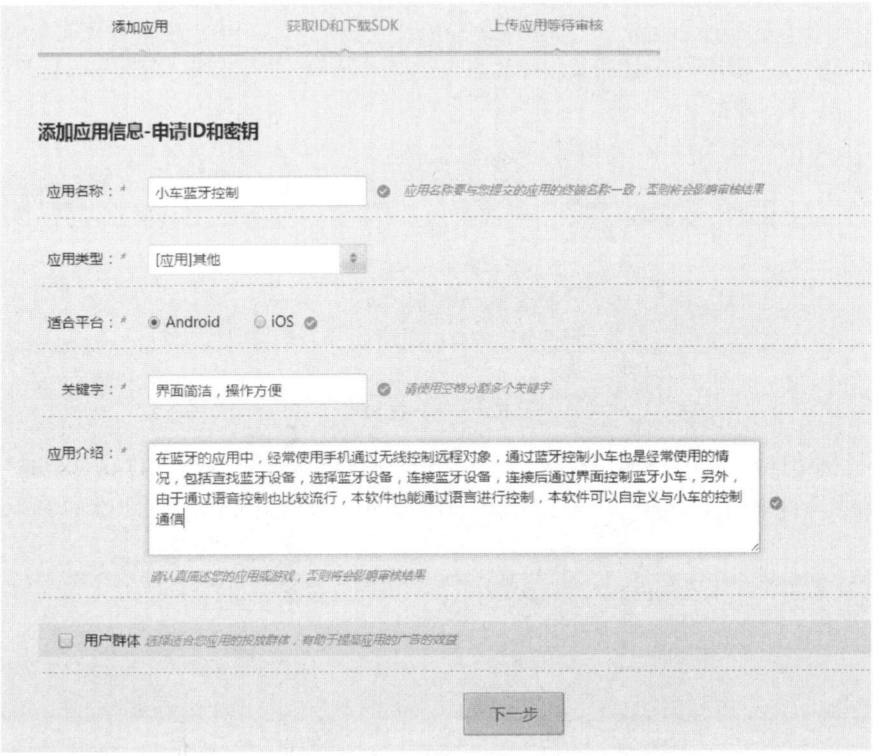

添加应用 获取ID和下载SDK 上传应用等待审核

添加应用信息-申请ID和密钥

应用名称：* 小车蓝牙控制 应用名称要与您提交的应用的终端名称一致，否则将会影响审核结果

应用类型：* [应用]其他

适合平台：* Android iOS

关键字：* 界面简洁，操作方便 请使用空格分割多个关键字

应用介绍：* 在蓝牙的应用中，经常使用手机通过无线控制远程对象，通过蓝牙控制小车也是经常使用的情况，包括查找蓝牙设备，选择蓝牙设备，连接蓝牙设备，连接后通过界面控制蓝牙小车，另外，由于通过语音控制也比较流行，本软件也能通过语言进行控制，本软件可以自定义与小车的控制通信

请认真描述您的应用或游戏，否则将会影响审核结果

用户群体 选择适合您应用的投放群体，有助于提高应用的广告的效益

下一步

图 12-54 填写应用信息

（4）进入如图 12-55 所示的界面，获取应用密钥。

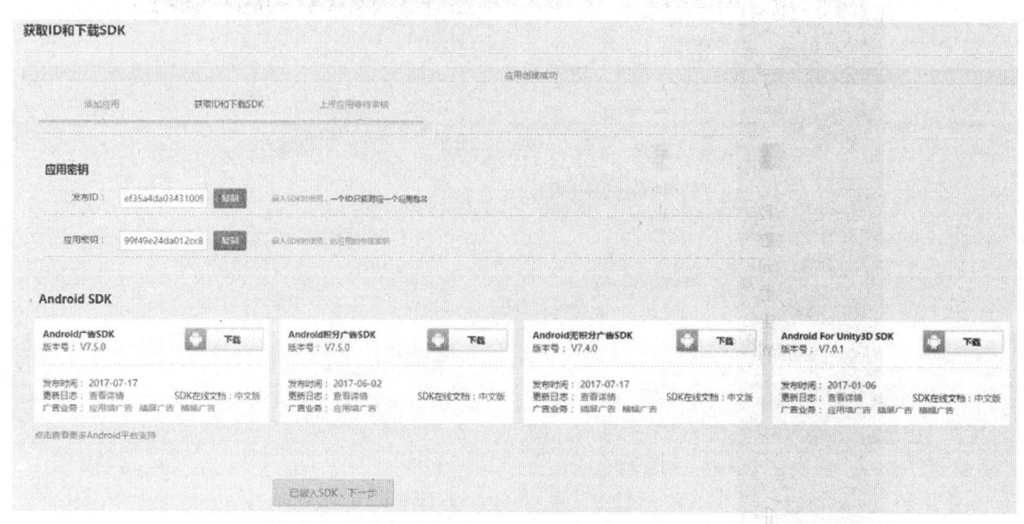

图 12-55　获取应用密钥

（5）在上面的界面中可选择下载的 SDK 有：Android 广告 SDK 、Android 积分广告 SDK、Android 无积分广告 SDK 和 AndroidFor Unity 3D SDK，选择需要的 SDK，单击“下载”按钮，如图 12-56 所示。

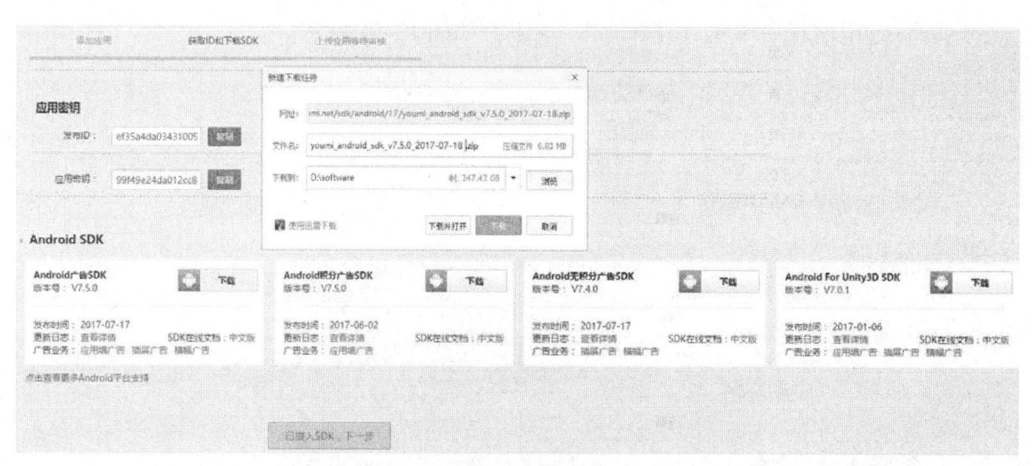

图 12-56　下载 SDK

（6）此例中选择下载的 SDK 文件名为“youmi_android_sdk_v7. 5. 0_2017-07-18. zip”，这是一个压缩文件，文件中包含了嵌入广告使用的库、Demo 项目和开发文档，如图 12-57 所示。

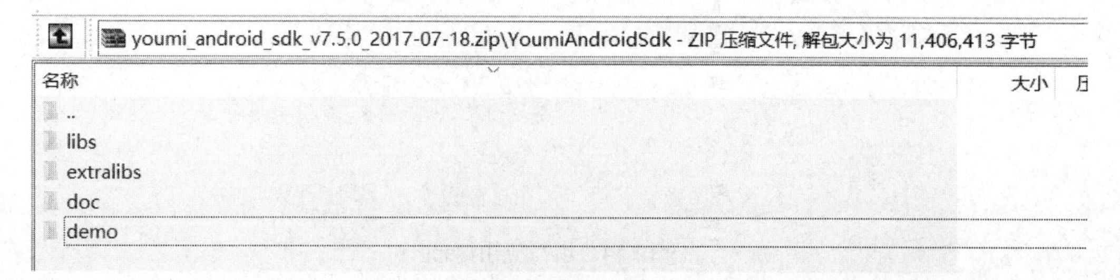

图 12-57　SDK 文件目录

（7）在 SDK 的开发文件目录 doc 中包含了开发者嵌入广告的使用方法，如图 12-58 所示。

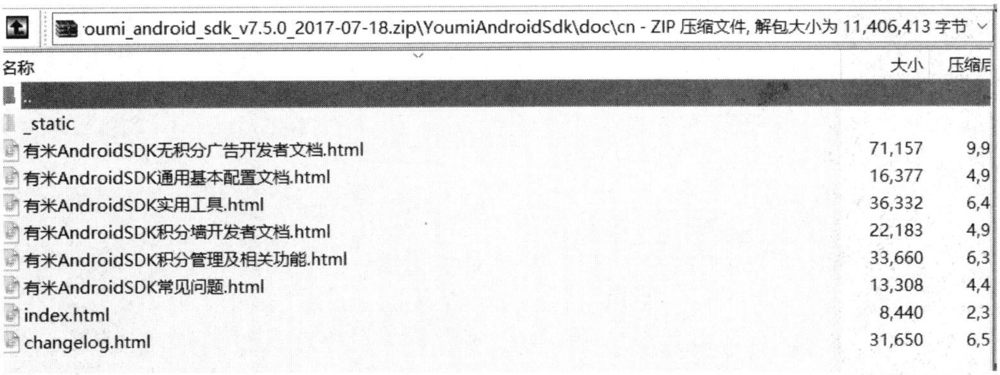

oumi_android_sdk_v7.5.0_2017-07-18.zip\YoumiAndroidSdk\doc\cn - ZIP 压缩文件, 解包大小为 11,406,413 字节

名称	大小	压缩后
..		
_static		
有米AndroidSDK无积分广告开发者文档.html	71,157	9,9
有米AndroidSDK通用基本配置文档.html	16,377	4,9
有米AndroidSDK实用工具.html	36,332	6,4
有米AndroidSDK积分墙开发者文档.html	22,183	4,9
有米AndroidSDK积分管理及相关功能.html	33,660	6,3
有米AndroidSDK常见问题.html	13,308	4,4
index.html	8,440	2,3
changelog.html	31,650	6,5

图 12-58　开发文档 doc 目录

（8）打开开发者手册“index. html”，如图 12-59 所示，其中详细介绍了使用方法，可参照开发者手册实现广告嵌入，也可参考其中的 Demo 程序。

图 12-59　开发者文档导航

（9）实现嵌入广告后，上传应用程序并等待审核，如图 12-60 所示。

图 12-60　上传应用程序并等待审核

12.3.4.2　发布渠道

具体的发布渠道有以下几种。

（1）安智市场：安智市场是目前国内装机量比较大的应用市场，国内品牌大多数的手机都没有携带 Google 市场，所以该市场是国内非常重要的一个渠道，审核时间一般为 1～2 个工作日。

（2）安卓市场：安卓市场的安装量在国内仅次于安智市场，也是开发者必不可少的一个渠道，审核时间一般为 1～2 个工作日。

（3）应用汇：应用汇的安装量也比较大，开发者应当考虑这个渠道，审核时间一般为 1～2 个工作日。

（4）腾讯手机应用平台：腾讯平台的安装量也比较大，虽然跟安智市场和安卓市场还有差距，但具有庞大的用户群体及有力的推广模式，因此也是开发者需要的一个渠道，其审核流程包括审核、测试和上架，一般至少需要 3～4 个工作日。

（5）91 手机商城：91 手机商城也是开发者不能忽略的一个渠道，审核时间一般为 2～3 个工作日。

（6）智汇云：智汇云是华为提供的市场平台，因为华为手机国内市场占有量在迅速提高，所以智汇云的用户量是比较可观的，该平台的审核时间会稍微久一些，通常需要 5 个工作日以上。

（7）N 多网：N 多网的应用量相对来说少一些，如果您有足够的精力，那么也可以利用这个渠道，审核时间一般为 1 ~2 工作日。

（8）机锋网：跟 N 多网类似，应用量也相对少一些。

（9）联想商城：联想商城要求提交固定的圆底图标，另外，其审核和测试非常仔细和严格，需要说明的是，联想的测试会给出一份详尽的报告，告知应用的功能缺陷、Crash 出现频率等，其内容会仔细说明具体步骤及结果，所以不失为一个很好的免费测试渠道，一般审核时间为 3 个工作日以上。

（10）其他平台，包括搜狐、网易应用、安智迷、三星 App（英文）、MOTO app（英文）、安卓星空、爱米吧、eoe 亿优等。

12.4 本章小结

本章介绍了应用程序的最后环节，即应用程序的托管和发布。由于大部分应用程序是团队开发的，所以需要把代码提交到一个平台上，GitHub 是一个免费的代码托管平台，代码的传送通过 Git 实现，本章介绍了具体的操作方法。本章还以 360 应用商店为例，介绍了应用程序的发布操作方法，最后介绍了嵌入广告赢利的方法。

Android（中文名为“安卓”）操作系统正在持续扩展市场，已经成为全球应用最广的操作系统之一，引领了终端智能化的浪潮。其在智能手表、智能电视、智能手机、智能眼镜、智能平板、电子书阅读器、游戏机，甚至是家居、家电、音响产品、汽车面板等设备的智能化方面表现出了卓越的功能效果。因此 Android 凭借着自身的优势，也得到了越来越多企业及开发者的青睐。

本书基于当前最新的 Android Studio 版本（稳定版 Android Studio 2.3）、Android SDK 和最主流的应用，以 Android 项目开发的视角，循序渐进地讲解并展示了 Android 项目开发过程的主要流程，依次介绍了开发环境的搭建、项目设计、界面设计、应用程序构成设计、高级界面设计、数据持久化方案、多媒体应用开发、网络开发、无线通信、开源库和开源项目，以及应用程序的托管和发布等内容。在讲解每项知识点时，都遵循了理论联系实际的讲解方式，配以实战演练，从而详尽剖析了 Android 项目开发的完整实现流程。

通过对本书进行学习，初中级开发者将极大地提高 Android 开发能力，向 Android 高级开发者迈进。而对于高级开发者来说，仍然可以从本书的知识体系中学习到更加规范的操作流程和并获得不少设计灵感。

本书适用于对 Java 编程有一定基础，并且已经有一定的 Android 开发经验，想进一步提高 Android 开发能力的读者，可作为高等院校信息类相关专业的教材，也可作为 Android 程序设计的培训教程，还可作为广大 Android 开发爱好者自学的参考手册。

图书在版编目（CIP）数据

新编 Android 应用开发从入门到精通/何福贵等编著．—北京：机械工业出版社，2017.12

ISBN 978-7-111-58810-8

Ⅰ．①新…　Ⅱ．①何…　Ⅲ．①移动终端—应用程序—程序设计　Ⅳ．①TN929.53

中国版本图书馆 CIP 数据核字（2017）第 330416 号

机械工业出版社（北京市百万庄大街 22 号　邮政编码 100037）
策划编辑：丁　伦　责任编辑：丁　伦
责任校对：丁　伦　封面设计：子时文化
责任印制：孙　炜
北京中兴印刷有限公司印刷
2018 年 3 月第 1 版第 1 次印刷
185mm×260mm · 25.25 印张 · 621 千字
0001—3000 册
标准书号：ISBN 978-7-111-58810-8
定价：85.00 元（附赠源代码）

凡购本书，如有缺页、倒页、脱页，由本社发行部调换

电话服务	网络服务
服务咨询热线：010-88361066	机 工 官 网：www.cmpbook.com
读者购书热线：010-68326294	机 工 官 博：weibo.com/cmp1952
010-88379203	金 书 网：www.golden-book.com
封面无防伪标均为盗版	教育服务网：www.cmpedu.com